BOSTON STUDIES IN THE PHILOSOPHY OF SCIENCE

VOLUME XXXIII

SCIENCE AND ITS PUBLIC: THE CHANGING RELATIONSHIP

SYNTHESE LIBRARY

VOLUME 96

BOSTON STUDIES IN THE PHILOSOPHY OF SCIENCE

EDITED BY ROBERT S. COHEN AND MARX W. WARTOFSKY

VOLUME XXXIII

SCIENCE AND ITS PUBLIC:
THE CHANGING RELATIONSHIP

Edited by

GERALD HOLTON AND WILLIAM A. BLANPIED

D. REIDEL PUBLISHING COMPANY

DORDRECHT-HOLLAND / BOSTON-U.S.A.

Library of Congress Cataloging in Publication Data
Main entry under title:

Science and its public.

(Boston studies in the philosophy of science ; v. 33)
(Synthese library ; v. 96)
    Includes bibliographical references and index.
    1.   Science—Social aspects.   2.   Science and state.
I.   Holton, Gerald James.   II.   Blanpied, William A.,
1933–   III.   Series.
Q174.B67   vol. 33   [Q175.5]   501s   [301.24'3]   75-41391
ISBN 90-277-0657-3
ISBN 90-277-0658-1 pbk.

Published by D. Reidel Publishing Company,
P.O. Box 17, Dordrecht, Holland

Sold and distributed in the U.S.A., Canada, and Mexico
by D. Reidel Publishing Company, Inc.
Lincoln Building, 160 Old Derby Street, Hingham,
Mass. 02043, U.S.A.

Printed in The Netherlands by D. Reidel, Dordrecht

Editorial Foreword

To STUDY the philosophy of science has always been a complex task, reaching to the methods and achievements of the sciences, to their histories and their contexts, and to their human implications. Sometimes favored by their social environment, sometimes dissenting from their *Zeitgeist*, the scientists have taken varying roles in the social spectrum, allied with differing interests, classes, powers, religions, evaluative outlooks. Philosophers should be interested as much in the changing social situations of science and of scientists as in the changing empirical findings and explanatory conceptions; recognition that rationality, experience, and inquiry have a history is no longer novel. Moreover the historical development of scientific perceptions of nature is linked—whether loosely or tightly—by the development of perceptions of science itself. Perceptions of science are located not only in the self-awareness of scientists but also in the critical awareness of their fellow human beings. No doubt some friends or critics are more articulate than others, but the context for science has not been bland or neutral. Plaything, weapon, savior, hireling, magician, devil, priest, the stereotypes of science and scientist are neither the simple result of plain ignorance nor the obvious reflection of some successes and some failures of the scientific enterprise. Public perceptions of science have great importance for understanding both the public in society and the sciences at the stage perceived. For the public, when understood, may lead us to subtlety, may underscore ambiguity, may stress dangerous omissions and incompleteness, and all to the good; and understood, or not, the public perceptions may lead either to stifling constraints or warm encouragement. If the public perceptions were themselves understood, education might be undertaken in the widest humane sense to enlighten the publics of whom Gerald Holton speaks so wisely.

So we take studies of the public perceptions of science to be of serious concern to philosophers, but they are not philosophical alone; they must be broadly social-scientific, empirical at times, reflective and critical, and sympathetically listening to what may only be the tacit content of the public judgment. Tacit or plain, the public view of science must not be evaded.

Center for the Philosophy and History of Science   ROBERT S. COHEN
Boston University   MARX W. WARTOFSKY
November 1975

Table of Contents

)

Preface

THE CHIEF AIM of this volume is to examine some of the serious critiques of twentieth-century science, at least in the more industrialized countries, and to stimulate research into the significance of arguments made from both within and outside science. The debates concerning the legitimacy and utility of scientific endeavor impinge upon scientists from both sides, from their traditional "clients" no less than from their traditional critics. The possibility they have to face again, as in certain periods in the past, is that the bulk of criticisms is not simply the ephemeral reaction of isolated individuals or small groups to specific political and moral dilemmas, but that it may be the sign of a more basic discontent—possibly even a harbinger that our world view, which until recently has been generally supportive of science, may be about to turn, if scientists, educators, philosophers, and other scholars do not make a serious effort of understanding and wrestling with the legitimate core of the questions being raised today.

Another related issue is the extent and validity of the understanding of science, both research oriented and applied, by the public. (Perhaps the term "publics" would be more appropriate, since they are of many different kinds.) While many sections of the public would agree that science is an immensely successful and practical pursuit, as remarkable for the quality of its theoretical contributions as for its utility in a wide range of fields, the evidence of public ambivalence about science and technology are too numerous to ignore.

This ambivalence, again, is not characteristic only of our time or of any single country; hence it seemed to me necessary, in planning this volume, to recruit scholars from a wide variety of disciplines, and to encourage at least some of them to go considerably beyond a study of the situation in North America in the 1970's. Both the historical and the international dimensions of the volume are there by design, as is the wide range of sciences and technology being considered.

These motivations were incorporated into the commissioning letter to the authors, a portion of which will help readers to see what tone we hoped to set for our enterprise:

At least until recently, the relationship between man and nature which has evolved from modern science has been more or less tacitly understood (at least in the western world) in about the same way among those who have received any substantial education. Briefly, man has come to be seen as a part of nature (rather than, as before Darwin, outside and above nature), linked to and limited by nature is several ways: by his dependence upon the chemical, physical, and biological processes within and around him; by his historical, biological debt to the chain of "lesser" creatures from which he evolved; and by his behavioral responses and potentials.

In this relationship, science was thought to have two largely beneficial functions: first, to provide a method for gaining a deeper understanding of the links between man and wider nature, and in this way to make him intellectually (and perhaps spiritually) more exalted; and second, to help him transcend his physical limitations, e.g. by paving the ground for medical and other technological developments. Throughout the present century, and increasingly during the years after World War II, there has been a considerable debate on whether the tangible by-products of the scientific endeavor have been on the whole beneficial or destructive; until recently, however, pure science was usually excused from participation in the supposed sins of technology. Indeed, the image of man within nature and yet in control of it, and the value-free, objective, and rational ethic commonly thought to be inherent in and supported by modern science, have come to be regarded as triumphal end results of the scientific revolution.

Regardless of its philosophical soundness, the old model of value-free science unlocking the secrets and powers of nature for man's benefit has had profound social and intellectual consequences. But several ancient flaws and new counterclaims have now become far more evident than ever before, at least in the "developed countries." Many feel these will become increasingly intense as scientific and technical advances reveal ever more effective methods for intervening directly in on-going biological and psychological processes within the person—rather than just indirectly (as physics did) through the more long-range processes of changing ideas and the external landscape of technological capabilities. In addition, frustration, disillusionment, and righteous anger have been consequences of some of the exaggerated or unfulfilled expectations that were associated with the compelling social visions of plenty and justice which purposeful, rational enlightenment seemed to offer a few short decades back. We seem thus to be in a crisis of reason, in which commitment to rational knowledge as a source of human freedom is being seriously challenged. It is clear that concern with this situation goes far beyond science, and involves also scholars and practitioners in social and humanistic studies.

Searching thinkers both in science and outside are troubled by the new images which at their fashionable extremes project science as almost mindlessly giving, to unprepared or unwise people, uncontrolled powers over nature and human life. They see a significant kernel of truth inherent in these objections, one that is not diminished by the fact that the most rapidly spreading alternative to the rational world view often appears to be a mindless searching for immediate experience and unreasoned, instant truth.

Essays and books by prominent writers with wide readership broadcast the doubts concerning the explanatory power that was thought to be resident in science itself. The counter-culture essayists, with large audiences of their own, would drastically change the predominant balance favoring rational elements. Matters are only made worse by some reactions from within the scientific community, which often seem, and sometimes also *are*, self-serving. These frequently confuse the obligation to take seriously the state of public understanding of science with the short term benefits that would accrue from better public relations.

Little headway will be made in countering these attacks either by refusing to take them seriously at all or by starting from the assumption that the methods of science and the image of nature offered by modern science must be accepted as the unchallengeable basis for any debate. On the contrary, it is wiser to assume that at least the more seriously meant current challenges to scientific rationality have a philosophical basis that invites interdisciplinary study, and that, in any case, these challenges may have profound philosophical and sociological implications which we should try to understand and be prepared for.

Acknowledgments

It is a pleasure to acknowledge the indispensable help received from numerous individuals and institutions in planning and assembling this volume. At every step I had the benefit of the advice and active work of the coeditor of this volume, Dr. William A. Blanpied, who, as Executive Director of the Harvard University Program on Public Conceptions of Science, incorporated the preparation of this collection into our Program as one of its important products.

The period of gestation stretched over two years, and included two conferences, one for planning and one for the discussion of the drafts of the papers among the authors.* These conferences were held under the joint auspices of the Harvard Program and of *Daedalus*, the quarterly journal of the American Academy of Arts and Sciences; the latter published the major part of the essays in this volume as a special issue, but additional essays have been incorporated for this book publication. The Editor of *Daedalus*, Professor Stephen R. Graubard, not only helped throughout to achieve this publication, but encouraged it and participated in the planning of it from the very outset. His taste and perception shaped the enterprise throughout. Another debt of gratitude is owed to Geno A. Ballotti, Managing Editor of *Daedalus*, and to his staff.

Last but not least, it is a great pleasure to acknowledge the financial support given by the two chief financial sponsors of his issue and of the Harvard Program, the National Science Foundation's Office of the Public Understanding of Science, and the Commonwealth Fund. Theodore Wirths and Richard Stephens of the former, and Quigg Newton, Robert Glaser, and Reginald Fitz of the latter are indeed to be thanked for their encouragement throughout.

Jefferson Physical Laboratory   GERALD HOLTON
Harvard University

---

* The persons involved in these conferences, or in major correspondence, and who thereby formed an informal Advisory Committee, were:
James Ackerman, William A. Blanpied, David Z. Beckler, Peter S. Buck, Saul G. Cohen, Emilio Q. Daddario, A. Hunter Dupree, Amitai Etzioni, Arthur Galston, Loren Graham, Stephen R. Graubard, Daniel S. Greenberg, Gerald Holton, Kenneth Keniston, Franklin Long, Leo Marx, André Mayer, Jean Mayer, Ernst Mayr, Russell McCormmach, Everett I. Mendelsohn, Leonard B. Meyer, John A. Moore, Clyde Nunn, David Perlman, Don K. Price, Gardner C. Quarton, Marc J. Roberts, David J. Rose, Barbara Rosenkrantz, Theodore Roszak, Brian Schwartz, Edward Shils, Richard E. Stephens, Jeremy Stone, Stephen Toulmin, Steven Weinberg, Charles Weiner, Robert R. Wilson, and Harriet Zuckerman.

WILLIAM A. BLANPIED

Introduction

DURING THE LATE 1960's and early 1970's, concern was voiced by several leading scientists and their supporters about the ability of science to survive the attacks that were being launched against it—at least in the industrially developed, Western countries—from a number of directions. By the mid-1970's the virulence of these attacks had largely abated; however, there remained as a potentially positive residue the perception that the relationship between science and its public was in a state of flux. Thus although it is now generally agreed that science and technology will continue to play a central role in contemporary society, there is widespread recognition of the fact that many if not most of the earlier tacit assumptions that were made about their relationships to other important societal institutions can, and indeed must, be subject to searching critical examination.

The essays that appeared in the Summer 1974 issue of *Dædalus*, the journal of the American Academy of Arts and Sciences, provide the core of the present volume. The issue was intended to examine a number of problems relating to the public understanding and appreciation of science in contemporary society. Of course that society is characterized by the existence of many different subgroups; and of course the word "science" has more than one legitimate connotation. If, then, the relationship between the sciences and their publics are in a state of flux, both the scholar and the policy maker urgently need detailed analyses of the major components of this kaleidoscopic situation. Each of the original essays in the *Dædalus* volume was regarded as a significant and timely contribution in its own right, as well as a contribution to the total pattern that is emerging. The additional amount of space available here in book form has allowed us to include also the three further essays, by George Basalla, Sally Kohlstedt and Dorothy Nelkin; these are published for the first time in this volume.*

Re-reading all of these essays, I am now particularly struck by the repeated references to issues of scientific freedom and responsibility, or, more broadly, to the public perceptions of the ethical and human value implications of science and technology. It also seems evident that many of the authors regard these issues as central to the relationship between science and its public, as we have come to do more and more since the start of the decade. Indeed, today there is

---

* The essays of Basalla and Kohlstedt are based on papers they presented at a conference at the Radcliffe Institute in April 1974. A paper by Dora Weiner, entitled "Public Health under Napoleon: *The Conseil de Salubrité de Paris*, 1802-1815," which was also presented on that occasion, has since been published in *Clio Medica*, IX (1974), 271-84.

G. Holton and W. A. Blanpied (eds.), Science and its Public, XIII–XXV. All Rights Reserved.
This article Copyright © 1976 by D. Reidel Publishing Company, Dordrecht-Holland.

an increasingly sharp series of debates over questions of science and values, both within the scientific community and among the publics with which it interacts. By way of fleshing out the discussions in several of the essays that touch on that aspect of the relationships between science and its public, it may therefore be useful to sketch out a few aspects of the broad intellectual framework in which many contemporary problems in science and values are being discussed.

Concerns Largely Intrinsic to the Scientific Community

Much of the debate on science and values within the scientific community focuses on questions such as these: what, if any, are the values inherent in the conduct of science? What are the moral rights and responsibilities of practitioners of science, of their professional associations, and of the scientific community in general? Traditionally, such issues have been perceived in terms of a largely private ethos based on the conscience of the individual investigator. Greatly simplified and idealized, this private, traditional ethos would argue that if a search for truth may be regarded as being an absolute good, then the primary moral obligation of a person who searches for truth should be to facilitate that quest. Indeed, the traditional ethical duties of a scientist have been described with adjectives such as "objective", "disinterested", "universal".

In his candid essay, "Reflections of a Working Scientist", Steven Weinberg provides, in a largely autobiographical defense of the traditional ethos, one of the most eloquent statements of that position available. He has evidently listened to the holistic declarations of the neo-romantic critics who regard scientific rationality as *the* primary dehumanizing influence in contemporary society, but concludes that science would destroy itself if it sought a compromise with any of the "alternate cognitive systems" proposed by these critics. Having made the decision to enter into a compact with nature for the purpose of discovering its underlying laws, says Weinberg, the scientist becomes nature's pupil. That decision is itself a value judgment. Thereafter, he cannot flinch from the realization that the evidence provided by nature indicates that those laws are impersonal and free of human value. Yet, Weinberg remarks, he sees immense compensations. By becoming nature's pupil the scientist can be rewarded with a grand, holistic vision of universal order and harmony.

Not all contemporary scientists regard the compensations Weinberg writes about as sufficient, for during the past five or six years significant numbers of them have raised important questions relating to the scientist-as-truth-seeker ethos. For example, scientific professional societies used to be regarded as essentially passive organizations functioning to facilitate communications between scientists by arranging meetings and publishing journals. Yet many associations of scientists and engineers have recently either adopted or propsed formal codes of professional conduct.

In addition, both official and ad-hoc science-and-society groups have emerged within these professional associations. Many of them are pressing their parent societies to abandon their neutral roles as communicators, and institutionalize what they regard as meaningful social roles for the scientific and engineering communities. A related program would also provide mechanisms within the societies for recognizing and protecting the rights of those members who elect to engage in politically sensitive scientific advising or "whistle-blowing" activities outside their traditional truth-seeking roles. The Report of the Committee on Scientific Freedom and Responsibility of the American Association for the Advancement of Science, for example, discusses many of the ethical problems faced by contemporary scientists and engineers in considerable detail, and concludes by advocating that the AAAS should no longer take a neutral stance with regard to these issues.[1]

Most of the scientists who have become involved in non-traditional pursuits regard their involvement as extra-curricular, and would probably subscribe to those parts of the scientific ethos which place the highest premium upon the intrinsic value of a search for truth for its own sake. Yet it is not at all obvious that the traditional private ethos is completely consistent with the new position.

Edward Shils' essay, "Faith, Utility and the Legitimacy of Science", seems to suggest that it might not be. He agrees with Weinberg that a value judgment is implied in the decision to give the search for truth a high priority. More than that, the support for science as an intrinsically valuable human activity must ultimately be regarded as an act of faith. The conviction that other dedicated individuals share that faith provides an important element of cohesion for the community of scientists. If the public advocacy thrust seriously alters the priorities of outstanding scientists, the enterprise itself, he believes, might be seriously distorted. Indeed, public perceptions of the legitimacy of science are largely based not only on the presumed utility of scientific knowledge, but also on the degree to which the practitioners of science exhibit faith in its intrinsic merit.

Should Research Be Subject to Regulation?

Shils' arguments also imply that attempts to regulate scientific investigation are inimical to the most fundamental values intrinsic to science. Yet others have asked whether research in particular areas ought not to be expressly forbidden.

In this regard, the meeting of geneticists at Asilomar during February 1975 which resulted in a set of voluntary guidelines for research on recombinant DNA may well be a milestone, and as such is worthy of critical scholarly examination.[2] The fact that such a meeting even took place is indicative of a growing realization that unless the scientific community is prepared to regulate itself, additional external constraints will very probably be imposed upon it. Guidelines regulating the conduct of experiments with human subjects were promulgated by the U.S. Public Health Service as early as 1966, and continue to be enforced and expanded by the U.S. National Institutes of Health.[3] More recently, the social and behavioral sciences have come under public scrutiny. There is evidence of growing resistance to sociological surveys and educational experiments, particularly by various minority and ethnic communities.[4] These and other related issues are currently being considered by the National Commission on the Protection of Human Subjects, which submitted its recommendations on fetal research to the Secretary of the U.S. Department of Health,

Education and Welfare during May 1975, and has scheduled the filing of a complete, far-ranging report in 1976.[5]

If the desirability or at least the practical necessity of placing limitations on scientific activity is accepted, are consistent means available for deciding how to do so? For example, could the social sciences provide broad guidelines both for themselves and for the natural sciences? Marc J. Roberts' essay "On the Nature and Condition of Social Science" argues that on the contrary his professional group has accomplished less than it might have, primarily because it has tried to imitate certain characteristics it perceived in the natural sciences. Foremost among these characteristics has been the presumption that science must be value-neutral. Yet the assumption of a set of values must preceed the gathering of data, says Roberts, otherwise no means would be available to decide which types of data should be gathered and which ignored. Rather than attempting to suppress their ethical differences, natural and social scientists should recognize explicitly the important role they play in leading to divergent scientific positions. It follows, in Roberts' view, that the value relativity of social science can and must be accepted, and its implications examined, if social science is to make an effective contribution to contemporary society. If this is so, then the scientific community can legitimately limit or even proscribe certain areas of research. Indeed, its failure to do so might in some instances be interpreted as unethical.

As the search for guidelines and institutional mechanisms to regulate allegedly undesirable activity is being sought, there has also been a growing realization that new ethical guidelines and institutions may have to be developed if scientists and engineers are to involve themselves in the largely desirable activity of seeking solutions to critical contemporary problems. These problems are familiar ones and include the search for environmental and consumer protection ethics; the ethics of population growth and population control; the effects of science and technology on development; and the role of scientists and technologists in "designing the future".

David Rose's essay, "New Laboratories for Old", provides an illuminating case study on the failure of an attempt to convert a laboratory devoted primarily to the development of nuclear reactors into one that would be capable of carrying out research on a wide variety of problems related to energy and the environment. He traces that failure in part to the fact that the scientific and technical communities have not yet learned how to build institutions that accurately reflect their value systems.

It has also been argued that although scientists and engineers often seem to regard themselves as having similar social responsibilities, the conventional wisdom which strongly couples science with technology is not only seriously flawed, but, from a societal viewpoint, can be destructive. According to this point of view, scientists are engaged in activities that are primarily cognitive, whereas the activities of engineers and physicians are more properly described as professional. Thus, there is no particular reason why scientists and engineers should try to adopt similar ethical postures towards society, or that scientists should attempt to act within traditional ethical guidelines when they engage in non-traditional activities.[6]

Science and Public Policy

Much of the societally related activity that is being promoted by groups within the professional societies has crystallized in the emergence of the public interest science field. (In Great Britain, public interest science is frequently referred to as Critical Science, to emphasize the role of the scientist as a critic.[8])

Advocates of public interest science are concerned with using scientific knowledge and methodology to protect the public as a whole from what they regard as possible harms and dangers of emerging technologies. They are also seeking to institute scientific adversary mechanisms at the state and local as well as at the national level, and to make critical advice readily available to private advocacy organizations.

The forms that such adversary mechanisms should assume is by no means clear. Indeed, although it has been taken for granted for some time that more appropriate strategies might well be devised for getting pertinent scientific information into the Executive Branch of the Federal Government, the recognition of the desirability for a more visible legislative approach to science policy making has come only recently. Emilio Q. Daddario's "Science Policy: Relationships are the Key" argues that part of the genius of a legislature lies in its institutionalization of adversary procedures. As members of the Congress have become more sophisticated about science and technology, several have come to recognize that there can be honest differences of interpretation regarding science policy between groups of equally competent scientists. It seems likely to Daddario that if Congress were to establish the means for evaluating responsible scientific advice from many diverse quarters, then a good beginning would be made in at least one branch of the government toward institutionalizing scientific adversary procedures.

In view of the fact that the bulk of the financial support for scientific research in the United States now comes from the Federal government, it is surprising that not more critical attention has been given to a range of issues concerning the extent to which academic scientists can or indeed *do* continue to behave according to guidelines derived from the traditional ethos even when they interact strongly with various agencies of the government.[9] The massive influx of Federal funds has made possible the rise of "big science", the practice of which departs significantly in many respects from the idealized model which involves a single investigator and a graduate student working in a small university laboratory. Novel, value-related problems created by "big science" include the changing relationships between scientific colleagues, and between teacher and graduate student, as well as the ethics of "empire building" with public funds in a field in which individual accomplishment rather than the control of resources is still regarded as the primary criterion for professional recognition[10]

Don K. Price's "Money and Influence: The Links of Science to Public Policy" examines the broad spectrum of relationships that have grown up between the scientific community and the Federal government. Price reminds us that criteria other than pure scientific merit have come to be accepted as a basis for the disbursement of Federal funds in support of even basic research.

He also points out that the scientific community has been notably reticent about devising ethical or professional restraints to help political authorities know when scientific advice is valid, and when the limitations of science make it milseading. Most members of the President's Scientific Advisory Committee (PSAC), which was abolished in January 1973,[11] made the assumption that they could retain their status in private scientific institutions while serving as presidential advisors. That is, they sought to combine their roles as neutral seekers for truth with novel, highly political roles.

In his "The Precarious Life of Science in the White House", David Beckler takes explicit note of the tension between the values implicit in these separate roles. He also suggests that the unwillingness of the PSAC members to recognize distinctions between the values intrinsic to science and to the political decision making process, as exemplified primarily in the debates over the ABM and the SST, may have been largely responsible for the downgrading and ultimate elimination of the Presidential science advisory system during the Johnson and Nixon administrations.

A somewhat different set of value-related questions comes into focus when international rather than domestic science policy is emphasized. Many of these questions are related to the effects of the transfer of high technology to less developed countries. For example, what special strategies are available for balancing environmental protection with developmental needs? Is it necessary or desirable to transfer the epistemological content of Western science along with technology? What effects, if any, is the former likely to have on traditional, non-Western value systems? Significantly, perhaps, most of the concern with these questions seems to be emanating from the developed countries, the intellectuals in the developing countries remaining on the whole in a state of untroubled euphoria regarding the benefits of science and technology.

However, the serious and well-publicized food shortages that were prevalent in large areas of Asia and Africa during 1974 and early 1975 raise serious questions regarding the appropriateness of the technology that has been transferred to the less developed countries. In their eloquent and prophetic essay, "Agriculture, The Island Empire" André and Jean Mayer chronicle the divorce of agriculture from the mainstream of American science and of the intellectual tradition during the 19th century, with the result that few scientists would now regard agriculture as *the* model science, or indeed as a science at all. Yet in view of the authors, the success or failure of science as a whole will, by the end of the present century, be judged by the success or failure of agriculture. Partly as a result of the separation of agriculture from the academies, U.S. scientists, as well as many of their elite Western-trained counterparts in the less developed countries, seem ill-equipped to deal with the growing world food crisis.

Changes in the Perceived Values of Science

The recent upsurge of interest in issues of science and values has led to the tendency on the part of several commentators to write about such issues as if they were unique to contemporary society; others sometimes appear to make the implicit assumption that the values intrinsic to science are immutable. However, historians of science generally recognize that the now-familiar status hierarchy that reserves the greatest rewards for the research professor and ranks pure research above applied research and engineering originated in Germany during the mid-19th century. Yet, two generations later, as Russell McCormmach shows in his intriguing essay, "On Academic Scientists in Wilhelmian Germany" many scientists felt quite uncertain of their roles in their rapidly changing society. German scientists saw themselves in part as bearers of a universal culture that transcended national boundaries. Yet at the same time they were proud of the parts they conceived for themselves as creators of a unique national science. During much of the era they attempted to steer an uncertain middle course between too close an identification with a quickening scientific technology on the one hand, and an avant-garde culture imbued with artistic, mystical and nature-worship values on the other. While the dilemmas confronting those scientists were different from those facing Western scientists in the late 20th century, the debates over the realtionships between values intrinsic to science and to other societal institutions have striking parallels and may have been equally agonizing to the protagonists.

The intrinsic values perceived by American scientists also shifted considerably during the 19th century. As André and Jean Mayer note, agriculture was widely regarded as the most important American science during the early years of the century. Yet by the end of the century many had ceased regarding it as a science at all. Likewise, as Sally Kohlstedt's essay "The Nineteenth-Century Amateur Tradition: The Case of the Boston Society of Natural History" reminds us, much of the science that was done early in the century was carried out by amateurs. Indeed, the activities of non-academic groups in American cultural centers such as Boston and Philadelphia were marked by remarkable vigor and diversity. Presumably most of these dedicated amateurs would have subscribed to the prevailing scientific ethos. Yet the professionalization and institutionalization of science that took place during the century made it increasingly difficult for them to continue to contribute significantly to science.

Kohlstedt's chronicle of the increasing frustrations of the amateur scientists suggests that in the process of restricting the ranks of scientific experts to those who qualified as professional guild members, the American scientific community made a conscious value choice which has made its ability to relate to the larger society increasingly difficult. That the role of scientific expert in the U.S. has not been completely accepted by large segments of the public is made abundantly clear in Dorothy Nelkin's "Science or Scripture: the Politics of 'Equal Time'." To most modern scientists it is incomprehensible that they still have to convince some school boards that creationist ideas have no place in public school biology courses. But Nelkin argues persuasively that the current creationist-evolutionist controversy should not be regarded primarily as a battle between enlightenment and obscurantism. Instead she shows that those groups who have been pressing to give creationism equal time with evolution in the schools are, in part, seeking to re-interpret the openness of science in terms of a democratic egalitarian ideal derived from the forum of politics. From this perspective the efforts of scientists and science-educators to devise and make widespread use of standard curricula in the schools with the aid of Federal funds can be regarded as an attempt by a distant authority to uproot traditional and deeply cherished values.

John A. Moore's companion essay, "Creationism in California", suggests that scientists and educators may have to learn how to protect the integrity of their message by building defenses on a sector of the front with which they have been rather unfamiliar. As a biologist and educator Moore would like to believe that the evidence in favor of evolution is so persuasive that any rational person must of necessity agree with Huxley that the acceptance of "Darwinism" as a working hypothesis is "the only rational course for those who have no other object than the attainment of truth." However it appears that many educated persons, including high-school biology teachers, are prepared to accept or reject evolution not because they comprehend the scientific processes that provide such overwhelming evidence in its favor, but according to the degree of their faith in scientists. If this is so, says Moore, the scientific community has been remiss in its efforts to convey to the public a genuine under standing of science.

Epistemology and Values

The essays of Kohlstedt, Nelkin and Moore touch on what may be the central issues in the relationship between science and its various publics: who should determine the development of scientific knowledge and its applications? For whom should science be done? Who should be able to tell scientists what methodologies to employ?

Gerald Holton's essay "On Being Caught Between Dionysians and Apollonians" suggests that in order to approach these questions it is desirable to examine in detail the nature of the serious and frequently acrimonious epistemological debates that are currently taking place in scholarly circles. Several historians of science and science educators have been concerned for some time with questions relating to the origins of scientific creativity. Certainly a better understanding of these processes might be central to the intrinsic value system of science. In addition, such an understanding could conceivably be translated into strategies for facilitating more fruitful interactions between science and its various publics. Yet, Holton notes, today's investigators are caught between the anvil of the neo-Dionysians—who would deny the value of rationality as embodied in classical scientific methodology and who advocate a totally passive, experiential approach to nature—and the hammer of the neo-Apollonians, those philosophers of science who assert that the admission that there can be any non-rational elements in the scientific process seriously undermines science itself. Indeed, a close reading of the neo-Apollonian literature indicates that more is at stake than academic epistomology, for the advocates of this viewpoint seem to have cast themselves in the roles of the defenders of Western civilization against intruding barbarians. Remarkably enough, all three groups —Dionysians, Apollonians, and the majority of scientists caught in the middle— seem to agree on the questionable proposition that the process of creativity itself cannot be a proper subject for research.

Steven Weinberg's elegant exposition of classical scientific methodology in the present volume implicitly denies the neo-Apollonian claim that scientific objectivity and rationality must necessarily be equated with a completely non-emotional approach to nature. In contrast, Theodore Roszak's "The Monster and the Titan: Science, Knowledge and Gnosis" is a moving and articulate defense of a neo-Dionysian viewpoint. Although Roszak does not deny the value of science as one means for attaining knowledge, he argues that such knowledge is reductive rather than augmentative and therefore can never be regarded as complete. The single fire of Prometheus, he says, cannot be equaled by the million individual candles now being lighted continuously by modern scientists. Thus the assumption that science must be *the* primary cognitive system ultimately brings humanity face to face with a fearsome monster called meaninglessness.

Roszak's essay touches upon several of the mystical and visionary roots of modern science, and calls upon contemporary scientists to reassert their Promethean roles in seeking for wonder and ultimate belief in the universe. Indeed, coming from the opposite direction, more than one scientist has urged a self-awareness of the "charismatic" element in science. They and Roszak would at least agree that it is precisely the perception by the public that scientists hold the key to the wonder of the universe that largely legitimizes scientific activity.

The Public Understanding of Science

Roszak in effect is summoning scientists to accommodate their values to a neo-romantic view of nature, and in so doing to win favor with the followers of the counter-culture. During the early 1970's it seemed evident to many serious commentators that a large number of college undergraduates had become committed, lifetime members of that culture, and that their hostility constituted the most serious of all the many perceived threats to the continued health of science.

How serious was the alleged erosion of faith in science during the late 1960's and early 1970's? In their essay on "The Public Appreciation of Science in Contemporary America" Amitai Etzioni and Clyde Nunn review the available survey data (to early 1974), and conclude that while there may well have been a shift in public attitudes towards science, the data do not support a crisis theory. For example, between 1966 and 1971 the percentage of the population in a Harris Poll sample that professed to have "great confidence" in science decreased from 52% to 32%, then rose to 37% in 1973. During these same years other American institutions such as medicine, education, and religion also experienced downward shifts in public confidence levels of similar magnitudes, whereas institutions such as the Federal government, major U. S. corporations, and the media suffered larger erosions in confidence. Significantly, the relative position of "science" among the institutions eliciting relatively great confidence rose during these years. In 1966 and 1971 it ranked fifth, behind medicine, finance, the military, and education. By 1973 science and education tied for second place behind medicine.

A secondary analysis of these data belies the contention that science is distrusted primarily by rebellious youth, for the age group between 18 and 29 had substantially more confidence in science than the national average. In addition, the scientific community ranked higher in confidence among this group than it did among any other. In fact, the major distrust for science as for other institutions is discernible among the so-called "forgotten American", lower-status, less-educated groups, and those who live in less affluent parts of the country.

Etzioni and Nunn's analyses lead one to a conclusion that is so obvious that it is frequently ignored: the perceptions of science by any particular group are conditioned by the values it perceives, not only in science, but also in a host of other societal institutions with which science presumably is coupled. In addition, the attitudes of a particular group towards these institutions depends on how strongly and in what ways they intrude on the lives of its members. Analyses of data on public attitudes toward technology reach a similar conclusion.[12] Indeed, it seems likely that each individual possesses a complex and in some respects a logically inconsistent set of images about science. Etzioni and Nunn and their colleagues are currently conducting an indepth survey in the U.S. on these questions, and expect to have publishable results by the end of 1976.

George Basalla's essay, "Pop Science", examines and contrasts popular images of science, technology and medicine as reflected by vehicles of pop culture such as best selling novels and commercial television serials. Even though there exists the underlying assumption that science and technology are useful and even desirable attributes of modern and, indeed, of futuristic society, scientists are most often pictured in pop culture as distant, unemotional, and frequently absent-minded or even mad. If permitted to pursue their investigations without being restrained by others who possess more understanding of "real" human values, these scientists may well bring disaster upon themselves and their fellows. In contrast, the physician is usually wise and serious, yet manages to strike a perfect balance between scientific competence and human understanding. Indeed, those few doctors in pop culture who do turn out badly most often do so because they permit the desire for scientific glory to outweigh their human instincts.

It seems entirely possible, then, that the public conception of the "good" scientist may well be conditioned by its perceived images of medicine. It is worth asking whether the growing public interest in bioethics can be generalized to encompass a broader range of issues. For if many contemporary concerns relating to value questions resulting from the effects of science and technology on society dwell on largely negative issues, there is also a considerable body of contemporary literature which argues that scientific and technological change have illuminated traditional ethical dilemmas, and/or can make significant new contributions to a broader ethic. It is tempting to subsume the entire range of biomedical issues into this category: population control, genetic counseling and engineering, child care, behavior control, and death. For example, since the advance of technology has not only made the definition of death ambiguous in certain cases, but has also multiplied the options for delaying death, it has multiplied the range of ethical choices made available to address questions relating to the continuation—or termination—of life. In addition, it has brought to the fore age-old questions relating to the sanctity and quality of life.

Even though the recent upsurge of concern in the ethical and human value implications of science and technology may well result from the development of new technologies, many specific ethical problems of current interest are related to a search for means for reasserting traditional values. In particular, it would appear that a good deal of the concern with ethical problems expressed by members of the medical profession, and made explicit through courses in ethics at medical schools and hospitals, as well as through inter-intsitutional medical school-divinity school programs, can best be understood in this context.

It seems clear from the essays in the present volume that the public understanding and appreciation of science is largely conditioned by the intersections between values intrinsic to science, and values that characterize other societal institutions. In his essay "Science and the Mass Media", David Perlman argues from his experience as a science writer on a large-circulation daily newspaper that there is a high level of public interest in genuine science. However, he takes issue with those members of the scientific community who regard the science writer as a passive conduit for information from the laboratory to the public. On the contrary, Perlman believes that the implicit ethics of the journalistic profession require that the science writer should be a responsible critic, subjecting scientific developments to the same close scrutiny that has come to be expected in the journalistic coverage of political and economic developments. Better communication between scientists and journalists requires that the value differences between practitioners of the two crafts be understood and appreciated. Since openness has long been one of the most important idealized attributes of science, it can be argued that scientists cannot be completely true to their own intrinsic ethics if they find it unacceptable to allow their work to be examined by conscientious members of the press.

Concluding Notes

Perlman concludes by asserting that all scientific inquiry must ultimately serve the public, for it is the whole of society that endows science with its charter. Few would argue with that assertion. Yet, as the essays in the present volume indicate, it is frequently difficult to reconcile the ideal of "science for humanity" either with the ethical system that has traditionally guided the practice of scientific research or with the values underlying the activities of other important sectors of contemporary society.

Given the rapidly increasing interest in so many diverse topics relating to the ethical and human value implications of science and technology, the existence of several *lacunæ* in the present volume and other such collections should not be surprising. Nor is it surprising that these white spots on the map are most readily discernible in the newer areas of concern, and on the borderlines separating various categories of issues.

For example, it would be desirable to encourage critical investigations on the linkages between the traditional scientific ethos and the new public-interest thrust. Recent advances in the sociology of knowledge, as well as case studies which have been completed or are in progress on the structure of the scientific professions, indicate that such lines of analysis can yield important insights into these problems.[13] The development of a few case studies of the processes of decision making in the public sector is also encouraging, and, one hopes, signifies a growing scholarly interest in this field.[14]

Obviously, there is also a need to subject the whole range of relatively new, pressing social problems that are now dealt with in largely pragmatic terms to historical, philosophical, and sociological analysis. The fact that concern with these problems has, in fact, led to a recognition of their ethical content suggests that such analyses are necessary and may be feasible. Several of these issues appear to be generically related to problems which are usually regarded as being peculiar to the biomedical fields. If this is the case, an analysis which proceeds from the assumption that technological change sharpens old ethical dilemmas more often than it raises new ones may perhaps be fruitful.

A good deal of contemporary activity relating to the ethical and human values implications of science and technology is now being pursued by scientists and engineers who bring a deep human concern to bear on these problems, but who have a layman's view of ethics and may be naive concerning the methodologies of the social sciences and the humanities. Conversely, philosophers and humanists frequently seize with fervor upon a particular set of issues, but are all too often led astray because of their failure to understand the sophisticated conceptual basis of, and the quantitative constraints imposed by, science and technology. Be that as it may, it now appears imperative to encourage strong, lasting transdisciplinary research in the ethical and human value implications areas, and to inject a concern for them at all stages of the educational process. The sixteen critical essays that comprise the present volume can, one may hope, make significant contributions to the further development of fruitful scholarly activity along these lines.

REFERENCES

1. *Report of the AAAS Committee on Scientific Freedom and Responsibility*, prepared for the committee by John T. Edsall (Washington, D.C.: American Association for the Advancement of Science 1975), *Science*, 188 (May 1975), pp. 687-93.

2. Nicholas Wade, "Genetics: Conference Sets Strict Controls to Replace Moratorium", *Science*, 187 (March 1975), pp. 931-35. Christine Russell, "Biologists Draft Genetic Research Guidelines", *Bioscience*, 25 (April 1975), pp. 237-40; 277-78.

3. National Institutes of Health, *The Institutional Guide to DHEW Policy on Protection of Human Subjects*, (Department of Health, Education and Welfare, Publication No. NIH/72 (December 1971), pp. 72-102. National Institutes of Health, "Protection of Human Subjects: Policies and Procedures," Federal Register, 38, No. 221 (November 1973), Part II, pp. 31738-749. See also Jay Katz (ed.), *Experimentation With Human Subjects* (New York: Russell Sage Foundation, 1972).

4. See, e.g., Henry K. Beecher, "Human Studies", *Science*, 164 (June 1969), pp. 1256-58; Irwin Katz and Patricia Gurin (eds.), *Race and the Social Sciences* (New York: Basic Books, 1969); Margaret Mead, "Research with Human Beings: A Model Derived from Anthropological Field Practice," *Dædalus* (Spring 1969), pp. 361-86; Edward Shils, "Muting the Social Sciences at Berkeley," *Minerva*, 11 (1973), pp. 291-95.

5. Peter Steinfels *et al.*, "The National Commission and Fetal Research", *The Hastings Center Report*, 5, No. 3 (June 1975), pp. 11-46.

6. Allen B. Rosenstein, "A National Professions Foundation," *Engineering Education*, 61, No. 2 (November 1970), pp. 100-104; "The Applied Humanities and the Institutionalization of Change in Higher Education", *Engineering Education*, 63, No. 7 (April 1973), pp. 523-27; and No. 8 (May 1973), pp. 599-602.

7. Martin Perl *et al.*, "Public Interest Science", *Physics Today*, 27, No. 6 (June 1974), pp. 23-45. The entire issue of the journal is devoted to this topic.

8. Jerome R. Ravetz, "A New Science: Critical Science", *Intellectual Digest*, 2, No. 9 (April 1972), p. 56.

9. Joel Primack and Frank von Hippel, *Advice and Dissent: Scientists in the Political Arena* (New York: Basic Books, 1974).

10. Jerome R. Ravetz, *Scientific Knowledge and Its Social Problems* (Oxford: Clarendon Press, 1971).

11. President Ford announced his decision to recreate an Executive Office position of science and technology advisor to the President on May 22, 1975. See, e.g., Robert Gillette, "White House Science Adviser: You *Can* Go Home Again", *Science*, 188 (June 1975), pp. 995-97.

12. Todd LaPorte and Daniel Metlay, "Technology Observed: Attitudes of a Wary Public". *Science*, 188 (April 1975), pp. 121-27.

13. See, e.g., André F. Cournand and Harriet Zuckerman, "The Code of Science: Analysis and Some Reflection on Its Future", *Studium Generale*, 23 (1970), pp. 941-62; Joseph Ben David, *The Scientist's Role in Society: A Comparative Study* (Prentice Hall, 1971).

14. See, e.g., Primack and von Hippel, *op. cit.* (ref. 9); Dorothy Nelkin, *Jetport: The Boston Airport Controversy* (New Brunswick, NJ: Transaction Books, 1974); "The Political Impact of Technical Expertise", *Social Studies of Science*, 5, No. 1 (January 1975), pp. 35-54.

EDWARD   SHILS

Faith, Utility, and the Legitimacy of Science

ECCLESIASTICAL antagonism to science has now pretty well expired. Religious thinkers have made many concessions to science as a body of knowledge, and have attempted to find a practicable division of "spheres of interest." They seek areas of harmony with science and possible bases for alliance, and rebuff only scientistic excesses. Popular religions, healing cults, spiritistic practices, and astrological beliefs which are regarded as antithetical to science have had a very long history, and they retain a considerable following at present. Their devotees, however, do not have much contact with science. They do not form a major part of the public which interests itself in one way or another in science, and the low opinion in which they are held by scientists is not reciprocated. They are neutral or even admiring, claiming that what they believe has behind it the authority of science. Even though on particular points of medical technology, such as blood transfusion, inoculation, or surgery, they might resist strongly, on the whole they do not conduct a campaign against science.

Such active antagonism toward science as exists in the present century is, on the face of it, secular, although perhaps it is deeper than that. It certainly does not employ arguments drawn explicitly from Christian doctrine. Secular antagonism to science is a relatively new phenomenon in the world. It was at its height in the nineteenth century. It was associated then with literary romanticism, political conservatism, and philosophical idealism. Literary romantics thought science tore the veil of beauty from nature and replaced the warm vision which came from the heart by the cold calculations of reason. They hated its ratiocinations and its dissolution of primary qualities. Political conservatives thought that science would dissolve the traditional ties of master and man and the religious beliefs which legitimated those ties. They feared it would introduce a rationalistic, experimental attitude toward social and political institutions and would thereby fail to perceive what was essential in them. Philosophical idealists denied claims which asserted that rational-empirical procedures were the sole paths to reliable knowledge; they also emphasized their ethical insufficiency and their powerlessness to understand the mind and its works. On the other side, political radicalism always regarded science as its great ally against the forces of clerical and worldly authority. Science and reason were at one in their implacable opposition to the traditional, the arbitrary, and the supernatural. Progressives and liberals regarded science as their ally in the campaign to erode the superstitions of traditional beliefs and hierarchical institutions.

The dispute over the value of science subsided, in educated circles, by the beginning of the twentieth century. Philosophical idealism now satisfied itself with restricting the area in which the methods of the natural sciences were valid. The romantic tradition in literature resigned itself to a world of which it passionately disapproved and of which science was an integral part; science was no worse than industrialism and the entire apparatus of impersonal, quantitative, rationalistic ways of dealing with man and the world. It was part of a world gone wrong, about which nothing could be done. Political conservatism withdrew, for the most part, from its earlier intellectual positions, although, where it remained intellectually alive, it censured the amorality of the scientific attitude, which it thought injurious to moral discipline. Radicalism and liberalism continued to place themselves firmly on the side of science. Liberals regarded one of their tasks to be the defense of the freedom of scientific research and of the scientific outlook on the world against reactionaries and obscurantists; radicals regarded science as a liberating and revolutionary force which would come into its own only when society was reorganized through revolution. For those who were more or less content with the existing order of things, science, through its accomplishments in medical and agricultural technology, seemed to improve the condition of man. It was the appropriate instrument of a hedonistic view of life. These were the views of science, put forth with varying shades of enthusiasm, which prevailed for much of the first two-thirds of the present century throughout much of the population of Western societies.

In the perturbation of spirits which came to the surface in the 1960's, marked changes took place. Although science is more deeply embedded in our societies now than it has ever been, it has not been under such vehement attack with such publicity for a long time. Although individual scientists were troubled and humiliated by authority in the past, the wider public was not witness to their tribulations; criticism was not widespread. Now, although there is little or no persecution by authorities in Western countries, criticism is public and widespread. Scientists are still relatively richly supported, and generally, except in a few fields, well employed. Nonetheless, in intellectual circles few speak unapologetically on behalf of science.

I do not think that, in the long term, scientific research and the expansion of scientific knowledge are in serious danger from its current conservative and radical critics, since both are utilitarian and view science as a milch-cow for their various practical ends. Except for their slighting attitude toward pure or basic science, those who accept the utilitarian standpoint also accept an empiricist conception of nature and its knowability. The numinous penumbra of scientific activity is a very thin fringe as far as they are concerned, but they do not go out of their way to deny or discredit it.

They usually act deferentially toward established scientists, but for them scientific research is valuable insofar as it is an instrument in the service of substantive ends other than an end itself, and its practical reputation is generally intact in these circles and the United States. Industrialists, military men, and legislators are not likely to be much moved by arguments about the failures of scientists to control the use of their discoveries or by epistemological arguments against the objectivity of scientific knowledge. Intellectually deeper and morally graver criticisms of science are not likely to discourage these practical patrons of science.

Yet the matter is not as simple as it seems. Support for science is based not only on purely practical grounds, but also on general beliefs about its future efficacy and about the value of the way of life sustained by scientific technology and the scientific view of the world. And these beliefs are affected by the tides of mood and opinion which rise and fall in the ocean of the larger society. Moods change imperceptibly, and beliefs follow them. The movement is not random; the changes can occur only within the potentialities of previously held beliefs. The recent salience of egalitarianism in social policy in the United States is an instance of such a change—one which can, in fact, injure the condition of science. It can do so by fostering antipathy against the major universities which produce so much outstanding scientific work and so many outstanding scientists, but which are, for that very reason, disparaged as "elitist" in the contemporary egalitarian idiom. Adherence to the acknowledged tradition of the modern type of university will become difficult if "elitist" universities are forced to diminish the scale of their fundamental research and of their advanced training of doctoral and post-doctoral students—and this could occur even while science is being praised as the basis of economic and social progress!

The growth of science, insofar as it is a function of the financial support for the conduct of research and teaching, depends, to a large degree, on the belief of a relatively small number of persons in any society that science is valuable—both instrumentally and intrinsically. The beliefs of these rather small elites of businessmen, legislators, party politicians, and civil servants are not, however, self-contained, and they are not inevitably self-perpetuating. In any case, they constitute only one part of a more complex whole.

The continuity and growth of science is a function both of the internal traditions and institutions of science, and of external support in the form of financial resources and affirmative beliefs. Financial support is given in a matrix of opinion. Whether the decisions to support science are made by private individuals, princes, legislators, or civil servants, they rest on beliefs—the beliefs, in the first instance, of patrons of science, and, at a further remove, of the public to whom patrons are responsible and on whom they depend for the medium in which they act and think. Without these beliefs, there would be no support.

Outside the scientific community itself, readiness to support science rests in part on the belief that science contributes to the material well-being of society. If the belief in this link between science and material well-being were broken within the influential circles of the lay public of science, support would diminish. If those circles come to think that science does harm to the society in which it is supported—or to mankind in general, although for the present this is a less prominent consideration—then science, or at least the present arrangements for conducting science, will be less amply supported.

This condition could come about if "evidence" becomes available which shows that fundamental science does not contribute to technological progress in medicine, agriculture, etc., or that it does not contribute sufficiently to pay for "investment" in it. At present, the evidence that fundamental scientific research contributes to material well-being is very uneven and not by any means rigorously conclusive. The conclusion is accepted because there is a mood to accept it and because, according to prevailing standards of proof in the lay circles which count, the evidence is good enough. But it is largely a matter of faith—a faith resting on fragments of substantial evidence of the practical productivity of scientific research, and on other fragments susceptible to multiple and divergent interpretations and derived from a profound and diffuse "will to believe" in the efficacy of science.

The "will to believe" in science is deep in our cultural inheritance. It derives not only from a utilitarian ethic, but also from beliefs in the intrinsic value of truth, in the merit of cognitive activity and in the necessarily beneficial consequences of cognitive activity. These elements, however, can exist in different combinations, and the combinations are susceptible to change. If the belief in the intrinsic value of cognitive activity fades, the burden for continued munificence toward science would rest preponderantly on its utility. It is possible that, except for spectacular cases, this utility could not be readily demonstrated. Thus, to some indeterminate degree, the support of science is an act of faith.

II

Faiths have changed, and the faith that truth is valuable and that cognitive activity of the scientific sort is an intrinsic good could also change. This faith is widespread at the center of our society, and it has been widely diffused in the periphery. It gave birth to outstanding private and state universities considerably before the time when scientific research and higher learning generally were thought of as "investments" in economic growth. This faith, generally and widely shared in the lay public, is a necessary condition of the financial support of science anywhere—above all in liberal-democratic societies where opinions and votes count.

Scientists are sustained by this faith. Without it, an important part of their motivation would disappear. The faith of the laity in the intrinsic value of scientific knowledge is sustained by the adequate and visible embodiment of that faith in the practice of scientists. If scientists in sufficient numbers cease to believe that the truths attained through disciplined research and rational analysis are intrinsically good, and show it by their actions, then the faith of the laity will also falter. If scientists are seen as hypocritical in their assertion that scientific activity is good, and antagonistic to the application of scientific knowledge to the ends desired by nonscientists, then faith in science and in scientists would become weaker. Much belief in science, among the various circles of the laity, depends on their faith in the integrity and disinterestedness of scientists.

There is no doubt that faith in science exists, not just among the powerful in state and economy, but also among quite ordinary citizens. Human beings have a need for faith in certain serious truths and for the embodiment of these truths in certain human figures. There is less faith in priests nowadays than there used to be, and less faith in earthly authority as well. Scientists are the recipients of some of the faith no longer attached to objects which are religious in the traditional sense. That faith is not always salient or tangibly articulated, but it does exist.

The acceptance of science by the various circles of the laity in the modern period has rested partly on the beliefs that scientists had no "interest" other than truth, that they had a method for discovering truth, and that they applied that method without reservations, qualifications, or distortions. They seemed to be superior to churchmen who quarreled, often acrimoniously, with each other, and whose quarrels appeared to be motivated by their own earthly interests.

Scientists have a great advantage over churchmen in that the structure of the scientific community keeps them on the strait and narrow path. Their conclusions are scrutinized by the use of consensually accepted techniques for testing the validity of the evidence, so that the risk of being discredited sustains their integrity. The traditional ethos of the scientific community makes scientists scrupulous about what they present to their peers, but the ethos alone might not be sufficient without the reinforcement of the assessment of other scientists.

The faith in science which rests on the belief that scientists embody the ideal of selfless devotion to truth is reinforced by the belief that what they discover also contributes to practical improvement. Not every scientific assertion need have this practical merit, but scientists benefit by the tendency of the public to generalize their practical effectiveness. Thus, the faith in science among laymen rests on the belief that scientists have a method which can produce practically and intellectually reliable results. The laity sees the detachment of scientists as lying largely in the application of this method which expels bias, passion, and partisanship, and which results in the actual attainment of important objective truths.

The laity's faith is confirmed by the public bearing of scientists. Laymen could come to see scientists not as disinterested but as concerned to promote their own material advantage, to enhance their own status among the laity or their fellow scientists, to indulge themselves in their own pleasures, and to pursue their own partisan political ends by knowingly untruthful statements which they claim to be scientific. The line between being engaged in and passionately committed to a higher form of disinterested activity, and strenuous partisanship in pursuit of partisan ends, under the guise of scientific disinterestedness, is a very thin and faint one. Similarly, a slight change in the moral mood of a person who admires and believes in science and scientists could change him into one who disparages and distrusts science because he no longer believes in the probity of scientists.

The laity know few scientists at first hand. They only hear about them. They occasionally read about them in the press and in books; they see them on television. Some know more than others, but images are formed and they are of consequence. The images formed by some parts of the laity are of greater consequence than those formed by other parts, even in a democracy.

III

Alongside this faith in science as a body of important truths objectively established, there is another faith in science which is expressed in the utilitarian attitude. This faith rests on anterior beliefs that the material side of life is of great importance and that one of the very first obligations of government is to protect and promote the material well-being of the citizenry Hedonism is not a new thing in the world; it has almost always been appreciated by those who could afford it, and even those who could not have always sought their pleasures where they could find them. The rulers of societies have always been known—and often hated—for their desire to enjoy the things of this world; usually they have also alleged that they were concerned to promote such enjoyment among their peoples. In fact, however, they did little. There were neither the resources to enable them to do so, nor the constraints of opinion or voting power to force them to do so. Nowadays, however, the resources and the constraints do exist, and rulers—especially rulers in liberal democratic countries—are under great pressure to look after the material well-being of the electorate. Of course they do so very unevenly and some parts of the electorate are more looked after than others; but almost all are clamorous for material well-being, or have some one who is clamorous on their behalf. The politicians and higher civil servants are more or less attentive to their demands. Their attentiveness to particular demands is derived from a general hedonistic belief that a way of life replete with comfort and secure from pain and early death is one of the highest goods man can seek. Science benefits from this belief because it is thought that science contributes to its fulfillment.

It is not, however, hedonism as such which underlies much of the laity's support of science; it is a progressive populistic hedonism. It is a belief that all human beings—or at least all or most of the members of a given national society—should be gratified in their material wants and that this gratification should be continuously extended. Technological innovation would make no sense, except to engineers, if there were not the prior desire for a more economical use of resources to produce more goods with the same resources. The goal is a continuous increase in the supply of products available for the gratification of needs. The pressure for technological innovation postulates an insatiable demand—insatiable at least as far as the productive capacity of existing economies is concerned. If human wants were satiated, there would be no demand for technological innovation beyond the existing "state of the art." (I omit here such independent pressures as international economic competition and military rivalry and conflict.) If the pattern and intensity of individual wants in the richer societies were to go into attrition, if a more puritanical or ascetic outlook on life replaced the present hedonism, one important base of the readiness to support science at anywhere near its present level as a precondition of technological innovation and increased productivity would be weakened. Such an attrition of wants is a possibility, but it is not very likely in the foreseeable future. In fact, wants, even in the most affluent countries, are far from the point of satiation.

Still, if it were demonstrated "scientifically" that certain parts of fundamental science had, and could have, no bearing whatsoever on technology even if they were of great theoretical importance, and that their cost represented a very poor investment of resources, the incentive to support them would be reduced. It might, of course, be possible to show that every field of fundamental scientific research contributes to some technological innovation in an economically productive way, even if not immediately or directly. This is the contention of those who wish to have fundamental research amply supported. It may well be true, and it may be demonstrated by rigorous research, although this has not yet occurred. If it does turn out to be true, the situation for the support of science will be even better than when the proposition was believed without any rigorous evidence. If it turns out not to be true, or at least not to be demonstrable by rigorous research, then the support of fundamental research, in whatever fields, will be damaged, even if faith in the intrinsic value of the pursuit of truth persists, and if higher education, including higher education in science, continues to be conducted on a large scale. Scientific knowledge would grow in such a situation much less rapidly than it has in recent decades. Privately supported research might still persist on a considerable scale. Research for monumental purposes, such as landing on the moon, might occur from time to time, and so would biomedical research, which is clearly connected with those practical problems which dominate concern. Chemistry, linked as it is with industry and medicine, would continue to receive ample support. On the other hand, certain branches of physics, which have no evident application at present and which are very costly, might be markedly restricted. The scale of the scientific enterprise as a whole in advanced countries would be much smaller. In some fields, scientific research might become more of a handicraft than it has been in recent decades.

Some persons who believe that scientific research is a practical necessity for the growth of the economy, the military, personal security and health, etc. assert that in view of the various defects of the universities, it would be better to withdraw science from them, thereby saving science and leaving the universities to flounder as best they could. Such a separation of research from higher education would deprive fundamental research of still another of its protective alliances. Under the conditions such persons recommend, science would lose the stimulus of graduate students; the stream from which ensuing generations of scientists flow would dry up. The training of prospective scientists could, of course, be confined to independent research institutes, but that would only separate scientific work from undergraduate education by putting them into two different types of institutions in separate locations. Such an arrangement might well be practicable, but it might have the roundabout effect of cutting off the limb on which scientific research has been growing. Even more important, it would weaken the institutional bases of fundamental research. These have been located very preponderantly in universities ever since scientific pursuits became institutionalized and professionalized. The consequent weakness of fundamental research left in the academic setting would, in the course of time, change the public image of the scientist. It would also weaken the ethos of the scientific enterprise, which has been most effectively transmitted from one generation to the next in universities.

IV

The dispositions of scientists themselves also help to determine the security of the position of science in our contemporary societies. Scientific work which makes a positive difference to the state of knowledge requires strong propensities—driving and sensitive curiosity, unceasing persistence, and, of course, great ratiocinative and imaginative powers. It requires fluent mastery over a vast and complex tradition, which is, in differing degrees, in a perpetual process of obsolescence and growth. It requires a subtly poised position between deferential assimilation of the tradition and a daring readiness to cut loose from it. Even given the existence of the propensities, it takes a long period of severe and stimulating discipline to become a scientist. The complex tradition must be assimilated; the propensities must be aroused and kept in a persistent state of strain; the sensibility must be cultivated; the imaginative and ratiocinative powers must be aroused, focused, and kept at a high pitch. This is a very demanding mode of life, and not many human beings have been capable of it through much of human history.

For more than two centuries, there have been many such persons. There were great scientists in the world before science began to be taught in schools and universities. Persons of powerful intellectual impulse could, following leads picked up in a scattered literature, begin to reflect on the problems which attracted them; they were able to find masters to whom to apprentice themselves; and they established, on their own initiative, contacts with scientists scattered over the earth's surface. There were very few opportunities to gain a livelihood through the pursuit of science. Under these conditions, being a scientist was not only entirely voluntary, it had to be done under very difficult circumstances and at a high cost to the scientific aspirant or practitioner. It required self-consecration to lead the life of a scientist.

It has become much easier to become and to be a scientist than it once was, at least in terms of external things. The "professionalization" of science has made it possible for science to attract great numbers of votaries who would never have found their way to it under the conditions which prevailed until the nineteenth century. The historically fortuitous circumstance of the association of higher education with research has, by combining teaching with training, also fostered the recruitment of young persons to make careers in academic and applied science. These and many other factors have made it easier to find a way into science and to carry it on as a career.

Yet there are other respects in which science has become more difficult. There has been an increase in the intensity of competitiveness. There has been an increase in extremely expensive widely used machines which require close scheduling, and this imposes the strain of an external discipline. The need for speedy publication adds to the strain of the scientific career. In addition, the prominence of the greatest scientists and the large rewards they receive make being a more mediocre scientist, especially in an undistinguished institution, somewhat disagreeable.

We have heard in recent years of the "flight from science" in the United States and Great Britain. There has been a certain anxiety about this lest it be a first step in the desertion of science by its potential recruits. It is true that science would come to a halt if there were no first-class minds willing to devote themselves to it. It would become like Protestant theology in the United States, where a great tradition has fallen into disregard and has ceased to develop because not enough first-class minds have been willing to cultivate it. Is it entirely out of the question that this could happen to science?

For such a change to occur, sports, business, administration, sensual gratification, artistic creation, politics, religious contemplation, the cultivation of friendship, or the duties and pleasures of family life would have to become so absorbing to the minds best endowed with the potentiality for outstanding scientific work that the strenuous, strength-consuming, soul-dominating practice of science would seem, by comparison, not worth the unceasing, nerve-racking exertion which its ethos demands. It would be necessary for those minds which reach out toward the comprehension of the ultimate powers which guide the universe and control human life, either to lose their interest in the universe or to be satisfied by mystical and traditional religious experience.

How probable are such changes? Science, the ethos of science, and "scientism" have never shared a monopoly over the human mind in Western countries; they certainly do not do so in the twentieth century, which is the time of their greatest ascendancy. From the depths of superstition to the heights of spirituality, there have been reluctances and resistances to the acceptance of the results, the procedures, and the modes of thought of the natural sciences. Astrological and magical beliefs have never completely disappeared; they certainly have not done so during this period when the ascendancy of the natural sciences has seemed unquestioned. The divinely revealed truths of the great religions yielded to the truths of science only slowly, even among the educated; they have not abdicated completely even now. Throughout the nineteenth and twentieth centuries, there have been very serious and learned efforts to define the limits of the natural sciences as well as to delineate other procedures for dealing with social behavior and intellectual works. In some respects, the work of Rickert, Dilthey, and Weber could be interpreted as delimitations of the domain in which the natural sciences could legitimately prevail. The outlooks of Stefan George and D. H. Lawrence, the spiritual exercises of Gurdjeff and Ouspensky, the spiritual instruction and guidance offered by Krishnamurti, Kahlil Gibran, and Count Keyserling had many followers among the highly educated in the period which preceded the present radical rancor against science.

The existence of all these religious and spiritual movements, at various levels of sophistication, has not prevented the growth of science. But might not these ways of approaching the world expand their following in the same way the scientific approach to the universe has in the course of the past four centuries? It is possible that the sensibilities which regard nature as an object of aesthetic contemplation or try to understand man's nature through his imagined myths might come to be more appreciated and sought after than they are today. During the past three decades, there has certainly been a marked expansion of aesthetic sensibility—often very crude and unsophisticated—and of a mythological outlook. In consequence, the cast of mind required for science might become rarer, and science might cease to attract and recruit young persons who still possess it.

The appreciation of scientific discoveries might also shrink in the public which forms the "catchment area" of the scientific profession and the foundation of its spiritual and material support. Financial support for science might then diminish markedly, and spectacular scientific accomplishments might become less common. The institutional machinery for the making of scientists might also shrink because it would not have enough outstanding recruits. Our unprecedently elaborate institutional machinery for the making of scientists would have to become much more efficient. The undergraduate years would have to be used more effectively for bringing young persons, better trained in secondary schools, into a more concentrated and demanding scientific course. A process of detecting young persons of outstanding promise as scientists earlier and "force-feeding" them to the point where they could begin research earlier might offset shrinkage in the reservoir of young persons inclined to live through science.

It is, however, more probable that, as the size of this reservoir shrank, the machinery for the making of scientists would also deteriorate. The mathematical and scientific component of primary and secondary education would be reduced or lightened. Training at the advanced undergraduate and graduate levels would be adulterated to the point where new entrants into science would have mastered neither the substantive traditions necessary for doing research nor the ethos of science—its tacit rules, its standards for the selection of problems for research, and its moral discipline. Had the "student revolution" in its extreme forms extended and won all it sought, such an outcome might have been expected. Even in defeat, it has won notable victories in furthering the slackness of intellectual requirements in certain parts of a considerable number of universities in the United States and in the Federal Republic of Germany.

Although these developments could reduce the proportion of a generation seeking to become scientists, they are not likely to divert those with the strongest propensities for scientific work. It is even conceivable that by discouraging those of infirm will, it will raise the average and make for a purer scientific vocation. The actual practice of science at the level of fundamental and original discovery can never be an affair engaging great numbers of persons. As long as science is taught in primary and secondary schools and higher educational institutions, it is likely that there will be enough scientists to form a "critical mass" in the major centers of teaching, training, and research, and in the international scientific communities.

V

Why should not an indifference toward science in the larger public be just as bearable by contemporary scientists as it was by their great seventeenth-century predecessors? Would indifference or outright hostility toward science on the part of the laity make any serious difference to contemporary science—apart from the reduction in funds for equipment, salaries, etc.? It seems to me that it might.

The corps of scientists of the present day might become demoralized if they were to see their numbers shrink toward what they were proportionately in the scientifically most creative countries of the sixteenth and seventeenth centuries. The maintenance of scientific morale depends on a fairly constant flow of outstanding young recruits to the subject. It needs them not only for the ideas, skills and spirit they bring but because they give evidence of the legitimacy of the scientific enterprise. A pronounced shrinkage in numbers of new entrants might strike at the self-confidence of science.

The self-confidence of scientists depends on their sense of collective accomplishment as well as on each individual scientist's sense of his own personal accomplishment. The scintillating progress of science on many fronts at once has a great deal to do with the buoyancy of spirit of many scientists in the period from the end of the Second World War until the late 1960's. If this rate of progress slows down, scientists will be strongly affected. Both in the individual's own sphere of work and in that of the scientific community as a whole, the contemplation of old truths is much less to the scientists' taste than the discovery of new ones. A decline in numbers, especially of first-rate scientists, in resources, and in spectacular discoveries might have a dispiriting effect on scientists. It might make them think that science is becoming "exhausted." This would tarnish their image of science and weaken the self-esteem which is necessary not only for the actual conduct of scientific research but equally for the public's esteem for science.

The great scientists of the seventeenth century believed in science as a profoundly religious person believes in the divine order of the universe. The rationale for the religious understanding of the world was still acceptable and scientific understanding had, in the eyes of the great scientists, the same standing. But to most scientists today, even if they are believers, their acceptance of the religious understanding of the world lacks conviction and their scientific understanding gains no support from any association with religious belief. On the whole, scientists nowadays are lacking in the ability to feel that their scientific work has transcendent significance. This does not mean that they are not unswervably committed to science, but rather that they cannot justify it in public terms acceptable to themselves and to a nonscientific audience. Their deepest convictions are not utilitarian, but their idiom is. Utilitarianism provides, in various forms, the idiom of intellectual circles in our time; to speak of transcendent or intrinsic values which are not just matters of idiosyncratic preference is considered old-fashioned. As a result, scientists, lacking an appropriate way of describing the value of what they are doing, often give a poorer impression of themselves than they need to give. And because scientists cannot express their belief in the transcendent value of science, their belief in its transcendent value is made more uncertain—an uncertainty which has further negative effects on the public's esteem for science.

VI

A separate factor which has come into existence in recent years merits mention. In the nineteenth century and in the early twentieth century, the assessment of science and scientists was done by scientists, philosophers, theologians, industrialists, politicians, and civil servants. The specialist press which dealt with science was the scientific press, which published scientific papers and, on the margin, factual information about the scientific community. Speeches by leading scientists on general issues were also included. Journalists who dealt with science were popularizers of the results of scientific research. They did not comment on the moral or political aspects of science. With the growth of scientific enterprise, a new specialty has arisen; it might be called "science policy journalism." Alongside it has grown up an academic interest in science which some call the "science of science," and others "science policy studies" or "liberal studies in science." Sociologists, political scientists, economists, journalists, historians, and philosophers have all contributed to the formation of the new discipline or at least to a new body of opinion. It began in the late 1940's in a state of adulatory enthusiasm for the potentialities of science; it has now turned a little sour. What was accepted with unthinking acclamation fifteen or twenty years ago has now become the object of stricture, doubt, and condemnation.

Recently economists have been extending their jurisdiction and spreading the light of their analysis to subjects which they hitherto left untouched, including science. Applying their dichotomy between producer's and consumer's goods, some economists analyze science either as a producers' good—an investment in future production, to be assessed in terms of the return which it offers in comparison with alternative objects of investment—or as a consumer's good—a form of self-indulgence for scientists at the cost of the lay community. These economists put scientists into an awkward position because they refuse to discuss the claim that science is an intrinsically valuable cognitive activity worthy of appreciation and support on that ground alone. Their implicit criticisms are that basic research must justify its claims on resources by its technological fruitfulness, and that basic research which has no technological application is nothing other than a consumer's good provided at public expense.

There is something too tough-minded about the economic analysis of science for scientists to shrug it off. Indeed, since so many scientists themselves argue, without rigorous evidence, that science should be supported because of its technological benefits, they are particularly vulnerable to economic analysis. The fact that empirical analysis has not determined the economic profitability of investment in research does not enhance the status of scientists, but contributes to their "demythologizing."

The publicists of science policy have a much broader audience because they are less sophisticated intellectually than economists, because they are in more crucial positions in certain scientific publications, and because a considerable number of alienated scientists like what they have to say. They wish to show that scientists are selfish and hypocritical, rushing wherever money is to be had. The publicists of science policy are against governmental authority, and see science as entwined in authority; they are against "isolation from society," and imply that science, because of its service to government, has become isolated from the "real needs of society"; they intimate that the scientists' claim that they need autonomy is only a demand for privilege. The publicists of science policy are against "elitism" and the "establishment," but science is "elitist" and has within itself an "establishment"; they are egalitarian, while science is stratified in accordance with acknowledged achievement. These views in their most radical form are expressed in demands for "people's science" and "critical science."

Such criticisms of "science policy," in and of themselves, will crumble only the edges of the faith of the laity in science. They can, however, damage the broader public's conception of scientists if they affect the self-image of scientists and cause them to deny their calling. The publicistic disclosure of the shortcomings of scientists and the radical demand for the reform of the scientific enterprise accept the utilitarian view of science, but express considerable animosity against its leadership and its prevailing ethos. They seem to give evidence that there has been a loss of collective self-esteem within the scientific community. Some embittered disillusionment is to be heard expressed in fairly prominent positions within the scientific community, but it is probably more common in its lower strata.

VII

Although practically no one thinks that science will fall back into the position it held in the Middle Ages, the animosity and lack of enthusiasm shown about science nowadays raise questions about the possibility of a decline of science as a continuously expanding body of knowledge. Is it possible for science to fall from the important position it has had in the present century? And, if so, how far will it fall? Will it fall everywhere in the world, or only in the United States? And, what would a fall in the United States do to the position of science in other countries?

Let us ask whether it is conceivable that science could recede everywhere in the scientifically advanced countries and make no further progress in the backward ones, that what is already discovered might be retained, but without further significant additions to the stock of knowledge.

One possibility is that the mysteries of nature and of man's existence might be exhausted in the foreseeable future. Then science would be finished, although technological innovation could go on—at first exploiting hitherto unused scientific achievements and then becoming empirical as it was before it became scientific.

However, the very imperfections of man's mind—its powerlessness combined with its great powers—suggest that in every answer given to a particular fundamental question, a later arrival of similar intellectual powers and curiosity will discover some imperfection—some gap to be filled by further scientific research. Thus, even short of the unanswerable problem of the meaning of universal and human existence, which some human minds will always feel themselves deeply impelled to ponder, every step forward in understanding eventually discloses questions previously unasked and unanswered. Thus, the universe and man's nature hold in store for us a limitless supply of questions, only a very few of which we are even empowered to ask at present. This being so, there will always be tasks for science, and the human mind is too questioning and too expansive not to respond to them.

But will powerful minds be content to operate within the mode of thought which has become established over more than four centuries? The great scientific tradition was not established arbitrarily and it did not grow from strength to strength by accident. Notwithstanding numerous imperfections, it has a great coherence and is persuasive to many very powerful minds. Great minds can certainly think in other frameworks. Nonscientific modes of thought exist in our civilization, and have predominated in other civilizations. Nonetheless, unless we postulate a collapse in the genetic properties of the species, it seems utterly unlikely that the scientific mode of thought will cease to offer a framework congenial to many very imaginative and intelligent minds.

Nature can appear to be exhausted only if the human mind loses its capacity to inquire. If the properties of the human race remain what they have been, nature will always withhold some answers from and raise further questions in the most pressing and penetrating minds. The question is whether those minds will have available to them the powerful and costly instruments they will need, and whether they will be provided with the discipline to discern the problems and seek their answers. The answers depend on whether those who dispense the resources needed for equipment, institutions, and remuneration "believe" in science as instrumentally useful and intrinsically good. Given hedonism, it does not seem probable that they will come to see science as instrumentally useless. The growing demands of a world population of three to four billions of human beings create a demand for science to meet even basic needs. Recent ways of using technology may be severely criticized, but this does not lessen the need for a better technology more responsibly used. The evident superiority of scientific technology to empirical technology is an assurance that the technological side of science will continue to receive substantial support. As long as our societies operate on a large scale with complex institutions, as long as the mass of the population makes many demands for material gratifications that only improved technology can provide, and as long as a belief in the technological potentiality of science remains part of "faith in science," there will continue to be support for science.

VIII

Much of the recent skepticism about the enterprise of science cannot be characterized as "irrationalist"—even though some of it is unjustifiable and wrong—for it shares the rationalistic and utilitarian postulates of modern science. But there is also an irrationalist criticism of science which is deeper and more far-reaching in its implications. It denies the value of the knowledge gained through science, and praises categories of experience with which science has nothing to do. If it found a very widespread following, it would surely weaken the incentive to do science and the willingness to support it. At present, it is a marginal phenomenon, espoused by fewer prominent intellectuals than in the nineteenth and early twentieth centuries. Although it has gained some attention because it has coincided with certain other criticisms of science, I myself doubt whether it is likely to have a massive effect on the appreciation and motivation of scientific work.

The disparagement of the ordered and disciplined experience of scientific study, and the belief that through the cultivation of "states of intense consciousness" which transcend and rise to a "higher" level than individual consciousness through the aid of drugs and spiritual exercises are not likely to have a very wide or lasting appeal. Mysticism, the direct experience of the divine, has never been very common in any major religion, compared with organized and disciplined worship. Minor religious enthusiasms have always existed alongside the organized religious life. Although they have occasionally intruded into it, they have soon receded and a less enthusiastic form of worship, mediated through priests and institutions, has reasserted itself. Mankind in the mass cannot live by enthusiasm—by possession by spirits or the "secular" equivalents thereof. The criticisms which enthusiasts make of established institutions usually maintain their force only for short periods. The present irrationalist current will therefore not ignite the mass of society, which, although disillusioned about much in today's society, does not seem to be disillusioned about the value of earthly gratification, social stability, and a relatively ordered existence formed around respectable norms, persons, and institutions.

More important for the future of science than the "irrationalist" criticism of science by nonscientists is the bearing of scientists themselves. The morale of the scientific community is crucial. It is the state of opinion in the scientific community which sustains individual scientists, and it is the outlook and bearing of the leading scientists which set its tone. Their bearing toward their own work and their relationships with their sustaining laity—politicians, civil servants, industrial managers and enterprisers, military men, journalists, and other members of the intellectual classes—will be decisive. If scientists appear to be diabolical, cynical, self-serving and self-indulgent, or if they are disparaging toward science, the faith of the laity will waver more than it has. If that rupture of the tacit contract, which binds them to these other sectors of society's center, and the abdication of their own dignity become enduring traditions among scientists, science might lose a good deal of its support. Still, I think that support for science cannot disappear short of a cataclysm which destroys man's intellectual powers and his curiosity about the mysteries of the universe as well as his economic capacity to support the working of that curiosity.

The human race did not take up science lightly a half a millennium ago. It had been preparing for at least two millennia. It was making no mistake, any more than it was when it took up religion or art. Science has roots deep in what is deepest in man's mind and it is not likely to be dislodged by a decade of bitter criticism by academic humanists and journalists, skepticism on the part of politicians, and self-derogation by some scientists and their friends. Nonetheless, it is not irrelevant to bear in mind the influence of the wavering faith of clergymen on the present situation of the Christian religion in Protestant and Roman Catholic countries.

THEODORE ROSZAK

The Monster and the Titan: Science, Knowledge, and Gnosis

THE TITLE OF the book was *Frankenstein*. The subtitle was *The Modern Prometheus*.

An inspired moment when Mary Shelley decided that a maker of monsters could nonetheless be a Titan of discovery—one whose research might, in our time, win him the laurels of Nobel. She claimed the story broke upon her in a "waking dream." It may well have been by benefit of some privileged awareness that one so young fused into a single dramatic image the warring qualities that made Victor Frankenstein both mad doctor and demigod. A girl of only nineteen, but by virtue of that one, rare insight, she joined the ranks of history's great myth makers. What else but a myth could tell the truth so shrewdly, capturing definitively the full moral tension of this strange intellectual passion we call science? And how darkly prophetic that science, the fairest child of the Enlightenment, should find the classic statement of its myth in a Gothic tale of charnel houses and graveyards, nightmares and bloody murder.

Asked to nominate a worthy successor to Victor Frankenstein's macabre brainchild, what should we choose from our contemporary inventory of terrors? The bomb? The cyborg? The genetically synthesized android? The behavioral brain washer? The despot computer? Modern science provides us with a surfeit of monsters, does it not?

I realize there are many scientists—perhaps the majority of them—who believe that these and a thousand other perversions of their genius have been laid unjustly at their doorstep. These monsters, they would insist, are the bastards of technology: sins of applied, not pure science. Perhaps it comforts their conscience somewhat to invoke this much muddled division of labor, though I must confess that the line which segregates research from development within the industrial process these days looks to me like one of gossamer fineness, hardly like a moral *cordon sanitaire*.

I realize, too, that there are some—those who champion a "science for the people"—who believe that mad doctors are an aberration of science that can be wholly charged to the account of military desperados and corporate profiteers. Their enemies are also mine; I write in full recognition of how the wrong-headed power elites of the world corrupt the promise of science. But I fear there are more unholy curiosities at work in their colleagues' laboratories than capitalism, its war lords, and hucksters can be made the culprits for. Certainly they must share my troubled concern to see the worst excesses of behavioral psychology and reductionist materialism become unquestionable orthodoxies in the socialist societies.

I will grant to both these views some measure of validity (less to the first, much more to the second). But here and now I have no wish to pursue the issues they raise, because I have another monster in mind that troubles me as much as all the others—one who is nobody's child but the scientist's own and whose taming is no political task. I mean an invisible demon who works by subtle poison, not upon the flesh and bone, but upon the spirit. I refer to the monster of meaninglessness. The psychic malaise. The existential void where modern man searches in vain for his soul.

Of course there are few scientists who will readily accept this unlovely charge upon their paternity. The creature I name wears the face of despair; its lineaments are those of spiritual desperation; in its bleak features scientists will see none of their own exhilaration and bouyant morale. They forget with what high hopes and dizzy fascination Victor Frankenstein pursued his research. He too undertook the adventure of discovery with feverish delight, intending to invent a new and superior race of beings, creatures of majesty and angelic beauty. It was only when his work was done and he stepped back to view it as a whole that its true—and terrifying—character appeared.

The pride of science has always been its great-hearted humanism. What place, one may wonder, is there in the humanist's philosophy for despair? But there is more than one species of humanism, though the fact is too often brushed over. In the modern West, we have, during the past three centuries, run a dark, downhill course from an early morning humanism to a midnight humanism; from a humanism of celebration to a humanism of resignation. The humanism of celebration—the humanism of Pico and Michelangelo, of Bacon and New-ton—stems from an experience of man's congruency with the divine. But for the humanism of resignation, there is no experience of the divine, only the experience of man's infinite aloneness. And from that is born a desperate and anxious humanism, one that clings to the human as if it were a raft adrift in an uncharted sea. In that condition of forsakenness, we are not humanists by choice, but by default—humanists because there is nothing else we have the conviction to be, humanists because the only alternative is the nihilist abyss.

If I say it is science that has led us from the one humanism to the other, that it is science which has made our universe an unbounded theater of the absurd . . . does that sound like an accusation? Perhaps. But I intend no condemnation, because I believe that, at every step, the intentions of the scientists have been wholly honest and honorable. They have pursued the truth and followed bravely where it took them, even when its destination became the inhuman void. In any case, I say no more than thoughtful scientists have themselves recognized to be true—in some cases with no little pride. Thus, Jacques Monod:

By a single stroke [science] claimed to sweep away the tradition of a hundred thousand years, which had become one with human nature itself. It wrote an end to the ancient animist covenant between man and nature, leaving nothing in place of that precious bond but an anxious quest in a frozen universe of solitude.[1]

Or, as Steven Weinberg puts it elsewhere in this volume:

The laws of nature are as impersonal and free of human values as the rules of arithmetic. We didn't want it to come out this way, but it did. . . . The whole system of the visible stars stands revealed as only a small part of the spiral arm of one of a huge number of galaxies, extending away from us in all directions. Nowhere do we see human value or human meaning.

Our universe. The only universe science can comprehend and endorse. "A universe," Julian Huxley has called it, "of appalling vastness, appalling age, appalling meaninglessness." But not, for that reason, an uninteresting universe. On the contrary, it is immensely, inexhaustibly *interesting*. There is no reason, after all, why what is wholly alien to us should not be wholly absorbing. Nor is there any reason why, in such a universe, we should not make up meanings for ourselves—whatever meanings we please and as many as we can imagine. Is this not the favorite preoccupation of modern culture, the intellectual challenge that adds the spice of variety to our lifestyle? We may even decide to regard science itself as the most meaningful way of all to pass the time. All we need remember—if we are to remain scientifically accountable—is that none of these meanings resides in nature. They simply express a subjective peculiarity of our species. They are arbitrary constructions having no point of reference "Out There." Which is to say: the universe we inhabit—insofar as we let it be the universe science tells us we inhabit—is an *in-human* universe. We share some minute portion of its dead matter, but it shares no portion of our living mind. It is (again to quote Jacques Monod) an "unfeeling immensity, out of which [man] emerged only by chance. . . ." and where "like a gypsy, he lives on the boundary of an alien world, a world that is deaf to his music, just as indifferent to his hopes as it is to his suffering or his crimes."

Perhaps not every reader agrees with me that meaninglessness is a monster. If not, then our sensibilities are of a radically different order and we may have to part company from this point forward, for this is not the place to try closing the gap between us. But I believe more than a few scientists have looked out at times upon the "unfeeling immensity" of their universe with some unease. Note Weinberg's phrase, "We didn't want it to come out this way. . . ."

Perhaps not every reader regards the degradation of meaning in nature as a *moral* issue. But I do. Because meaninglessness breeds despair, and despair, I think, is a secret destroyer of the human spirit, as real and as deadly a menace to our cultural sanity as the misused power of the atoms is to our physical survival. By my lights at least, to kill old gods is as terrible a transgression of conscience as to concoct new babies in a test tube.

But even if scientists should agree that their discipline buys its progress at a dear price in existential meaning, what are they to do? Steven Weinberg faces the question squarely in his essay and offers an answer which would, I suspect, be endorsed by many of his colleagues. He tells us that "other modes of knowledge" (the example he gives is aesthetic perception) might be accommodated alongside science in a position of coexistence, but they cannot be given a place *within* science as part of a radical shift of sensibilities.

. . . science cannot change in this way without destroying itself, because however much human values are involved in the scientific process or are affected by the results of scientific research, there is an essential element in science that is cold, objective, and nonhuman. . . . Having committed ourselves to the scientific standard of truth, we have thus been forced, not by our own choosing, away from the rhapsodic sensibility. . . . In the end, the choice is a moral, or even a religious, one. Having once committed ourselves to look at nature on its own terms, it is something like a point of honor not to flinch at what we see.

"The univserse," Weinberg insists, "is what it is." And science, as the definitive natural philosophy, can have no choice but to tell it like it is, and "not to flinch."

One cannot help admiring the candor of such an answer—and grieving a little for the pathos of its resignation. But it is, in any case, a Promethean answer, one that reminds us that the free pursuit of knowledge *is*, after all, a supreme value, a need of the mind as urgent as the body's need for food. However much one may upbraid science for having disenchanted our lives, sooner or later one must come to grips with the animating spirit of the discipline, the myth that touches it with an epic grandeur. Call up the monster, and the scientist calls up the Titan. Press the claims of spiritual need, and the scientist presses the claims of mind as, in their own right, a sovereign good.

Any critique of science that challenges the paramount good of knowledge risks becoming a crucifixion of the intellect. If Prometheus is to stop producing monsters, it must not be at the sacrifice of his Titanic virtues. The search for knowledge must be a free adventure; yet is must not choose, in its freedom, to do us harm in body, mind, or spirit. One no sooner states the matter in this way than it seems like an impossible dilemma. We are asking that the mind in search of knowledge should be left wholly free and yet be morally disciplined at the same time. Is this possible?

I believe it is, but only if we recognize that there are *styles* of knowledge as well as *bodies* of knowledge. Besides *what* we know, there is *how* we know it—how wisely, how gracefully, how life-enhancingly. The life of the mind is a constant dialogue between knowing and being, each shaping the other. This is what makes it possible to raise a question which, at first sight, is apt to appear odd in the extreme. *Can we be sure that what science gives us is indeed knowledge?*

Plato, Don Juan, and Gnosis

For most Western intellectuals that might seem a preposterous question, since for the better part of three centuries now science has served as the measure of knowledge in our society. But to raise it is only to recall the Platonic tradition, within which our science would have been regarded as an intellectual transaction distinctly beneath the level of knowledge. There is no telling for sure how highly Plato might have rated the spectacular theoretical work of the modern world's best scientific brains, but I suspect he would have respected it as "information"—a coherent, factually related account of the physical structure and function of things: a clever scheme for "saving the appearances," as Plato liked to characterize the astronomy of his day. Here we have a demanding and creditable labor of the intellect; but on Plato's well-known four-step ladder of the mind, science would be placed somewhere between the second and third levels of the hierarchy—above mere uninformed "opinion," but distinctly below "knowledge."

Easy enough to dismiss Plato as backward or plain perverse for refusing to rate science any higher on the scale. But how much more interesting to let the mind follow where his gesture takes it when he invites us to look beyond experiment, theory, and mathematical formulation to a higher object of knowledge which he calls "the essential nature of the Good . . . from which everything that is good and right derives its value for us."

Significantly, when Plato tried to put this object of knowledge into words, his habit, like that of many another mystic, was to enlist the services of myth and allegory, or to warn how much must be left unsaid. "There is no writing of mine about these matters," he tells us in the Seventh Epistle in a passage that might be a description of the Zen Buddhist Satori, "nor will there ever be one. For this knowledge is not something that can be put into words like other sciences; but after long-continued intercourse between teacher and pupil, in joint pursuit of the subject, suddenly, like light flashing forth when a fire is kindled, it is born in the soil and straightaway nourishes itself." No doubt, at first glance, such an elusive conception of knowledge is bound to seem objectionable to many scientists. But in light of all that Michael Polanyi has written about the "personal knowledge" and the "tacit dimension" involved in science, Plato's remarks should not seem wholly alien. Plato is reminding us of those subtleties that can only be conveyed between person and person at some nonverbal level; to force such insights into words or into a formal pedagogy would be to destroy them. If we are to learn them at all, there is no way around intimate association with a guru who can alone make sure that each realization is sensibly adapted to the time, and the place, and the person. So too in science, as in every craft and art. Is not much that is essential to the study left to be learned from one's master by way of nuance and hint, personal taste and emotional texture? And does this not include the most important matters of all: the spirit of the enterprise, the choice of a problem deemed worth studying, the instinctive sense of what is and what is not a reputable scientific approach to any subject, the decision as to when a hypothesis has been sufficiently demonstrated to merit publication? How much of all this is taught by the glint in the eye or the inflection in the voice, by subtle ridicule or the merest gesture of approval? Even the exact sciences could not do without their elements of taste and intuitive judgment, talents which students learn by doing or from the living example before them.

Plato is, of course, pushing the uses of reticence much further. He contends that, if only the tacit dimension of instruction between guru and student is exploited to its fullest, we can find our way to a knowledge, properly so-called, which grasps the nature and *the value* of things as a whole, and so raises us to a level at which intellect and conscience become one and inseparable in the act of knowing. "Without that knowledge," he insists, "to know everything else, however well, would be of no value to us, just as it is of no use to possess anything without getting the good of it."

Again, I suspect that Plato is not so far removed in his pursuit from a familiar scientific experience, one which comes in the wake of any significant discovery. It is the sense that, over and above what the particular discovery in question has shown to be factually so, this activity of the mind has proved itself *good;* it has, as a human project, elevated us to a level of supremely satisfying existence. One has not only found out something correct (perhaps that is the least of it, in the long run) but one has *been* something worth being. It is an experience many people have known, at least fleetingly, in their work as artists, craftsmen, teachers, athletes, doctors, etc. We might call it "an experience of excellence," and let it go at that. But what Plato wished to do was to isolate that experience as an object of knowledge, and to treat it, not as the by-product of some other, lesser activity, but as a goal in its own right. He wished to know the Good in itself which we only seem to brush against now and again in passing as we move from one occasional task to another. Nothing in modern science would have appalled Plato more than the way in which a professional scientific paper seeks, in the name of objectivity, to depersonalize itself to the point of leaving out all reference to that "experience of excellence"—that fleeting glimpse of the higher Good. For, I believe Plato would have objected, if no such experience was there, then the work was not worth doing; and if it was, then why leave it out, since it must surely be the whole meaning and value of science? Once you omit *that*, you have nothing left except . . . information.

If I invoke Plato here, it is not because I wish to endorse his theory of knowledge, but only to use him as a convenient point of departure. I recognize the logical blemishes that have dogged his epistemology through the centuries—and regard many of them as unanswerable within the framework Plato erected for his work. He is, however, the most renowned philosophical spokesman for a style of knowledge which is far older than formal philosophy; in his work we confront a visionary tradition which runs through nearly every culture, civilized and primitive. The prime value of Plato—so it has always seemed to me—lies not so much in the intellectual territory he occupied and surveyed, as in his stubborn determination to keep open a passage through which the mind might cross over from philosophy to ecstasy; from intellect to illumination. His dialogues stand on the border of a trans-rational sensibility whose charm seems a constant feature of human culture—a sensibility perhaps as old as the mind itself, and yet as contemporary as the latest bestseller list. Recall what the Yaqui Indian shaman Don Juan calls himself in Carlos Casteneda's recent popular reports: "a man of *knowledge*." And, for all the differences of personal style and lore that part the two men, the old sorcerer means knowledge in exactly the way Plato meant it, as an ecstatic insight into the purpose and place of human existence in the universe, a glimpse of the eternal.

What both Plato the philosopher and Don Juan the sorcerer seek as knowledge is precisely that *meaningfulness* of things which science has been unable to find as an "objective" feature of nature. To follow where such a conception of knowledge takes us is not to denigrate the value or fascination of information. It is to be neither antiscientific nor antirational. It leads us not to an either/or choice, but to a recognition of priorities within an integrated philosophical context. Information can be exciting to collect; it can be urgently useful: a tool for our survival. But it is not the same as the knowledge we take with us into the crises of life. Where ethical decision, death, suffering, failure confront us, or in those moments when the awesome vastness of nature presses in upon us, making us seem frail and transient, what the mind cries out for is the meaning of things, the purpose they teach, the enduring significance they give our existance. And that, I take it, is Plato's knowledge of the Good.

To call this *another kind* of knowledge may seem a convenient compromise or a generous concession. But I submit that either as compromise or concession, this policy of Cartesian apartheid is treacherous. At best, it asks for the sort of schizophrenic coexistence that divides the personality cruelly between fact and feeling. At worst, it is the first step toward denying the "other knowledge" any status as knowledge at all—toward considering it a sort of irrational spasm devoid of any claim to truth or reality, perhaps an infantile weakness of the ego that is only forgiveable because it is so universally human. At that point we are not far from treating the need for meaning as a purely subjective question for which there is no objective answer—as an unfortunate behavioral trait which we leave psychologists or brain physiologists to stake out for investigation. Once it ceases to be the basis for knowledge, it may finish as an occasion for therapy.

My purpose here is to call back to mind the traditional style of knowledge for which the nature of things was as much a reservoir of meanings as of facts—a style of knowledge which science is now aggressively replacing in every society on earth. Let us call this knowledge "gnosis," borrowing the word not to designate a second and separate kind of knowledge, but an *older* and *larger* kind of knowledge from which our style of knowledge derives by way of a sudden and startling transformation of the sensibilities over the past three centuries. My contention is that this process of derivation has been spiritually impoverishing and psychically distorting. It has resulted in a narrowing of our full human potentialities and has left us—especially in science— with a diminished Titanism that falsely borrows upon the myth it champions. When the modern Prometheus reaches for knowledge, it is not the torch of gnosis he brings back or even searches for, but the many candles of information. Yet not a million of those candles will equal the light of that torch, for these are fires of a different order.

Augmentative Knowledge

I will not try to characterize gnosis here as an "alternative cognitive system" in any programmatic way—as if to offer a new methodology or curriculum. Rather, I want to speak of gnosis as a. different sense of what knowledge is than science provides. When we search for knowledge, it is a certain texture of intelligibility we first and most decisively seek, a *feeling* in the mind that tells us, "Yes, here is what we are looking for. This has meaning and significance to it." Though it may work well below the level of deliberate awareness, this touchstone of the mind is what makes the persuasive difference in our thinking. Indeed, science itself arose in just this way, when men of Galileo's generation came to feel, with an uncanny spontaneity, that to know was to measure, that all else was subjective and unreal, a realm of "secondary qualities."

In the broadest sense, gnosis is *augmentative* knowledge, in contrast to the *reductive* knowledge characteristic of the sciences. It is a hospitality of the mind that allows the object of study to expand itself and become as much as it might become, with no attempt to restrict or delimit. Gnosis invites every object to swell with personal implications, to become special, wondrous, perhaps a turning point in one's life, "a moment of truth." Paul Tillich has called gnosis "knowledge by participation . . . as intimate as the relation between husband and wife." Gnosis, he tells us, "is not the knowledge resulting from analytic and synthetic research. It is the knowledge of union and salvation, existential knowledge in contrast to scientific knowledge."

It is the guiding principle of gnosis that only augmentative knowledge is adequate to its object. As long as we, at our most open and sensitive, feel there is something left over or left out of any account we give of any object, we have fallen short of gnosis. Gnosis is that nagging whisper at the edge of the mind which tells us, whenever we seek completeness of understanding or pretend to premature comprehension, "not yet. . . not quite." It is our immediate awareness, often at a level deeper than intellect, that we seem not to have done justice to the object—not because there remains quantitatively more of the object to be investigated, but because its essential *quality* still eludes us.

I speak here of the experience many people have known when faced with some brutally reductionist explanation of human conduct. We feel the explanation "reduces" precisely because it leaves out so much of what we spontaneously know about humanness from inside our own experience. We look at the behaviorist's model and we know—as immediately as our eye would know that a circle is not a square—that this is not *us*. It may not even be an important part or piece of us, but only a degraded figment. Even if such knowledge "worked"— in the sense that it allowed others to manipulate our conduct as precisely as an engineer can manipulate mechanical and electrical forms of energy—would we not still protest that *knowing how* to dangle us like a puppet on a string is not *knowing* us at all? Might we not insist that such "knowledge" works in the very opposite direction—that it is an ignorant, insulting violation of our nature? As Abraham Maslow once observed of his own experience in behavioral psychology: "When I can predict what a person will do under certain circumstances, this person tends to resent it. . . . He tends to feel dominated, controlled, outwitted."[2] Between "knowing" and "knowing how" there can be a fearful discord—like Bach being played on skillets and soup kettles: more mockery than music.

That discord shows up readily enough when we ourselves are the specimens under study. In that case, the standard of adequacy is provided by the object of investigation. We can speak for ourselves and fend off the assault upon our dignity. But what about the nonhuman objects of the world? Does it make any sense to say that our scientific knowledge of them may be *qualitatively* inadequate?

To answer that question, let us begin with a familiar comparison: that between art and science. The coincidence of the two fields has been observed many times, especially in so far as they share a common fascination for form and structure in nature. Yet, while there is an overlap, it is, from the scientist's point of view, an overlap of interest only, not of intellectual competence. Both art and science find an aesthetic aspect in nature (though of course many scientists have done significant research without pausing over that aspect). But for the scientist, the aesthetic appearance is a *surface;* knowledge stands behind that surface in some underlying mechanism or activity requiring analysis. What the artist sees is not regarded by science as knowledge of what is *in* the object as one of its constituent properties. Instead, what preoccupies the artist is called "beauty" (though often it would better be understood as awe, conceivably mixed with fear, anxiety, dread). Beauty is, for science, a sort of subjective supplement to knowledge, a decoration the mind supplies before or after the act of cognition, and which can or even ought to be omitted from professional publication. Aesthetic fascination may attract us to the object; it may later help flavor popularized accounts of research. From the scientist's viewpoint, however, only further study (dissection, deep analysis, comparison, experiment, measurement) allows us to find out something about the object, something demonstrable, predictive, useful. Compared to such hard fact, the artist's perception is merely dumb wonder, which, apparently, artists have not the intellectual rigor to go beyond. Jacob Bronowski has, for example, referred to the artist's response to nature as "a strangled, unformed and unfounded experience." But, he goes on, "science is a base for [that experience] which constantly renews the experience and gives it a coherent meaning."[3]

If this were not the supposition, we might imagine an entire specialization in science devoted to studying the nature poets and painters: biologists sprinkling their research with quotations from Wordsworth or Goethe . . . neophyte botanists taking required courses in landscape painting . . . astronomers drawing hypotheses from Van Gogh's "Starry Night" . . . theoretical physicists pondering the bizarre conceptions of time and space one finds in the serial tone row, cubism, constructivism, or Joyce's *Finnegans Wake*. Of course, nothing forbids scientists from wandering into these exotic realms, but what curriculum *requires* that they do so?

From the viewpoint of gnosis, however, what artists find in nature is decidedly knowledge of the object, indeed knowledge of a uniquely valuable kind. It is not repeatable or quantitative, nor is it open to experimentation or utilitarian application. It is usually not logically articulable; that is why special languages of sound, color, line, texture, metaphor and symbol have been invented to carry the message, in much the same way that mathematics has been developed as the special language of objective consciousness. But that message is as much knowledge as when, in addition to knowing your chemical composition, I discern that you are noble or base, lovable or vicious. So artists discover the communicative mood and quality that attach to form, color, sound, image. They teach us those qualities, and these become an inseparable part of our total response to the world.

Of course, these qualities can be screened out if our interest is directed to something less than the whole, but this does not mean the sensuous and aesthetic qualities are not really there as a constituent property of the world—a property that is being artfully *displayed* to us. Would it not, in fact, be truer to our experience to conceive of the world about us as a *theater*, rather than as a mechanism or a randomized aggregation of events? It is surely striking how often science quite naturally presents its discoveries as if it were unfolding a spectacle before us, thus borrowing heavily on sensibilities that have been educated by the dramatists and story-tellers. All cosmology is talked about in this way, and even a good deal of high energy physics and molecular biology. Everything we have lately discovered about the evolution of stars is, quite spontaneously, cast in the mode of biography: birth, youth, maturity, senility, death, and at last the mysterious transformation into an afterlife called "the black hole." Or, take the classic example of aesthetic perception in science. Can there be any doubt that much of the cogency of Darwin's theory of natural selection stemmed from the pure drama of the idea? Natural selection was presented as a billion-year-long epic of struggle, tragic disasters, lucky escapes, triumph, ingenious survival. Behind the sensibility to which Darwin's theory appealed lay three generations of Romantic art which had pioneered the perception of strife, dynamism, and unfolding process in nature. Behind Darwin stand Byron's Manfred, Goethe's Faust, Constable's cloud-swept landscapes, Beethoven's tempestuous quartets and sonatas. All this became an integral part of the Darwinian insight. I doubt there is anyone who does not still bring to the study of evolution this Romantic taste for effortful growth, conflict, and self-realization. The qualities are not only in the idea, but also in the phenomenon. It is not that these dramatic qualities have been "read" into nature by us, but rather that *nature* has read them into *us* and now summons them forth by the spectacle of evolution we find displayed around us.

We should by now be well aware of the price we pay for regarding aesthetic quality as arbitrary and purely subjective rather than as a real property of the object. Such a view opens the way to that brutishness which feels licensed to devastate the environment on the grounds that beauty is only "a matter of taste." And since one person's taste is as good as another's, who is to say—as a matter of *fact*—that the hard cash of a strip mine counts for less than the grandeur of an untouched mountain? Is such barbarism to be "blamed" on science? Obviously not in any direct way. But it is deeply rooted in a scientized reality principle that treats quantities as objective knowledge and qualities as a matter of subjective preference.

The Spectrum of Gnosis

Now to push the point a little further. If art overlaps science at one wing, it overlaps visionary religion at the other. If artists have found the cool beauty of orderly structure in nature, they have also found there the burning presence of the sacred. For some artists, as for the Deist scientists of Newton's day, God's imprint has appeared in the rhythmic cycles and stately regularities of nature. For other artists—Trahern, Blake, Keats, Hopkins—the divine grandeur of the world appears all at once, in an ecstatic flash, a jolt, a "high." Here we find the artist becoming seer and prophet. For such sensibilities, a burning bush, a storm-battered mountaintop can be, by the sheer awesomeness of the event, an immediate encounter with the divine.

To know God from the order of things is a deduction, a shaky one perhaps in the eyes of skeptical logicians, but at least remotely scientific in character. To know God from the power of the moment is an epiphany, a knowledge that takes us a long way from scientific respectability. Yet here is where gnosis mounts to its heights, becoming knowledge willingly obedient to the discipline of the sacred. It does not close itself to the epiphanies life offers by regarding them as "merely subjective." Rather, it allows, it *invites* experience to expand and become all that it can. After all, if Galileo was right to call those men fools who refused to view the moon through a telescope, what shall we say of those who refuse Blake's invitation to see eternity in a grain of sand? Gnosis seeks to integrate these moments of ecstatic wonder; it regards them as an advance upon reality, and by far the most exciting advance the mind has undertaken. For here is the reality that gives transcendent meaning to our lives.

Perhaps the best way to summarize what I have said so far is to conceive of the mind as a spectrum of possibilities, all of which properly blend into one another—unless we insist on erecting barriers across the natural flow of our experience. At one end, we have the hard, bright lights of science; here we find information. In the center we have the sensuous hues of art; here we find the aesthetic shape of the world. At the far end, we have the dark, shadowy tones of religious experience, shading off into wave lengths beyond all perception; here we find meaning. Science is properly part of this spectrum. *But gnosis is the whole spectrum.*

If, in the past, gnosis has been more heavily weighted on the side of meaning than information, it should not be difficult to understand why. Our ancestors saw fit to put first things first. Before they felt the need to know how fire burns or how seeds germinate, they needed to know the place and purpose of their own strange existence in the universe. And this they found generously offered to them in the nature of things. Yet, I know of no visionary tradition that has ever refused to agree that natural objects possess a structure and function worthy of study. Certainly none of these traditions has been as adamantly closed to the technical level of knowledge as our science has been closed to gnosis. Plato may have wanted the mind to rise to a level of ecstatic illumination, but he never said there was no such thing as information or that its pursuit was a sign of madness or intellectual incompetence. Similarly, the alchemists may have sought their spiritual regeneration in natural phenomena, but they never refused to examine the way nature works. Undeniably, where gnosis becomes our standard of knowledge, science and technology proceed at a much slower rate than the wild pace we accept (or suffer with) as normal. This is not to say, however, that gnosis is without its practical aspect, but rather that its sense of practicality embraces spirit as well as body, the need for psychic as much as for physical sustenance.

The most familiar examples we have of culture dominated by gnosis are in the world's primitive and pagan societies. Many of these societies have been capable of inventing agrarian and hunting technologies every bit as ingenious as the machine technics of modern times. But, in stark contrast to the culture of urban-industrialism, their technology blended at every step with poetic insight and the worship of the elements. The tools and routines of daily life normally participated in the religious sensibility of the society, functioning as symbols of life's higher significance. From the viewpoint of the modern West, such a culture may look like a hodge-podge of wholly unrelated factors. In reality, it is an ideal expression of gnosis, for it expresses a unitary vision bringing together art, religion, science, and technics. Our habit, in dealing with such cultures, is to interpret their technics as lucky accidents and their aesthetic-religious context as an encumbrance. But by at least one critical standard, these "underdeveloped" cultures have proved more technically successful than our own may. They have *endured*, in some cases a hundred times longer than urban industrialism may yet endure. Surely that is some measure of how well a culture understands its place in nature.

Most of the world's mystic and occult traditions have been worked up from the gnosis of primitive and pagan cultures. At bottom, these traditions are sophisticated, speculative adaptations of the old folk religions, which preserve in some form their antique wisdom and modes of experience. Behind the Cabbala and Hermeticism, we can still see the shadowy forms of ritual magic and fertility rites, symbols of a sacred continuum binding man to nature and prescribing value. In all these mystic traditions, to know the real is to know the good, the beautiful, and the sacred at the same time.

This is not to say that all who followed these traditions achieved gnosis. The human mind goes wrong in many ways. It can go mad with ecstasies as well as with logic. Discriminating among the levels and directives of transrational experience is a project in its own right—one I do not even touch upon here, for the discussion would be far too premature at this point. There are disciplines of the visionary mind as well as of the rational intellect, as anyone will know who has done more than scratch the surface of the great mystic traditions. All I stress here is the difference between a taste for gnosis and a taste for knowledge whose visionary overtones have been systematically stilled as a supposed "distortion" of reality.

The Visionary Origins of Science

Our science, having cut itself adrift from gnosis, contents itself to move along the behavioral surface of the real—measuring, comparing, systematizing, but never penetrating to the visionary possibilitites of experience. Its very standard of knowledge is a rejection of gnosis, any trace of whose presence is regarded as a subjective taint. Yet, ironically, the scientific revolution of the sixteenth and seventeenth centuries was in large part launched by men whose thought was significantly colored by lingering elements of gnosis in our culture, most of which survived in various subterranean occult streams. Copernicus very nearly resorted to pagan sun worship as a means of supporting his heliocentric theory, the sheer aesthetic beauty of which seems to have been as persuasive for him as its mathematical precision. Kepler's astronomy emerges from a search for the Pythagorean music of the spheres. Newton was a life-long alchemist and student of Jacob Boehme. The scholarship on early science finds more and more hidden continuities between the scientific revolution and the occult currents of the Renaissance. Frances Yates has gone so far as to suggest that science only flourished in those societies where there had been a strong, free influx of Hermetic and Cabbalistic studies.[4] From this origin came the number magic and nature mysticism which were to be assimilated into science as we know it. These historical links have yet to be fully traced, but certainly the key paradigm of "law"—that mysterious sense of natural right order without which early science could never have gotten off the ground—carried with it in the thinking of early physicists unmistakable moral and theological reverberations. It was the concept of universal law that made the study of nature as a celebration of the grandeur of God compatible with the Christian doctrine.

What this confabulation with occult tradition suggests is that many lively minds of the seventeenth century, including some founding fathers of modern science, looked forward to seeing the New Philosophy become a true gnosis, possibly to replace the rigid, decaying dogmatism of Christianity. The trouble was that their exciting new approach to nature progressively screened out the very dimension of consciousness in which gnosis can alone take root: visionary insight. In seeking to externalize gnosis by raising it to a wholly articulate and mathematical level of expression, the New Philosophers left behind the mystic and meditative disciplines which might have taught them that introspective silence and transcendent symbolism are necessary media of gnosis. It was as if someone had invented an ingenious musical instrument with which he hoped to replace the full orchestra, with the result that thereafter all orchestral music had to be scaled down to the capacities of his instrument. And once that had been done, he and his audience began to lose their ear for the harmonies and overtones that only the orchestra can achieve. Quantification is just such an instrument of severely reduced resonance.

There is a haunting and troubling strangeness about this interval in our history. One might almost believe that perverse forces which baffle the understanding were at work beneath the surface of events turning science into something that did not square with the personalities of its creators. What was it, for example, that inspired Descartes to regard mathematics as the new key to nature? An "angel of truth" who appeared to him in a series of numinous dreams on three successive nights. But in his writing, he never once mentions the epistemological status of dreams or visionary experience. Instead, he turns his back on all that is not strict logic, opting for a philosophy of knowledge wholly subordinated to geometrical precision. Yet that philosophy purchases its apparent simplicity by an appalling brutalization of the very existential subtleties and psychic complexities that are the living substance of Descartes' own autobiography. Newton, a man of stormy psychological depths, spent a major portion of his life in theological and alchemical speculation; but all this he carefully edited from his natural philosophy and his public life. He even allowed himself to be talked out of attending the meetings of occult societies in London, lest he damage his reputation as a scientist. Arthur Koestler is not wide of the mark in calling the early scientists "sleepwalkers," men who unwittingly led our society into a universe whose eventual godlessness they might well have rejected vehemently.

This much of the problem stands out prominently enough: the mystic disciplines, on which gnosis depends, have never been as highly refined and widely practiced in the West as in the oriental cultures. In large part, they have suffered neglect because they cut across the doctrinal grain of conventional Christianity with its insistent emphasis on historicity and dogmatic theology. (I often think that few positivists realize how great a debt they owe to the peculiarly one-dimensional religious psychology of mainstream Christianity; its literalism and verbal rigidities paved the way for the secular skepticism of the religion's deadliest critics.) Still, in the Hermetic, Cabbalistic, Neo-Platonic, and alchemical schools of the Renaissance, at least a promising foundation existed for the building of a true gnosis. In these currents of thought we find an appreciation of myth, symbol, meditative stillness, and rhapsodic intellect that might, with maturity, have matched the finest flights of Tantric or Taoist mysticism.

But if these elements were mixed with early science in many exotic combinations, they were soon enough filtered out as violations of that strict objectivity which is the distinguishing feature of the Western scientific sensibility. It was Galileo's quantitative austerity and Descartes' dualism that carried the day with science, casting out of nature everything that was not matter in motion mathematically expressed. Here was the crucial point at which scientific knowledge ceased to be gnosis. Value, quality, soul, spirit, animist communion were all ruthlessly cut away from scientific thought like so much excess fat. What remained was the world-machine—sleek, dead, and alien. However much physics has, in our time, modified the mechanistic imagery of its classical period, the impersonality of the Newtonian world view continues to dominate the scientist's vision of nature. The models and metaphors of science may alter, but the sensibility of the discipline remains what it was. Since the quantum revolution, modern physics has ceased to be mechanistic, but it has scarcely become in any sense "mystical." The telling fact is that both in style and content it serves today as an ideal foundation for molecular biology and behavioral psychology, sciences which have of late become as mechanistic as the crudest reductionism of the seventeenth century. Almost universally these days, biologists regard the cell as a "chemical factory" run by "information-transfer" technology. And, at the same time, the arch-behaviorist B. F. Skinner suggests that since physics only began to make progress when it "stopped personifying things," psychology is not apt to gain a firm scientific footing until it likewise purges itself of "careless references to purpose" and ceases "to trace behavior to states of mind, feelings, traits of character, human nature, and so on"[5]—meaning, one gathers, that the way forward for psychology is to stop personifying people . . . and to begin mechanizing them.

The Suppression of Gnosis

Why has science taken this course toward ever more aggressive depersonalization? Perhaps the myth of Dr. Frankenstein suggests an answer—a tragic answer. Where did the doctor's great project go wrong? Not in his intentions, which were beneficent, but in the dangerous haste and egotistic myopia with which he pursued his goal. It is both a beautiful and a terrible aspect of our humanity, this capacity to be carried away by an idea. For all the best reasons, Victor Frankenstein wished to create a new and improved human type. What he knew was the secret of his creature's physical assemblage; he knew how to manipulate the material parts of nature to achieve an astonishing result. What he did not know was the secret of personality in nature. Yet he raced ahead, eager to play God, without knowing God's most divine mystery. So he created something that was soulless. And when that monstrous thing appealed to him for the one gift that might redeem it from monstrosity, Frankenstein discovered to his horror that, for all his genius, it was not within him to provide that gift. Nothing in his science comprehended it. The gift was love. The doctor knew everything there was to know about his creature—except how to love it as a person.

To find the cultural meaning of modern science, for *"Frankenstein's monster,"* read "nature-at-large" as we in the modern West experience it.

In the early days of the scientific revolution, Robert Boyle, convinced of the "excellency" of the new "mechanical hypothesis," insisted that nature, if it was to be mastered, must be treated like an "engine" or an "admirably contrived automaton." His argument prophetically relegated to the dustbin every lingering effort to personify nature, even by remote metaphor.

The veneration, wherewith men are imbued for what they call nature, has been a discouraging impediment to the empire of man over the inferior creatures of God. For many have not only looked upon it as an impossible thing to compass, but as something impious to attempt, the removing of those boundaries which nature seems to have put and settled among her productions; and whilst they look upon her as such a venerable thing, some make a kind of scruple of conscience to endeavor so to emulate her works as to excel them.[6]

Here was a deliberate effort—and by a devout Christian believer—to cut science off from every trace of Hermetic or alchemical influence, from every connection with animist sympathy and visionary tradition. Boyle—like Bacon, Descartes, Galileo, and Hobbes—realized that herein lay the promise of material power. From that point on, it became permissible for the scientist to admire the mechanical intricacy of nature, but not to love it as a living presence endowed with soul and reflecting a higher order of reality. A machine can be studied zealously, but it cannot be loved. By virtue of that change of sensibilities—which may of course have transpired at a subliminal level of consciousness—the New Philosophy could lay claim to power (at least short-term manipulative power) but it had lost the *anima mundi*, which, as an object of love, belongs only to gnosis.

Still, from time to time, something of the spirit of gnosis intrudes itself into scientific thought, if only as a passing reflection upon some aspect of design in nature which hints that there is indeed *something* more to be known than conventional research can reveal. Science is not without such moments. But they appear only as autobiographical minutiae along the margins of "knowledge," modest confessions of faith, personal eccentricities, a bit of subprofessional self-indulgence on the part of established great names. These ethical, aesthetic, and visionary aspects have long since become human interest sideshows of science, the sort of anecdotal material that never makes it into the textbooks or the standard curriculum, except perhaps as a whimsical footnote.

And yet, have scientists never noticed how the lay public hangs upon these professions of wonder and ultimate belief, seemingly drawn to them with even more fascination then to the great discoveries? If people want more from science than fact and theory, it is because there lingers on in all of us the need for gnosis. We want to know the meaning of our existence, and we want that meaning to ennoble our lives in a way that makes an enduring difference in the universe. We want that meaning not out of childish weakness of mind, but because we sense in the depths of us that it is *there,* a truth that belongs to us and completes our condition. And we know that others have found it, and that it has seized them with an intoxication we envy.

It is precisely at this point—where we turn to our scientists for a clue to our destiny—that they have indeed a Promethean role to perform, as has every artist, sage, and seer. If people license the scientist's unrestricted pursuit of knowledge as a good in its own right, it is because they hope to see the scientists yet discharge that role; they hope to find gnosis in the scientist's knowledge. To the extent that scientists refuse that role, to the extent that their conception of what science is prevents them from seeking to join knowledge to wisdom, they are confessing that science is not gnosis, but something far less. And to that extent they forfeit—deservedly—the trust and allegiance of their society.

Dr. Faustus, Dr. Frankenstein, Dr. Moreau, Dr. Jekyll, Dr. Cyclops, Dr. Caligari, Dr. Strangelove. The scientist who does not face up to the warning in this persistent folklore of mad doctors is himself the worst enemy of science. In these images of our popular culture resides a legitimate public fear of the scientist's stripped-down, depersonalized conception of knowledge—a fear that our scientists, well-intentioned and decent men and women all, will go on being titans who create monsters.

What is a monster? The child of knowledge without gnosis, of power without spiritual intelligence.

The reason one despairs of discussing "alternative cognitive systems" with scientists is that scientists inevitably want an alternative system to do exactly what science already does—to produce predictive, manipulative information about the structure and function of nature—only perhaps to do so more prolifically and more rapidly. What they fail to understand is that no amount of information on earth would have taught Victor Frankenstein how to redeem his flawed creation from monstrosity.

But there is, in the Hermetic tradition we have left far behind us, a myth which teaches how nature may, by meditation, prayer, and sacrifice, be magically transmuted into the living presence of the divine. That was the object of the alchemist's Great Work, a labor of the spirit undertaken in love whose purpose was the mutual perfection of the macrocosm, which is the universe, and the microcosm, which is the human soul.

> And what if all of animated nature
> Be but organic Harps diversely fram'd
> That tremble into thought, as o'er them sweeps
> Plastic and vast, one intellectual breeze
> At once the Soul of each and God of all?

Samuel Taylor Coleridge

REFERENCES

1. Jacques Monod, *Chance and Necessity* (New York: Knopf, 1971), p. 172.

2. Abraham Maslow, *The Psychology of Science* (New York: Harper and Row, 1966), p. 42.

3. J. Bronowski, *Science and Human Values* (New York: Harper Torchbooks, 1965), p. 95.

4. Frances Yates, *Rosicrucian Enlightenment* (London: Routledge, 1972). See also P. M. Rattansi, "The Social Interpretation of Science in the Seventeenth Century," *Science and Society 1600-1900* (Cambridge: Cambridge University Press, 1972).

5. B. F. Skinner, *Beyond Freedom and Dignity* (New York: Knopf, 1971), pp. 5-7.

6. Robert Boyle, "A Free Inquiry into the Received Notion of Nature," *Works* (1744), Ch. IV, p. 363.

STEVEN WEINBERG

Reflections of a Working Scientist

I ONCE HEARD the period from 1900 to the present described as "this slum of a century." Certainly the case could be made that the twentieth century fails to come up to the nineteenth in the grand arts—in music, in literature, or in painting. Yet the twentieth century does stand among the heroic periods of human civilization in one aspect of its cultural life—in science. We have radically revised our perceptions of space, time, and causation; we have learned the basic principles which govern the behavior of matter on all scales from the atomic to the galactic; we now understand pretty well how continents form and how the genetic mechanism works; we may be on the verge of finding out the over-all space-time geometry of the universe; and with any luck we will learn by the end of the century how the brain is able to think. It seems strange to me that of all the enterprises of our century, it should be science that has come under attack, and indeed from just those who seem most in tune with our times, with contemporary arts and ways of life.

I take it that my role in this issue is not so much to defend science—if science turns you off, then a scientist defending science must absolutely disconnect you—but rather to serve as an exhibit of the "genuine article," the unreformed working scientist. I will therefore simply list three of what I take to be the common current challenges to science, and react to each in turn.

These reflections arise from my own experiences as a theoretical physicist specializing in the theory of elementary particles, and I am not really certain how far they would apply to other areas of science. I intend most of my remarks to apply to the whole range of natural nonbehavioral pure sciences, but some of them may have a more limited validity. On the other hand, I explicitly do not intend my remarks to apply to the social or psychological sciences, which seem to me to face challenges of a special and different sort.

The Scientist as Dr. Frankenstein

I suppose that public attitudes toward science, favorable or unfavorable, are shaped far more by the expectation of good or evil technological developments, than by approval or disapproval of the scientific enterprise itself. This is much too big a problem to cover here in any but the most fragmentary way, and it can be logically separated from a judgment of science qua science, but it is a matter of such overriding public concern that it cannot be altogether passed over. I will discuss it briefly under the headings of five criticisms of the part that "pure" scientists have played in the creation of new technology.

1. *Scientists pursue their research, without taking due account of the harm that may be done by practical application of their work.*

This is in some degree true. There are even some scientists, though I think not many, who argue that it is their business to pursue knowledge wherever it leads them, leaving the question of practical application to businessmen, statesmen, and generals whose responsibility it is to worry about such matters. For example, many critics point to the nuclear weapon as the ugliest product of "pure" research. But this charge overestimates the degree to which the scientist can look into the future. The nuclear physicists who discovered fission at the end of the 1930's were not so much indifferent to the danger of nuclear weapons as they were unaware of it. (Meitner, Strassmann, and Hahn, for example, published their work in the open literature in 1938–1939.[1]) Later, of course, nuclear weapons were developed in the United States and elsewhere by scientists who knew perfectly well what they were doing, but this was no longer for the sake of pure research, but in the hope of helping to win World War II.

I do not see how my present work on elementary particles and cosmology could possibly have any applications, good or evil, for at least twenty years. But how can I be sure? One can think of many dangers that might arise from present pure research, especially research on genetics and the human mind, and I hope that the researchers will be able to hold back the most dangerous lines of research, but they will not have an easy time of it. For a scientist unilaterally to cut off progress along certain lines because he calculates that more harm than good will come out of it requires a faith in the accuracy of his calculations more often found among businessmen, statesmen, and generals than among natural scientists. And do the critics of science really want the scientist and not the public to make these decisions?

2. *In order to gain material support for their "pure" research or for themselves, scientists prostitute themselves to industry or government by working directly on harmful technological developments.*

Again, scientists being human, this charge is, in some measure, true. One has only to think of Leonardo's letter to the Duke of Milan offering his services in the construction of ingenious instruments of war. It seems strange to me, however, to single out scientists to bear the burden of this charge. Returning to the unavoidable example of nuclear weapons, Oppenheimer, Fermi, and the others who developed the nuclear fission bomb in World War II did so because it seemed to them that otherwise Germany would develop the bomb first and would use it to enslave the world. Since World War II a large fraction of the physicists whom I know personally have washed their hands of any involvement, part-time or full-time, in military research and development. I know of no other group, certainly not workers or businessmen, who have shown a similar moral discrimination. And what of those scientists who have not washed their hands? Admittedly, there are some who work on defense problems for money, power, or fun. There are a few others who are convinced on political grounds that any weapon that adds to military strength should be developed. However, most of the "pure" scientists in the U.S. who have been involved in military work have tried to draw a line at one point or another, and to work only on a limited class of problems where, rightly or wrongly, they felt that more good than harm could be done. My own experience has been mostly through work in the JASON group of the Institute for Defense Analyses, and more recently for the U.S. Arms Control and Disarmament Agency. Many of the members of JASON, myself among them, simply declined to do work in support of the U.S. effort in Vietnam. Others worked on the so-called "electronic battlefield," because they believed (as it happened, wrongly) that by laying an impassible barrier between North and South Vietnam, they would induce the U.S. to stop bombing North Vietnam. In recent years many of us have tried to switch our work over entirely to problems of strategic arms control, but it is not easy; the Nixon administration has recently fired or canceled the consultant contracts of many of those in the Arms Control Agency (including me) who had worked on SALT.

I would like to be able to argue that academic scientists have had a humane and restraining influence on military policy, but looking back, it is hard to find evidence that I, or even those much more active and influential than myself, have had any influence at all. However, I am convinced at least that the world would not be better off if we had kept our hands out.

3. *Scientific research of all types is oppressive, because it increases the power of the developed nations relative to the underdeveloped, and increases the power of the ruling classes relative to the ruled.*

This charge rests on such far-ranging political and historical assumptions that I cannot begin to do it justice. I am not convinced that new technology tends to support old power structures more than it tends to shake them up and put power in new hands. I am also not convinced that one should always support underdeveloped nations in conflict with more modern ones; for instance, it is the Arab states that threaten the existence of Israel, not the other way around. Furthermore, this argument for stopping scientific research logically requires a permanent general strike by everyone whose work helps to keep modern industrialized society going, not just by scientists. Perhaps some do reach this conclusion, but they must have more faith in their ability to look into the future than I have in mine. I would agree, however, that certain special kinds of technology are particularly liable to be used in an oppressive way, especially the modern computer with its capacity for keeping track of enormous quantities of detailed information. I would be in favor of cutting off specific kinds of research where specific dangers clearly present themselves, but decisions in this realm are always very hard to make. Usually, as in the case of computer technology, it is not possible, by closing off lines of research, to ward off the dangers of technology without at the same time giving up its opportunities.

4. *Scientific research tends to produce technological changes which destroy human culture and the natural order of life.*

I am more sympathetic to this charge than to most of the others. Even apart from what has been done with new weapons of war, a terrible ugliness seems to have been brought into the world since the industrial revolution through the practical applications of science. As an American, I naturally think of what I see from my car window: the great superhighways cutting cross the countryside, the suburban strips with their motels and gas stations, and the glittering lifelessness of Park Avenue.

I am not sure why this should have happened. Earlier new technology, such as the pointed arch and the windmill, created more beauty than ugliness. Perhaps it is a question of scale; so many people now have cars and electric appliances that the impact of highways, factories, and power stations is too great to be absorbed into the natural background—unlike an occasional windmill or cathedral.

If this diagnosis is correct, then a cure will be extraordinarily difficult. When industrialization offered cars and electric appliances to the general public, it offered a mobility and ease previously enjoyed only by the few who could keep carriages and servants, and people accepted with alacrity. Are we now going to ask them to go back to the status quo ante? I suppose that the only answer here, as before, is to make judgments as well as we can in favor of the civilizing technology and against the brutalizing. And there *are* examples of civilizing technology, like the bicycle, the LP record, and the railroad. As W. G. Hoskins, himself a bitter enemy of the superhighway and the jet airport, says in his wonderful book, *The Making of the English Landscape:*[2]

Indeed, the railways created as much beauty as they inadvertently destroyed, but of a totally different kind. The great gashes they inflicted on the landscape in their cuttings and embankments healed over, and wild flowers grew abundantly once more. Going down to the south-west in spring, the cuttings through Somerset and Devon sparkle with primroses. Even in Clare's own country, the railway has been absorbed into the landscape, and one can enjoy the consequent pleasure of trundling through Rutland in a stopping-train on a fine summer morning; the barley fields shaking in the wind, the slow sedgy streams with their willows shading meditative cattle, the early Victorian stations built of the sheep-grey Ketton stone and still unaltered. . . .

The problem of identifying the civilizing technology and of regulating society so as to suppress the rest is far too complicated to go into here. In any case, it is not a problem on which scientists' opinions are worth more than anyone else's.

5. *While serious human needs go unfulfilled, scientists spend large sums on accelerators, telescopes, etc., which serve no purpose other than the gratification of their own curiosity.*

There is no doubt that a great deal of scientific work is carried out without any expectation of practical benefit, and indeed would be carried out even if it were certain that no practical benefit would result. It is also true that some of this work is very expensive, for the simple reason that in any given field the experiments that can be done with string and sealing wax tend to have been done already.

I suppose that if one takes the strictly utilitarian view that the only standard of value is integrated public happiness, then scientists ought to be blamed for doing any research not motivated by calculations of how much it would contribute to public welfare. By the same reasoning, no one ought to support the ballet, write honest history, or protect the blue whale, unless it can be shown that this will maximize public happiness. However, anyone who believes that knowledge of the universe is, like beauty or honesty, a good thing in itself, will not condemn the scientist for seeking the support he needs to carry out his work.

This does not mean that the support must be granted; the public has to weigh the practical benefits that will be "spun off"—the teaching that most pure scientists do to earn their salaries and the general strengthening of technological capabilities that seems to accompany pure research. These are hard to calculate. As Julian Schwinger points outs,[3]

And one should not overlook how fateful a decision to curtail the continued development of an essential element of the society may be. By the Fifteenth Century, the Chinese had developed a mastery of ocean voyaging far beyond anything existing in Europe. Then, in an abrupt change of intellectual climate, the insular party at court took control. The great ships were burnt and the crews disbanded. It was in those years that small Portuguese ships rounded the Cape of Good Hope.

I do not want to argue here about whether the public gets its money's worth. My point is that, in seeking support for scientific research, scientists need not agree with the public as to why the work should be done.

The Scientist as Mandarin

There is a widespread suspicion that science operates as a closed shop, closed to unorthodox ideas or uncomfortable data, especially if these originate outside a small circle of established leaders. One recalls countless movies in which elderly scientists in white coats wag their grey goatees at the young hero and expostulate, "But what you propose is quite impossible, because. . . ." If the public is receptive to Sunday supplement stories about unidentified flying objects or quack cures for arthritis, it is in part because they do not believe the scientific establishment gives the possibility of such things a fair hearing. In short, not everyone is convinced that the scientists are as open-minded as they ought to be.

This is not one of the most important or profound challenges to science; nevertheless, I want to present some answers to it here, because this will give me a chance to explain some of my enthusiasm for the *process* of scientific research. Also, this is an easy challenge to meet, because it arises not so much from political or philosophical differences, as from simple misapprehensions of fact. For convenience I will discuss separately the questions of the receptivity of scientists to ideas from young or unestablished scientists; to ideas from outside the scientific profession; to unorthodox ideas from whatever source; and to uncomfortable data.

1. How open is science to new ideas from the young, unestablished scientist?

Of course, there is a scientific *cursus honorum*, and those who are just starting are less influential than their seniors. The fact is, however, the system of communication in science, probably more than that in any other area of our society, allows the newcomer a chance at influencing his field.[4]

In physics, my own field, the preëminent journal is the *Physical Review*. Almost all physicists at least scan the abstracts of the articles in their own specialties in each issue. The *Physical Review* has a panel of over a thousand reviewers who referee submitted papers, but in fact about 80 percent of all papers are accepted, and of the others a good proportion are rejected only because they are unoriginal. The *Physical Review* is an expensive operation, supported by subscriptions and page charges paid by the authors' institutions, but if an author cannot arrange to have the page charge paid, the paper is published anyway (though admittedly with a few months' delay).

There is also a more exclusive journal, *Physical Review Letters*, which publishes only short papers judged to contain material of special importance. As might be expected, there is a crush of authors trying to get their papers published in *Physical Review Letters*, and every year sees several editorials in which the editor wrings his hands over the difficulty of making selections. Nevertheless, *Physical Review Letters* does a good job of judging the paper rather than the author. (In 1959, when I was an unknown research associate, I had several papers accepted by *Physical Review Letters*; in 1971, as a reasonably well-known professor at M.I.T., I had one rejected.)

In addition to the *Physical Review* and *Physical Review Letters*, there are a great number of other physics journals in which it is even easier to publish. So well does this system work that it has become quite common for a physics department chairman who needs advice on the work of a young physicist in his own department to solicit comments from senior physicists in other universities who have never even met the young physicist, on the assumption that they will of course be familiar with his or her published work.

Of course, the humanities and social sciences also have widely circulated journals, but I have the impression that they do not provide anywhere near so effective a channel of communication for the young or unestablished scholar as do the natural science journals. The reason is that the natural sciences have more objective (though not necessarily more reliable) standards for judging the value of a piece of work. A young physicist who succeeds in calculating the fine-structure constant from first principles, or in solving any one of dozens of other outstanding problems, is sure of a hearing. For instance, my own subfield of theoretical physics was shaken up in 1971 by work of a previously unknown graduate student at Utrecht,[5] and then again in 1973 by a previously unknown graduate student at Harvard.[6] I suspect that a graduate student in history who has revolutionary ideas about the fall of the Roman Empire might have a harder time getting a hearing.

The less academic professions such as law, medicine, business, the military, and the church, are even less open. In these, a young person's work is, I believe, directed to a small circle of superiors rather than to an international community, and it is natural for their judgment of his ideas to be colored by subjective factors, such as the degree to which he accommodates himself to their preconceptions. Only a few, after getting over these hurdles, reach a level from which they can communicate to their whole profession.

None of this reflects any moral superiority in the scientists themselves. It is a natural outgrowth of the fact that they work in specialities small enough that a beginner has a chance to communicate with the whole international community of specialists, and with standards objective enough that they all can recognize the value of a piece of important research. However, it does seem peculiarly inappropriate to charge the sciences with being closed to new ideas from the young and unestablished.

For the sake of fairness, I should add here that these observations are strongly colored by my own experience as a theoretical physicist who works alone at his desk or at a blackboard with one or two colleagues. I concede that the scientific enterprise may look very different to experimental scientists, and most especially to those experimentalists in high energy nuclear physics who work in large research teams.

For instance, a recent paper[7] reporting the discovery of an important new class of neutrino interaction had no less than fifty-five authors from seven different institutions. I do not know to what extent a junior member of such a team can really get a hearing for an idea of his own.

2. *How open is science to new ideas from outside?*

My remarks so far only indicate the openness of the scientific community to ideas which are at least expressed in a language that is familiar to established scientists and deal with problems that they recognize as important. Otherwise, the work is unlikely to be published in a scientific journal or, if published, to be read. Then what about the prophet in the wilderness, the truly original genius outside the scientific community whose ideas cannot be understood by the pedants in university science departments?

I submit that there is no such person. I do not know of any piece of work in physics in this century which was originally generally regarded as crack-pot—as opposed to merely wrong—which subsequently turned out to be of value. It is true that Einstein was only a patent clerk when he invented special relativity, but his work was on a recognized problem, was duly published in the *Annalen der Physik*, and was received with respect, though not with instant acceptance by the physics community.

In reaching a judgment on the closed-mindedness of scientists to ideas from outside their ranks, it should be kept in mind that the system of scientific communication has evolved, not merely to transmit ideas and data, but to do so in a way that leaves the scientist time to get some of his own work done. If we had to struggle through every paper, even when the author did not accept the conventions of scientific language, we would literally have no time to do anything else. It may be that we miss a pearl of wisdom every century or so, but the price has to be paid.

3. *How open is science to truly revolutionary ideas?*

Even granting that the scientific communication system works as well as it ought to, are not scientists' minds closed to ideas, from whatever source, which challenge orthodox scientific dogma? (As Gershwin tells us, "They all laughed at Wilbur and his brother, when they said that man could fly.") Many laymen and some scientists seem to believe that any number of scientific revolutions would immediately become possible if only scientists would give up some of their preconceptions.

I believe that this is a mistake, and arises from a misconception as to the nature of scientific advance. The scientific principles which at any given moment are accepted as fundamental are like structural timbers which support a great superstructure of successful predictions. It is easy to imagine knocking down any of these timbers, but very hard to imagine what would then keep the roof from falling on our heads.

For a major scientific advance to occur, it must become clear not only that fundamental changes are necessary, but also how the successes of the previous theory can be saved. For example in 1957 T. D. Lee and C. N. Yang brought about a revolution in physics through their proposal that parity is not conserved—that is, that there is an absolute distinction in nature between left and right.[8] (It can be shown mathematically that if right and left are equivalent, then every physical

state can be classified as having odd or even parity, according to how it seems to
change when viewed in a mirror. It can also be shown that the parity is conserved—that is, it does not change with time.) It was quite easy to imagine that
parity is not conserved; what was hard to see was that parity conservation had to be
violated, and that it could be violated without losing the spectroscopic selection
rules and other consequences which had given rise in the first place to the idea of
parity conservation. As it happened, Lee and Yang were led to their proposal by a
puzzle in meson physics. Two different kinds of meson were identified as having
positive and negative parity respectively, through their decay into states of positive
and negative parity, and yet the masses and lifetimes of the two mesons were
observed to be identical. Many solutions were tried, including fundamental
changes in the principles of quantum mechanics. Finally, rejecting any such radical
solution, Lee and Yang proposed that the two different mesons were really only
one, that the meson had seemed like two because it could decay both into states of
the same and of different parity. This proposal would have gotten nowhere if they
had not pointed out at the same time that parity could be changed in these decays
because they were "weak" (that is, they have rates only of order 10^{10}/sec per particle), thereby leaving unchallenged the successful predictions of parity conservation in the much faster (say, 10^{20} to 10^{24}/sec) "strong" and electromagnetic
processes.

Even the greatest scientific revolutions show a similar conservatism. Einstein
changed our understanding of space and time, but he did so in a way which was
specifically designed to leave our understanding of electricity and magnetism intact. What the scientist needs is not a wide open mind, but a mind that is open just
enough, and in just the right direction.

4. How open is science to uncomfortable new data?

One often reads in popular histories of science that "So and so's data showed
clearly that this and that were false, but no one at the time was willing to believe
him." Again, this impression that scientists wantonly reject uncomfortable data is
based on a misapprehension as to the way scientific research is carried on.

The fact is that a scientist in any active field of research is continually bom-
barded with new data, much of which eventually turns out to be either misleading
or just plain wrong. (I speak here on the basis of my experience in elementary particle physics and astrophysics, but I presume that the same is true in other fields as
well.) When a new datum appears which contradicts our expectations, the
likelihood of its being correct and relevant must be measured against the total mass
of previously successful theory which might have to be abandoned if it were
accepted.

During the latter half of the nineteenth century, for instance, there were known
anomalies in the motions of the moon, Encke's comet, Halley's comet, and the
planet Mercury, all of which seemed to contradict Newton's theory of gravitation.
These anomalies might have caused a tremendous amount of effort to be wasted
looking for alternative theories of gravitation, but most physicists either ignored the
data or assumed that some less radical explanation would turn up.[9] As it happened,
they were 75 percent correct; simple explanations (such as an improvement in the
treatment of tidal forces) were later found for the anomalies in the motions of the

noon and the comets. The anomaly in the motion of Mercury did, in 1916, turn out to be of fundamental importance when Einstein showed how it arose from relativisitic corrections to Newtonian mechanics. But even this is an exception that proves the rule. If physicists had taken the anomaly in the motion of Mercury seriously from the beginning, presumably they would also have taken the anomalies in lunar and cometary motions seriously, and would thereby have been led away from rather than toward the discovery of general relativity.

Here is a simpler and more recent example. At a high energy physics conference in 1962, data were reported to the effect that neutral K mesons and their antiparticles can both decay into a positive pi-meson, an electron, and a neutrino. If true, this would have overturned a theory of weak interactions, the "current-current model," which had served as the basis of a great number of successes in other contexts. I remember Murray Gell-Mann rising and suggesting to the meeting that since the experiments didn't agree with the theory, the experiments were probably wrong. The next generation of experiments showed that this was indeed the case.

I realize that it may seem to the reader that the theorists in these examples were merely closed-minded and lucky. However, no scientist is clever enough to follow up hundreds of clues that lead in hundreds of different directions away from existing theories. (This is especially true of data of dubious provenance which would revolutionize scientific knowledge, such as evidence on unidentified flying objects, psychokinesis, and copper health bracelets.) What a scientist must do is to be open to just that piece of new data which can be integrated into a comprehensive new theory, and to file the rest.

Above all, in judging the openness of science, one should remember its unique capacity for discovering its own mistakes. Most natural scientists have the experience several times in their lives of being forced by new data or mathematical demonstrations to recognize that they have been seriously wrong about some important issue. (For instance, I was sure that Lee and Yang were wrong when they first proposed that parity is not conserved, and became convinced only by subsequent experiments.) On a larger scale, the physics community has many times been forced by new data to scrap large bodies of existing theory. If this takes away from our reputation for infallibility, it should also take away the impression that our minds are closed.

The Scientist as Adding Machine

The most profound challenge to science is presented by those, such as Laing and Roszak, who reject its coldness, its objectivity, its nonhumanity, in favor of other modes of knowledge that are more human, more direct, more rapturous.[10] I have tried to understand these critics by looking through some of their writings, and have found a good deal that is pertinent, and even moving. I especially share their distrust of those, from David Ricardo to the Club of Rome, who too confidently apply the methods of the natural sciences to human affairs. But in the end I am puzzled. What is it that they want me to do? Do they merely want the natural scientist to respect and participate in other modes of knowledge as well as the scientific? Or do they want science to change in some fundamental way to incorporate these other modes? Or do they want science simply to be abandoned? These three possible demands run together confusingly in the writings of the critics of science, with arguments for one demand often being made for another, or for all three. In accordance with my role here as a specimen of the unregenerate working scientist, I will try in what follows to keep the issues raised by these three demands logically distinct, and to analyze each in turn.

1. *We should recognize the validity of other modes of knowledge, more human and direct than scientific knowledge.*

Roszak expresses this view in terms of a metaphor he attributes to Stephen Toulmin:[11]

When we insist on making scientific expertise the arbiter of all knowledge, it is exactly like believing that cartographers know more about the terrain than the natives who live there, or the artists who have come to paint its beauties, or the priests who tend its holy places.

This does not seem to me to be an issue which raises any problems for science. Scientists, like other folk, are perfectly willing to respect and participate in various kinds of mental activity—aesthetic, moral, even religious. Perhaps the hang-up is with the word "know." For my part, since I view all epistemological arguments with perplexity anyway, I am willing to describe the perceptions of the Lake of Nemi experienced by Turner or the priests of Diana as "knowledge." For certain practical decisions, such as where to have a picnic, I would even be guided by this "knowledge" rather than by a contour map of the lake. Continuing Toulmin's metaphor, the real problem is whether maps should all be redesigned to incorporate aesthetic and moral information, or, if this is impossible, whether maps have any value at all? This is the problem I address below.

2. *Science should change so as to incorporate other modes of knowledge.*

To quote Roszak again,[12]

What should come of this ideally is not some form of separate-but-equal coexistence, but a new cultural synthesis.

And again,[13]

It is a matter of changing the fundamental sensibility of scientific thought—and doing so even if we must drastically revise the professional character of science and its place in our culture. There is no doubt in my mind that such a revision would follow. Rhapsodic intellect would slacken the pace and scale of research to a degree that would be intolerable by current professional standards. It would subordinate much research to those contemplative encounters with nature that deepen, but do not increase knowledge. And it would surely end some lines of research entirely out of repugnance for their reductionism, insensitivity, and risk.

My answer is that science cannot change in this way without destroying itself, because however much human values are involved in the scientific process or are affected by the results of scientific research, there is an essential element in science that is cold, objective, and nonhuman.

At the center of the scientific method is a free commitment to a standard of truth. The scientist may let his imagination range freely over all conceivable world systems, orderly or chaotic, cold or rhapsodic, moral or value-free. However, he commits himself to work out the consequences of his system and to test them against experiment, and he agrees in advance to discard whatever does not agree with observation. In return for accepting this discipline, he enters into a relationship with nature, as a pupil with a teacher, and gradually learns its underlying laws. At the same time, he learns the boundaries of science, marking the class of phenomena which must be approached scientifically, not morally, aesthetically, or religiously.

One of the lessons we have been taught in this way is that the laws of nature are as impersonal and free of human values as the rules of arithmetic. We didn't want it to come out this way, but it did. When we look at the night sky we see a pattern of stars to which the poetic imagination gives meaning as beasts, fishes, heroes, and virgins. Occasionally there is drama—a meteor moves briefly across the sky. If a correlation were discovered between the positions of constellations and human personalities, or between the fall of a meteor and the death of kings, we would not have turned our backs on this discovery, we would have gone on to a view of nature which integrated all knowledge—moral, aesthetic, and scientific.

But there are no such correlations. Instead, when we turn our telescopes on the stars and carefully measure their parallaxes and proper motions, we learn that they are at different distances, and that their grouping into constellations is illusory, only a few constellations like the Hyades and Pleiades representing true associations of stars. With more powerful instruments, the whole system of visible stars stands revealed as only a small part of the spiral arm of one of a huge number of galaxies, extending away from us in all directions. Nowhere do we see human value or human meaning.

But there are compensations. Precisely at the most abstract level, furthest removed from human experience, we find harmony and order. The enormous firmament of galaxies is in a state of uniform expansion. Calculations reveal that the rate of this expansion is not very different from the "escape velocity" which would just barely allow the expansion to continue forever. Furthermore, there seems to be a frame of reference in which the expansion is spherically symmetric, and we find that this cosmic frame is rotating at less than one second of arc per century.

The order we find in astronomy on the largest scale is only a small part of a much grander intellectual picture, in which all the systematic features of nature revealed by experiment flow deductively from a few simple general laws. The search for these laws forces us to turn away from the ordinary world of human perception, and this may seem to the outsider to be a needless specialization and dehumanization of experience, but it is nature that dictates the direction of our search.

When Galileo measured the frequencies of pendulums of varying lengths, Simplicio might have objected that this was a purely artificial phenomenon invented by Galileo himself, less worthy of attention than the natural bodies falling freely through the open air that had been discussed by Aristotle. However, Galileo perceived the existence of laws of motion which could more easily be approached through the nearly frictionless motion of a pendulum than through the study of bodies subject to the resistance of the air. Indeed, Galileo's great contribution to mechanics was precisely this perception, rather than the discovery of any particular law of motion.

In the same way, when we spend millions today to study the behavior of particles that exist nowhere in the universe except in our accelerators, we do so not out of a perverse desire to escape ordinary life, but because this is the best way we know right now to approach the underlying laws of nature. It is fashionable these days to emphasize the social and political influences upon scientific research, but my reading of history and my own experience in physics convince me that society provides only the *opportunity* for scientific research, and that the *direction* of this research is what it is to an overwhelming degree because the universe is the way it is.

We have, of course, a long way to go in understanding the laws of nature.[14] However, as far as we can now see, these laws are utterly cold, impersonal, and value free. By this, I don't at all mean that they are without beauty, or that there are no consolations in science. What I mean is that there does not seem to be anything in the laws of nature which expresses any concern for human affairs, of the sort which we, in our warm-blooded furry mammalian way, have happily learned to feel for one another.

Having committed ourselves to the scientific standard of truth, we have thus been forced, not by our own choosing, away from the rhapsodic sensibility. We can follow Roszak's lead only by abandoning our commitment. To do so would be to lose all of science, and break off our search for its ultimate laws.

3. *If science cannot be reformed, it should be abandoned.*

One must doubt that the world would be happier if we could forget all about the laws of nature. The prescientific mind peopled the world not only with nymphs and dryads, but also with monsters and devils; at least in one historian's view, it was only the triumph of science that put an end to the burning of witches.[15] But suppose for the sake of argument that the case could be made that we would be happier if science were driven into some obscure utilitarian corner of our consciousness. Should we let this happen?

In the end, the choice is a moral, or even a religious, one. Having once committed ourselves to look at nature on its own terms, it is something like a point of honor not to flinch at what we see. For me, and perhaps for others, the helplessness of man in the face of pain and death also gives a certain bitter satisfaction to the attempt to master the objective world, if only in the mind. Roszak and Laing point out what they see as the moral dangers of objectivity, fearing that it is likely to leave the scientist himself as cold and value free as an adding machine. I do not see this happening to my colleagues in science. But, in gurus and flower-children, I do see the danger of subjectivity, that the rejection of an external standard of truth can leave a person as solipsistic and self-satisfied as a baby.

Finally, I must emphasize again that the "coldness" I have referred to above only characterizes the discovered *content* of science, and has nothing to do with the wonderfully satisfying *process* of scientific research. In the last section I tried to show how scientists are joined together in a world society, fairer and more open than most. On an individual level, although we accept a discipline in testing our ideas against experiment, the generation of scientific premises is left to the scientist's imagination, guided but not governed by his previous experience. As Gerald Holton recently reminded us in citing Einstein's letter to Solovine, the method of scientific discovery often involves a logically discontinous leap upward from the plane of experience to premises.[16] For some scientists, in our time notably Einstein and Dirac, the aesthetic appeal of the mathematical formalism itself often suggested the direction for this leap. And even though scientific research may not fill us with the rapture suggested by a Van Gogh, the mood of science has its own beauty—clear, austere, and reflective, like the art of Vermeer. Or to use a different simile: if you accept the cliché that hearing a Bach fugue is like working out a mathematical theorem, then you ought also to realize that working out a mathematical theorem is like hearing a Bach fugue.

In the Science Museum in Kensington there is an old picture of the Octagon Room of the Greenwich Observatory, which seems to me beautifully to express the mood of science at its best: the room laid out in a cool, uncluttered, early eighteenth-century style, the few scientific instruments standing ready for use, clocks of various sorts ticking on the walls, and, from the many windows, filling the room, the clear light of day.

References

1. L. Meitner, F. Strassmann, and O. Hahn, *Zeitschrift für Physik*, 109 (1938), p. 538; O. Hahn and F. Strassmann, *Naturwissenschaften*, 26 (1938), p. 756.

2. W. G. Hoskins, *The Making of the English Landscape* (London: Hodder and Stoughton, 1955).

3. J. Schwinger, in *Nature of Matter—Purposes of High Energy Physics*, ed. L. C. L. Yuan (Upton, N.Y.: Brookhaven National Laboratory, 1965), p. 23.

4. The following remarks are based on my own observations, but the general conclusion, that the scientific communication system operates in a fair and open manner, is supported by detailed statistical studies. See H. Zuckerman and R. K. Merton, *Minerva*, 9 (1971), p. 66 and *Physics Today* (July 1971), p. 28. For comments on the reward system in science, see S. Cole and J. R. Cole, *American Sociology Review*, 32 (1967), p. 377.

5. G. 't Hooft, *Nuclear Physics*, B33 (1971), p. 173.

6. H. D. Politzer, *Physical Review Letters*, 30 (1973), p. 1346.

7. F. J. Hasert *et al.*, *Physical Review Letters*, 46B (1973), p. 121.

8. The original papers on this subject are conveniently assembled in *The Development of Weak Interaction Theory*, ed. P. K. Kabir (New York: Gordon and Breach, 1963).

9. The history of these problems is reviewed by S. Weinberg, *Gravitation and Cosmology* (New York: John Wiley, 1972), Sec. I.2. Also see E. Whittaker, *A History of the Theories of Aether and Electricity* (Edinburgh: Thomas Nelson, 1953), 2, Ch. 5.

10. For a bibliography and useful comments, see C. Frankel, *Science*, 180 (1973), p. 927.

11. T. Roszak, *Where the Wasteland Ends* (Garden City, N.Y.: Doubleday Anchor Books, 1973), p. 375.

12. T. Roszak, unpublished comment on an earlier version of the present article.

13. Roszak, *Where the Wasteland Ends*, pp. 374–375.

14. I have attempted to describe how far along we are now in coming to an understanding of this deductive order, in *Science*, 180 (1973), p. 276.

15. H. R. Trevor-Roper, *The European Witch-Crazes of the Sixteenth and Seventeenth Centuries* (Hammondsworth, England: Penguin Books, 1969), Ch. 5.

16. G. Holton, address at the Copernicus Celebration, National Academy of Sciences, Smithsonian Institution, April, 1973.

For help in the preparation of this article, I wish to thank M. Katz, E. Skolnikoff, L. Weinberg, and V. F. Weisskopf.

MARC J. ROBERTS

On the Nature and Condition of Social Science

The Imperfect is our Paradise.
Wallace Stevens

SOCIAL SCIENCE is still quite fashionable today. Public interest and support continue at high levels.[1] Yet various scholars, young and old, have expressed strong reservations about the meaning and orientation of current work. Two recent presidents of the American Economic Association have argued that current economic theory is unhelpful and too abstract.[2] A number of well-known sociologists urge that basic concepts in their discipline contain important, implicit biases.[3] And a distinguished political theorist writes that recent developments in political science will hinder students' intuitive grasp of the basic issues in his field.[4] At the same time, radicals proclaim that all these fields are mere tools for the maintenance of capitalism, and that their own work cannot be fairly judged by colleagues "corrupted" by conventional viewpoints.[5]

The experts were supposed to have the answers. Yet the persistence of crime and unemployment, poverty and inflation have made many uneasy. Is social science really useful after all? Indeed, should it even try to be useful? Some contend that a refusal to become immersed in pressing policy matters is immoral; others that detachment is a necessary correlative of the "scientific" attitude they consider essential to the long-run accumulation of human knowledge.

This questioning and debate, I will argue, is at least partially justified. Social science has accomplished less than it might because social scientists have inappropriately tried to imitate certain characteristics of natural science, especially physics. Social scientists have not understood that the nature of the particular phenomena they study has implications both for how they should proceed and what they can hope to find out.

I propose to begin by examining some fundamental epistemological questions. In order to understand what is unique about social science, we have to understand how science in general operates, and where the critical differences between social and natural science arise. I will argue that the social sciences can expect to discover fewer, less general, and less precise regularities than some natural sciences. Social science theories are relatively simplified and stylized, while the role of "craftsmanship" and "tacit" knowledge are correspondingly greater. Furthermore, although there is a role for values in all science, this role is more important in social than in natural science.

I will then sketch the implications of this situation for the current pattern of disciplinary organization and activity in social science. In particular, I will contend that some attractive, basic norms as to what science "should" do imply that less concern with "pure theory" in social science, and more concern with defining clear and relevant questions has much to recommend it. Such an orientation can better serve both practical and purely scientific ends than the vision of the social scientist as an unfettered, value-free seeker after the general laws of nature.

What Is Science?

Science is just one of many activities human beings have created to serve a series of related practical and psychological goals. Like crafts, religion, technology and art, science helps man to feel at home in the universe and enables him to improve his material lot by expanding his ability to explain, control, and predict the world around him. While all sciences share some features in common, the term "science" has been applied to a very varied collection of activities, institutions, methodologies and results.

To understand how sciences both resemble and differ from each other, we can view the conceptual systems which characterize science as hierarchical in nature. Any one area of study shares certain commitments, assumptions and definitions in common with all scientific areas—the commitment, for example, to employ data whose collection does not depend on the identity of the observer. Almost as fundamental—and often implicit and unself-conscious—are the epistemological postulates that delimit the admissible class of "facts" (and their legitimate interpretations). Less basic are the substantive conceptual systems of a particular science—its formulations of how the world works. These include specific, detailed definitions of relevant quantities or qualities, and the rules for measuring or observing them. These definitions are often interdependent with a quite complex conceptual apparatus which lends significance to the individual terms. Each science also includes assertions as to what research strategies are legitimate or useful, and likely to produce data that can be manipulated by approved methods. Most specifically, a science can include detailed statements about particular relationships and quantities. The existence of such systematic conceptual systems helps to distinguish science from "art,"[6] just as its emphasis on objectively verifiable data serves to distinguish it from religion. But all the boundaries between science and nonscience are ambiguous. Activities like engineering, alchemy, and history are hard to classify. They operate with mixed methods and on the basis of mixed commitments. There is no unambiguous or inherent essence to science which allows us to distinguish it precisely.

The substantive theories and concepts of science that tell us about the world are both the cause and the result of empirical work. Even the collection of data cannot proceed without some implicit or explicit set of categories for noticing and recording it. On the other hand, we can only know how well our concepts inform us about the world if we evaluate and try to improve them in light of actual experience.

Of course, an individual scientist typically does not set out to choose/construct all his conceptual equipment on his own. Instead, historically, logically, and psychologically, various kinds of notions have been linked together into more or less coherent and interdependent conceptual systems. Such conceptual clusters simultaneously suggest questions and answers, techniques and results. They are communicated to would-be scientists both explicitly and implicitly via accounts of "good" experiments or research projects.[7]

The much-debated question of whether the historical development of the conceptual structure of science does or does not involve "revolutions" does not have a general answer. My account makes it clear that there can be many kinds of differences among such structures, even when they claim to apply to the same phenomena. These differences may be profound or trivial, substantive or methodological. They may or may not be resolvable by an appeal to shared assumptions and information, depending on the particular situation.[8]

Such interlinked systems of methodological assumptions and empirical theory serve a deep psychological need. They provide both a statement about reality and a perspective from which to interpret reality which is very attractive, especially for individuals concerned enough about the order of the universe to become scientists. Such people often become very attached to their views of the world. Holders of a theory often ignore discordant data and may not accept a new theory despite the logically relevant evidence in its favor.[9] The personal psychological readjustment required is too great. The fact that a change in generations is often required to bring about a change in scientific beliefs reflects the psychological as much as the logical leap involved.

Psychology may be reinforced—consciously or not—by self-interest. When the conceptual system of a science changes, the professional position of a scientist and his past investment in training and experience, can be significantly endangered. Such changes can transform the skills and wisdom of older practitioners from valuable assets into constraining habits of mind that inhibit their use of newer concepts and techniques.

"Good" Science and the Structural Approach

Without making an explicit, elaborate moral argument, I propose to accept as appropriate goals for science the tasks of explanation, prediction, and control, to which much of science has in fact been devoted. I do not claim that these ends are somehow inherent in science; on the contrary, "science" is simply what we have come to call certain activities which have often served these purposes. Such general normative assumptions will not allow us to draw very specific conclusions about how science ought to be conducted. Nonetheless, they are strong enough to serve as the basis for a critical reassessment of social science.

While the notions of prediction and control are straightforward, there has been a great deal of discussion about the nature of explanation. I am using the term in the nontechnical sense: to "explain" something is simply to tell a story about what has happened that makes it plausible that events should have occurred as they did.[10] (What kinds of stories are "good" explanations for various purposes is a distinct issue.) In pursuing pure explanation for its own sake, science comes closest to serving a religious function—one which may be very important both to the scientist and to the society that supports him.

Explanation, prediction, and control are distinct phenomena. We can explain earthquakes, airplane crashes or the current inflation without being able to control or predict them very well. The Phoenicians and Babylonians could predict tides and planetary movements, but were quite unable to explain, still less control, them. And natural phenomena like electricity were controlled and put to practical use when no adequate explanations of them were available.

Like other activities aimed at explanation, prediction and control, science does not simply seek "truth." Instead, the usual problem of scientists is simply to find a *better* formulation of the phenomena being studied. The two-valued, true/false view of science fails to recognize both that approximate, hence untrue, generalizations can be very valuable and that such approximations constitute much of what we generally call "scientific knowledge."

In many sciences, the goals of explanation, prediction, and control have encouraged the use of a particular conceptual strategy which I will call the "structural" approach.[11] This tactic is not only common, but also, I contend, desirable, for it serves as an appropriate methodological maxim in many fields of study—including most social sciences.

The structural approach involves explaining phenomena by portraying the structure of the process which gave rise to them, often in terms of a physically less aggregate level of experience than that of the phenomena themselves. Thus one explains the decay behavior of uranium isotopes in terms of their atomic structure, and the physical properties of different metals in terms of their molecular structure. Such a tactic is not simply pure reductionism. Perhaps the "whole" is, in some sense, more than the sum of its "parts." All this approach says is that one should want to know what it is about the structure of the whole and the interactions of the parts that accounts for this result. At the same time, the process of giving structural explanations is a regression without a logically given end. Any account can be the starting point for further questions.

Conceptualizing the structure of a process is important for the aesthetic and psychological aims of science. Such theories provide what many people want when they ask for explanations or good explanations. For many, a tide table would not be a satisfactory explanation of tidal phenomena. In addition, from the viewpoint of prediction and control, structural knowledge can be essential. Suppose we want to manipulate a system or predict its response under circumstances beyond the range of our previous experience. Without structural insight, it is very difficult to know under what conditions past regularities will continue to hold up.[12]

This long-run goal is not always possible in the short run. Instead, we sometimes explain phenomena simply by portraying them within a larger system of relationships. This can happen when structural explanation is inappropriate because we do not yet have an account of the process in question. Before Newton could explain why the planets followed Kepler's laws, Kepler had to reveal those "laws." Before that, Copernicus had to discover that in fact the planets do go around the sun. In other cases, structural explanation is, for the moment, impossible because we have encountered the physically least aggregate objects to which we have access—scientists in the 1930's, for example, could not further break down electrons. In still other instances, we may be unable to relate superstructure to substructure because the system is too complex, as appears to be the case with human thought today.

In contrast to the structural approach, there is what I will call the "curve-fitting" view of science, which suggests that science simply looks for regularities in experience.[13] In this view, explaining a phenomenon involves merely setting forth the regularity of which it is a particular instance. Theories, models, or other conceptual systems are nothing more then logically irrelevant neumonics which help us to remember certain predicted regularities in observable phenomena. Unfortunately, few scientists seem to consider their theories as a mere arbitrary shorthand of this sort.

The central difficulty is that "curve fitting" ignores the role that our belief in the continuity of experience plays in all science. We tend to believe that similar situations will produce similar results—even if we cannot fully specify the dimensions of difference and resemblence.[14] Structural accounts automatically provide some clues as to where they will be most applicable—namely where the world most resembles the structure they portray. On the other hand, if the theoretical structure is considered imaginary and arbitrary, it can give us no hint about when and where it will apply. Instead we need to remember a series of separate "application theorems" to tell us where the "curve" will happen to "fit." In contrast, accepting the structural view means that science should attempt to construct the most accurate, realistic, and complete account it can of the actual structure of the process that generates the regularities it investigates.

Differences in the Role and Nature of Concepts

In some few areas of study—particle physics, for example—it has proved fruitful for scientists to adopt the stance that the mapping from concepts to reality should be exact and complete. A "theory" is seen as a detailed blueprint of phenomena. Any lack of correspondence between concept and data is an anomaly, warranting serious research. At the other end of the spectrum, in engineering, meteorology, and most social sciences, the conceptual system of a science is but a simplified model or even just a crude sketch.[15] Such formulations capture some of the features of the world and not others; they function primarily as heuristic and didactic devices to aid specific investigations of a statistical or case-study sort.[16] Of course much variety is possible in the adequacy even of crude models.

When concepts are stylized and inaccurate, the ability to explain/predict/control real events requires a large component of nonconceptualized "tacit" knowledge,[17] as Polanyi has argued. While we cannot fully formulate and write down what we know by means of such "skills" or "insights," such knowledge can be reliable and perfectable; in fact, it comprises much of what we know about the world. The distinction between science and craft is thus a matter of degree. In varying combinations, both tacit and conceptual components enter into all the activities we group under either heading. And almost all transmit the "craft" aspects of their knowledge via the traditional apprenticeship pattern. Of course, exact comprehensive theories are preferable to inexact, impressionistic models. But if the best theory available is only impressionistic, then one must use it.

This view sharply contrasts with the "covering law" notion of science which says that the business of science is to test and establish as true the laws of nature.[18]

In fact, strict tests would be largely beside the point when our model is a crude, stylized approximation which we know to be false to begin with.[19] Instead, we often use a more complex interactive process which simultaneously develops theory and our knowledge of the world.

In order for a theory to reproduce closely the behavior of phenomena, those phenomena must exhibit certain characteristics. First of all, they must be orderly; otherwise we will not be able to find empirical regularities to parallel conceptual ones. Second, we have to be able to relate the distinctions and concepts of the theory to clearly observable categories and characteristics in the real phenomena. Since the time of John Stuart Mill, it has been recognized that orderly, regular behavior, and clear, precise distinctions only exist where the world can be sorted into "real kinds."[20] That is, we must be able to divide objects of study into unambiguous classes whose members are more or less alike. The structural approach both assumes and implies that such classes exist when, first, all objects within each class are similar in internal structure, and second, the set of observed structures is discontinuous so that there are "gaps" between classes.

The point is that we can only seek to generalize about apples when we can tell apples from nonapples, and when all apples are more or less the same. Furthermore, apples will be the same if their internal structures are sufficiently similar, and they will be clearly distinguishable from nonapples if there are no fruits whose structure, and hence behavior, is similar to, but "just a little different from," apples in general.

This argument leads one to expect that theories will map experience most precisely where structures are simplest and structural variety most limited. Thus, as is in fact the case, we would expect theories to "fit" best in areas like atomic physics. In contrast we should not expect theories to do as well when we study complex heterogeneous objects that vary in both structure and behavior—like those of social science.[21] Business firms simply are not as much alike as electrons, and we should not expect to be able to say as much about them in general or in detail.

Highly varied and complicated objects are obviously difficult to conceptualize and categorize. But our limited intellectual capabilities imply that we must do it anyway in order to reduce the world to manageable proportions. Thus we often try to sort objects into a few distinct classes, despite the fact that their characteristics are scattered over a highly multidimensional continuum. This gives rise to theory as inexact sketch. Such conceptual schemes are often not really about the world as we know it. Instead they amount to parables about what might happen in a simpler and less ambiguous reality. Categories like "the steel industry" or "the middle class" represent the results of such a process. They are surrounded by an irreducible penumbra of ambiguous instances. The conceptual schemes from which they derive simply cannot resolve many real definitional problems in a nonarbitrary manner.

Of course, some classifications of complex, aggregate objects do reflect relatively clear distinctions, such as those among most species. But that alone does not mean that we can hope for exact generalizations about all members of each category. Structure and hence behavior within the class could still be quite varied. We may be able to tell gorillas from baboons, but that does not mean that all baboons will behave exactly alike.

The process of fitting exact models to experience finds its most complete expression in the use of mathematical models. Such models have obvious advantages. They force the scientist to make precise assumptions and allow him reliably to construct long and complex chains of reasoning. But this precision also has its drawbacks. A postulate is either assumed or it is not.[22] Mathematical notions assume *perfect* unambiguousness between categories and *perfect* homogeneity within them—assumptions which are a greater or lesser handicap, depending on how well the structure of reality corresponds to them.

The typical objects of study of social science (and many in natural science as well) are not "arithmomorphic," and mathematics must be used quite cautiously. Here the very precision of a formal statement is an oversimplification, and hence only an approximation.[23] As the physicist Percy Bridgman once argued, thought is simpler than reality, language simpler than thought, and mathematics simpler than language; thus only in special cases can mathematics closely mirror reality.[24]

Paul Samuelson argues, in contrast, that language and mathematics are strictly equivalent.[25] As an analogy, he suggests that everywhere you can get by railroad, you could also get by foot—and vice versa. I contend that the "vice versa" statement is false, although the initial analogy is perhaps more revealing than Samuelson realized. The train (mathematics) may run faster and more reliably, but it cannot take you everywhere. Instead it can only reach a simplified subset of all possible locations—those along the railroad tracks. Nevertheless, to get somewhere not on the track, it might well be advantageous to take the train (mathematics) at least as far as you can. Indeed, for some long and arduous journeys, going by foot (reasoning verbally) may not be a realistic possibility. For some other destinations, however, no train may be of any use whatever.

Mathematics has dangers as well as advantages. The lure of formalist virtuosity can seduce us away from the initial focus of our inquiry. If a specific real phenomenon, an actual empirical problem does not stand behind the formal model, it is always possible for theoretical efforts to become little more than recreational mathematics.[26] Under those circumstances, which I would argue are characteristic of contemporary economics, it is quite possible for a discipline to be pursued in a manner that does not most effectively achieve the goals of explanation, prediction and control.

The Role of Values in Science

I fail to see how science itself can be justified except on the basis of some prior norms. The goals of explaining, predicting, and controlling nature are, after all, only goals. The systematic, objective approach by which science both defines its activities and measures its progress is not the only conceivable approach to epistemological problems. A Zen mystic, for example, would accept neither its definition of the question nor its specifications as to what constitutes a good answer.

The relationship between facts and values is thus asymmetric. Logically speaking, values come first, for facts alone cannot serve to establish or justify values. The naturalistic fallacy is a fallacy. Even a discovery that certain modes of thought were due to man's biological structure and essential for species survival would not establish them as ethically desirable.[27] The discovery of facts, on the other hand, depends at least in part on concepts, assumptions, and inferences which can only be defended with reference to normative presumptions. All "knowledge" is ultimately dependent on some basic presumptions which cannot themselves be objectively established.

Values also enter into the choice of specific conceptualizations and theories. Criteria for choosing among conceptual schemes often conflict. Furthermore, the theoretical approach most useful in dealing with one aspect of the world might not be the most useful for dealing with another. Thus choosing an approach involves value judgments as to what is important to explain and predict, whether generality is to be sacrificed to simplicity, and so on.[28] The fact that on occasion we are able to choose among competing theories by appealing to the basic assumptions and norms which they share does not contradict this conclusion. We have to accept those norms and assumptions even to pose the issue.

We cannot avoid this difficulty by suggesting that "all" approaches should be developed because resources are both valuable and limited. When more resources are devoted to one theory, fewer resources are left to pursue another. The same constraint operates in the process of training new scientists. Time in class, in the library, and in the laboratory is limited. So is space in books and periodicals. In allocating them, some selection must be made as to what to emphasize. Even the suggestion that we devote equal time to every alternative does not avoid the problem, since that choice is no more value-free than any other.[29]

In economics today, for example, the standard approach to the study of business behavior spends a great deal of effort trying to explain variations in the margin between prices and costs in different industries. It accepts as given many features of these industries such as the distribution of their firms in terms of number and size, the types of products they offer, the technology they employ, and their relationships with the government.[30] To pursue this approach and not ask where these "exogenous" variables come from is a choice which cannot be adequately defended simply by claiming "That is what economics is." Similarly, there are any number of ways to measure and compute the unemployment rate or the price level. Choosing one as "better" than the others requires an answer to the question, "Better for what?"

Values also enter into the apparently neutral process of observation, since all inference is ambiguous in the presence of random elements. It does not matter whether the random elements result from imperfections in the process of observation or from the indeterminateness of the world itself; where they are present, what data can tell us is quite limited. In general, we cannot be sure what the truth is. In such cases, drawing conclusions about the real state of the world involves values. Suppose we must choose between two hypotheses. No matter which we select, there is always the possibility that the other is correct. Obviously the relative likelihood of making a mistake when we select one or the other matters—but so too do the costs of alternative mistakes, the costs of assuming A is true when in fact B is true or vice versa. We might well choose to risk a more likely small cost than a less likely large one. Yet the magnitude of the cost of being wrong in each case cannot be determined except on the basis of our values.[31]

Consider an extreme example: the view that there are genetic differences in the mental functioning of different races. Suppose society were to accept this view, and it proved false. I believe that very great evil would have been done. On the other hand, suppose society adopted the view that there are no differences, and that turned out to be incorrect. I would expect much less harm to result. Given these costs, I would want evidence which made the hypothesis of interracial similarity very unlikely indeed before I would reject it. My scientific choice depends on my values, not because I am uncritical or would just like to believe that there are no such differences, but because consistent choices under uncertainty can only be made by looking at the cost of making alternative kinds of errors.[32] In contrast, a would-be "value-neutral scientist" would presumably be willing to operate on the assumption that such differences exist as soon as evidence made it even slightly more likely than the reverse assumption.

These questions do not arise routinely in scientific work because traditional statistical methods typically subsume them under the choice of test criteria or of the particular technique to be used in estimating some magnitude. That choice is then made on conventional or traditional grounds, usually without discussion, justification, or even acknowledgement that value choices have been made.

Some might argue that a value-free scientist should not make any choice. Instead he should just present the probabilities and let each reader draw conclusions based on his own position. While I strongly favor presenting statistical results in such a way, this argument is beside the point when policy decisions or the allocation of scientific effort turns on the researcher's conclusions. Furthermore, such a presentation does not change the fact that the formulation of a statistical problem and the choice of a particular approach is itself value laden.

The practical importance of these difficulties varies among fields. In natural sciences, individuals with different *social* values will often evaluate the costs of making various kinds of *scientific* errors quite similarly. Liberals and conservatives could well have the same view of the cost of assuming that Jupiter had a solid core when in fact it did not. In the social sciences, however, those with different values will often see the costs of alternative errors quite differently. As a result, ethical differences can and should be important in leading scientists to divergent scientific positions. It is not necessarily a sign of unscientific behavior that conservative economists so often find reasons for less government action, while liberal ones find reasons for more. Even in the natural sciences, moreover, such differences can occur. Recent debates on pesticides, nuclear reactor safety, and the SST, for example, all found the experts divided, no doubt in part because they had different perceptions of the costs of making different kinds of mistakes.

Values then are involved in the choice to pursue any given question in a specific scientific manner. This statement has often been accepted, perhaps without full awareness of its implications.[33] Of course, one cannot consistently choose to believe anything. Choices are restricted by the "facts" as determined in accordance with one's basic epistemological and conceptual assumptions. That these most fundamental assumptions cannot themselves be derived from experience should not be worrisome to practical men. The lack of an objective or transcendental justification for science simply reflects the normatively empty character of the universe. All values and actions can ultimately be justified only in terms of those unprovable ends to which individuals choose to commit themselves.

The Difficulties of Doing Social Science

The subject matter of social science thus creates two basic difficulties. (1) The structural complexity and variety of the phenomena make detailed generalization difficult. (2) In addition, value differences among investigators are often relevant to scientific choices. As Planck is reported to have told Keynes, "Economics is harder than physics!"

A major aspect of this difficulty is our limited ability to give structural accounts of human behavior in terms of underlying biophysical processes. The social scientist faces several unsatisfactory choices. He can abandon all attempts at structural explanation and focus only on patterns of overt activity. Such "stimulus-response" psychology, however, does not explain complex individual choices very well.[34] Alternatively, he can portray actions as the result of an individual's conscious and unconscious decision processes—a strategy offering at least three specific options. One, the Freudian approach, was developed to explain abnormal behavior, and has not given rise to a fruitful model for analyzing much of what social scientists want to explain—why, for example, housewives buy less milk when the price of milk goes up.[35] A second option employs the simple utilitarian psychology of economics which amounts to little more than a set of consistency conditions.[36] This can be helpful in some circumstances, but it leaves many major issues utterly inaccessible, such as the origin of tastes and preferences. Furthermore, it assumes much more consistency than we observe in real choices. Finally, there is the cybernetic/information-processing approach. This may ultimately prove useful (and already has in limited cases), but it remains more a research program than a set of established results.

The problem is not just that research into what goes on inside a man's head is difficult, but that we lack an adequate conceptual framework to guide such research. The issue is not one of free will. If choices or behaviors are regular, for scientific purposes it does not matter whether or not they are free in one sense or another of that term. Yet human behavior which is extraordinarily complex and difficult to understand lies at the heart of all the processes studied by social science.

This fact brings at least four additional difficulties into play. First, our moral scruples; and the realities of social and political organization make it very difficult to experiment to obtain data on moral scruples. It is not impossible, of course, and we can use statistical techniques to analyze the experiments that nature and government policy provide for us. Often, however, these do not yield enough information for us to disentangle how the processes we are looking at actually operate. Second, interpreting the results of social science experiments is often quite difficult. It is always possible that the behavior of the subjects is influenced by the experimental situation. Although this problem is not unique to social science, it is certainly more common here than in, say, geology. Third, because human subjects learn, their awareness of a theory can affect the accuracy of that theory. The Keynesian view of business cycles leads to policy suggestions that are much more effective when most businessmen accept the Keynesian analysis. The fourth and most general difficulty is simply that social phenomena can and do change over time. What we learn today may not be accurate tomorrow. Big cars go out of fashion; party loyalty decreases. This is a problem that a student of the chemical properties of oxygen does not have to confront, although the meteorologist and population biologist face analogous, if more limited changes.

Interests and ideologies are more likely to intrude in the social sciences than in the natural sciences because of the link between social science and social policy. The logic of value-relative decision making imposes a great need for critical self-discipline. When much is at stake in the public arena and data are ambiguous, it can be difficult, even for a disinterested individual, to separate habits of mind from evidence, prejudice from insight. And not all participants in public policy discussions are disinterested. Organized socio-economic groups have explicitly supported or opposed various formulations or findings. These range from oil companies presenting statements on the energy crisis, to a group of conservative alumni of Harvard College who some years ago organized a society to oppose the diffusion of "socialistic Keynesianism" at their alma mater. Such pressures and their impact on patterns of funding do not have obvious parallels in the natural sciences.

Furthermore, the relationship of social science to policy problems makes it politically and psychologically difficult for an expert to take an agnostic position pending further research. Indeed, in the short run the experts know that on hard questions they cannot necessarily expect to learn much from additional research.

Because of its subject matter, social science has not discovered any highly general and detailed patterns. The regularities obtained by statistical methods are sharply limited in time and space. To understand the world at all, some conceptual structure has often seemed necessary. Since exact blueprint theories are not possible, social science disciplines have had to rely on models which are much simpler than the phenomena they are based on. They assume, for example, that there is a price for each commodity, or that the set of firms in a given industry is well defined. In reality, the dividing lines among products and producers are often irreducibly ambiguous. Statistics can only be collected by using arbitrary definitional criteria. These and other simplifications generally mean that we could not "test" the model in any simple sense, even if we wanted to. There is no way to relate the theory to a real world situation without doing violence to the theoretical structure. Instead, our models are "ideal types" in Weber's sense—obviously unrealistic presentations which function as a starting point, as an implicit list of questions, for a particular analysis. When we do want to understand a particular case in detail, we expect to face a significant burden in acquiring and analyzing relevant data.

Social science then is much closer to engineering or meteorology than it is to mathematical physics or astronomy, the disciplines that have been the models for many social scientists, including the pioneering mathematical economists of the 1870's. Alfred Marshall, the great British economist of the early years of the century, came closer when he compared economics to the problem of predicting the behavior of a new ironclad in a heavy sea. Experience was helpful, but the processes involved were too complex to write down with formal completeness. In a similar spirit, Keynes drew an analogy between economics and dentistry. Unfortunately, such modest images have not been widely accepted as a self-definition by social scientists.

The Current State of Social Science

While there are occasional interchanges over lunch or at cocktail parties, the day-to-day work of most social scientists does not adequately reflect these complexities. To some extent, this is only appropriate. Science only makes progress if scientists do science, and not philosophy or epistemology. However, the significance of unself-consciousness depends critically on what implicit presumptions are being ignored. In social science today, the results of this process are not totally desirable. Because social scientists have sought to be like natural scientists, they have tried (inappropriately and unsuccessfully) to develop highly general theories about broad classes of phenomena. Their failure to examine and broaden the normative basis of current work has left some very important questions unasked and unanswered. Controversial presumptions remain insufficiently explored. Without clearer norms and more realistic expectations, the cumulative development of what Kuhn has called "normal science" has not often been characteristic of social science. As a result, we can explain, predict, and control the world less well than we might have if scientific resources had been allocated differently.

As a first step, social scientists must recognize that all science is not physics. Physics has obtained equations that apply to all electrons because all electrons are, in the relevant sense, alike. All voters or consumers are not alike. When phenomena are heterogeneous, generality can only be gained at the price of content. One is forced to say less and less about each case in order to include all possible cases. Such abstract, non-phenomena-oriented theorizing in the social sciences most emphatically cannot be justified by analogy to basic research in natural science since the latter, unlike the former, is concerned with explicating real empirical events.

The failure of social scientists to make clear how little they know or can hope to know is understandable. Society seems to be most generous to and respectful of the "real sciences." Material well-being, power, status, and the scientist's ability to "fulfill his moral obligations" by influencing policy—all these depend on the acceptance by the wider society of his expertise. And when politics become involved, the chances increase that more will be promised than can be delivered, especially by the political actors in whose retinue social scientists are enlisted. In the short run, such claims both contribute to and are supported by the optimistic strain in American beliefs—the faith that all problems can be solved by sufficient hard work, money and expertise. However, when the program does not live up to its advertising, both taxpayers and supposed beneficiaries justifiably become a bit more suspicious of both politicians and experts.[37] This same pressure to find "results" also seems related to the frequently unsophisticated use of statistical methods in social science. Mechical, technically flawed studies are not at all uncommon as researchers with too little training or too little time rush to "publish or perish."

There are also pressures which provoke social scientists to work to control the distribution of society's resources within their own ranks—a practice, of course, which can produce technical conservatism and conformity. Organized scientific activity can only take place if scientists can acquire enough of society's resources to support themselves and their work. This influences both the distribution of scientific efforts and the development of scientific ideas. No one consciously has to choose a paradigm or viewpoint to please a potential source of patronage. Instead, the availability of support interacts with the interests of established scientists and new recruits. Some fields or lines of inquiry flourish and attract new adherents, while others decline. Both institutional and intellectual considerations play major roles in this evolutionary process. At any one time, practitioners may have more or less scientific discretion, and the external environment may be more or less coercive.[38] Fortunately or not, much support for social science remains problem oriented, a situation often decried by guild members on the grounds that it slights basic research.

Finally, the social scientist is under pressure not to make the role of values in social science more explicit, for this would tend to undermine his neutral status as an expert and his special professional role. This failure to clarify the ethical content implicit in social science theories is particularly important because the dominant conceptual systems in economics, sociology, and political science all share a tendency to legitimate the status quo. Much of what happens in society is taken as given, which makes it difficult to see how the world could or should be other than marginally different from what it is.

Economics operates on the model of a competitive economy which can be shown to produce an optimal outcome in terms of people's tastes. There is always a temptation to presume that the real economy is like the model, a conclusion that some economists urge quite explicitly. Similarly, in political science, it is difficult not to confuse the existing balance of group pressures with the correct outcome as given by the model of democratic pluralism. In sociology, the combination of a "value-free" perspective with structural-functionalism makes identification of what *is* with what is functional, and with what is desirable almost irrestible. In each case, the meaning and origin of the independent variables—tastes, influence, power—are left for someone else to study. Dissatisfaction with this state of affairs may well be the major reason for the resurgence of interest in Marxism as an alternative theoretical framework, especially among younger guild members.

Indeed, many economists simply ignore arguments which undermine the neutral posture. Long ago Paul Samuelson and other writers noted that the economist's usual recommendation that prices should equal costs—which is intended to guarantee economic efficiency—is not generally correct in an otherwise imperfect world.[39] Furthermore, efficiency may not be preferable to inefficiency, if the income distribution of the efficient state is unjust.[40] Yet time and again, an economist writing on a policy problem recommends that, to achieve efficiency, prices should be made equal to costs; they see such a measure as part of a neutral social policy which can appropriately be chosen purely on technical grounds.

This effort to maintain neutrality makes it very difficult for the professions to develop new lines of inquiry. Their internal agendas are limited and slow to expand. Graduate students, suddenly free to work on their own on a dissertation project, often become utterly unable to choose a topic. All of human interaction lies before them, yet they have rarely been encouraged to ask, "What would be worthwhile to study?" Thus it is disconcertingly difficult to make such decisions, and careerism, chance, inertia, and imitation often fill the gap.

For similar reasons, professional interest is more likely to follow social policy than to lead it. Work on Keynesian models of the economy grew greatly only after Congress, in the Full Employment Act of 1948, assumed responsibility for the level of employment. Studies of urban housing markets, poverty, and early childhood education all became much more extensive in scope after major federal programs were instituted in these areas. In part this pattern reflects the availability of outside funding, but not entirely. A failure of intellectual imagination is also at work. Because social science is a campfollower to public controversy, social and economic policy is often made at a time when only the slimmest scientific results are available.

Guilds adopt formal criteria of expertise in order to obtain support and fend off outsiders. Because it is so difficult to judge results, those who are in a position to do so often judge the techniques employed, or the educational backgrounds of the writers. But such means can take the place of ends. In my view, a significant proportion of recent theoretical work in economics has been of doubtful scientific value. Many papers explore questions posed not by the world itself, but by someone else's model. Whatever empirical puzzles or phenomena there are get lost in the shuffle. Consider the now quite fashionable work devoted to proving the existence of an equilibrium in various highly general models of the economy.[41] As the techniques employed have become more elegant, the point of the exercise has become less evident. And some very valuable human resources have been devoted to this effort.

Embedding an assumption within a larger mathematical formulation can help protect it from criticism. For many years, economists have posed policy problems in terms of how they affect the subjective happiness (utility) of individuals. Yet a goal of equality of utility—as opposed to equality of income—implies that individuals with expensive tastes should have an especially large share of society's resources. This assumption, when joined with another—that changes in economic arrangements do not alter the non-economic well-being of a community—facilitates the construction of an elegant mathematical apparatus. To me, both of these specifications seem quite controversial. Yet the obvious objections to them are seldom discussed.

Not all disciplines have suffered equally from these problems. Although economists do more than their share of pointless theorizing, they have apparently found it easier than, say, political scientists to construct a cumulative and steadily improving body of knowledge about at least some specific phenomena—about, for example, business cycles and the causes of different profit rates in different industries. In part this seems to be because economics has a normative framework that distinguishes between "good" and "bad" situations in terms of variables we can readily identify, such as prices, wealth and so on. Whether or not this framework is ethically definitive, it has provided the discipline with an indispensible ingredient to scientific progress: specific questions to ask.[42] Why does unemployment occur? Why are prices high and available quantities low in some markets? An explicit and sharply posed question, even if it derives from an unconvincing ethical argument, is a better starting place than no clear question at all.

Both sociology and political science lack easily determined and ethically appealing metrics by which outcomes can be characterized. The largely implicit normative basis of these disciplines just does not give rise to a series of specific distinctions among phenomena, and hence of specific questions about them. *If all outcomes are ethically equivalent, we cannot decide what events are worth noticing or explaining.* If everything that is, is functional, if every political outcome reflects, by definition, the balance of group pressures, our conceptual system does not pose specific problems. In this situation, disciplines arise in which the questions behind specific studies are often not clearly formulated, and in which they vary from one study to another. Successive studies do not check and build upon previous work.

We should note the possibility, of course, that the behavior studied by sociology and political science may be less regular than that explored by economics. The demands of the marketplace and the effective socialization of individuals within our society to the goals of material prosperity help to make the behavior of many business firms quite orderly. The processes and structures which shape political participation or family life may just be less restricted.

Social science is about highly aggregate objects whose structure and behavior vary. As a result, social scientists can realistically expect their ability to define, control, and predict to be limited. There are no numerically exact, far-reaching regularities like Newton's laws waiting to be discoverd. The structure of the system is so difficult to grasp that we have to build very stylized conceptualizations to be able to think about it at all. Such models must be employed with a good deal of craftsmanlike skill if we wish to explain, predict, and control actual phenomena at all. Furthermore, there are many ways to conceptualize social processes. We must choose among them, and our choices are inevitably based on values. Indeed, the whole enterprise is dependent on values in a number of ways, from our acceptance of the scientific approach to our interpretation of detailed statistical results. Economics may indeed be harder than physics; certainly the methods of physics are harder to apply to economics, or to sociology and political science, than to physics.

The pressures on contemporary social scientific guilds have prevented a full and frank assessment of this situation. Accomplishments have often been oversold. The lack of a clear and self-aware position has put social science in the position of a lady of easy virtue, living off social controversy and, at the same time, serving as an apologist for the status quo. Important questions go unasked and unanswered. The debate in sociology over whether "conflict" or "system" should be the key theoretical notion misses this point entirely.[43] Trading in one set of stylized answers for another will not provide sharper questions.

Social scientists need to construct their own notions of "good science," their own methodological approach appropriate to their particular subject matter. A critical reappraisal of what is known and can be known is the first step. We need a more modest view of the explanatory exactness and generality we are likely to obtain and a better understanding of the nature and use of theoretical constructs in social science. Second, we need to clarify the role of values in these disciplines. Are the questions currently being explored the ones we really wish to study? If not, what other issues should we pursue? How adequate are our current concepts for this task? Such stock-taking could provide a basis for groups of similarly inclined professionals to develop more explicitly focused research programs and new conceptualizations. To provide the necessary normative baseline, we appear to need a resurgence to substantive, as opposed to historical, efforts in philosophy and political theory.

Accepting the value relativity of social science, however, does *not* mean that the society should rapidly shift research resources about, as the focus of political attention wanders. The problems of social science are too hard for such faddishness to produce reliable results. But in order to establish research efforts which transcend the merely fashionable, each proposed investigation must have a coherent and consistent justification.

If the social sciences develop in this way, there is no reason to believe that less mathematical training would be required. Indeed, the opposite is likely to be the case. Social scientists will need to be more sophisticated in their approach to statistics. Theoretical activity will continue, and if anything become more difficult. The problem with current models, after all, is that they are often too simple to be helpful. An increase in the use of computer simulation models is one possible response. What is required is not an end to theoretical efforts but their redir ction toward *real* questions.

This view implies giving up the notion that there is some close analogy in the social sciences to basic research in the physical sciences. With complex heterogeneous objects that have many characteristics, we can hope to discern only limited regularities. The analogy is to engineering, not quantum mechanics. This makes the typical task of social science less glamorous, less general, and more expensive then it has generally been considered.

Finally, I want to urge the demystification of social science. Communication with nonexperts has to become a responsibility of all scientists, social and natural—a responsibility which is shirked when experts are not explicit about the role of values and judgments in their work, and use jargon and mathematics as professional status symbols. There is no reason to put something in a complex and inaccessible manner when the argument is inherently simple and potentially easy to understand. The problems of democratic policy making are difficult enough without this added complication.

REFERENCES

1. The analysis here derives from that in the author's doctoral dissertation of 1969. The vociferous discussions in the interfaculty seminar on the Social Role of Science, organized at Harvard in 1972-1973, contributed much to the current presentation, as have the comments of many of my colleagues, teachers and students, unfortunately too numerous to mention. The criticisms and psychological expertise of my wife, Ann S. Roberts, are also gratefully acknowledged.

2. W. Leontief, "Theoretical Assumptions and Nonobserved Facts," *American Economic Review*, 61, No. 1 (March 1971), pp. 1-7; J. K. Galbraith, "Power and the Useful Economist," *American Economic Review*, 63, No. 1 (March 1973), pp. 1-11.

3. A. W. Gouldner, *The Coming Crisis of Western Sociology* (New York: Basic Books, 1970); R. W. Friedrichs, *A Sociology of Sociology* (Glencoe, Ill.: Free Press, 1970).

4. S. Wolin, "Political Theory as a Vocation," *American Political Science Review*, 63 (December 1969), pp. 1062-1082.

5. "Symposium: Economics of the New Left," *Quarterly Journal of Economics* (November 1972).

6. For an account of conceptualization in art which emphasizes its conceptual structure, see E. H. Gombrich, *Art and Illusion* (New York: Pantheon, 1961).

7. T. S. Kuhn has called such conceptual clusters "paradigms." See T. S. Kuhn, *The Structure of Scientific Revolutions*, 2nd ed. (Chicago: University of Chicago Press, 1970), esp. "Postscript 1969," and "Reflections on My Critics"; *Criticism and the Growth of Knowledge*, eds. I. Lakatos and A. Musgrave (Cambridge: Cambridge University Press, 1970), pp. 231-278. For similar notions see also

S. Toulmin, *For Sight and Understanding* (Bloomington, Ind.: Indiana University Press, 1961); R. G. Collingwood, *An Essay on Metaphysics* (London: Oxford University Press, 1940); M. Polanyi, *Personal Knowledge* (Chicago: University of Chicago Press, 1958). An interesting examination of Kuhn's view is given by M. Masterman, "The Nature of a Paradigm," in Lakatos and Musgrave, *Criticism*, pp. 59-89.

8. For contrasting views, see Kuhn, *Structure*, pp. 111-135. Kuhn argues that a paradigm change is often a "conversion experience." The view presented here more closely resembles that of S. Toulmin, *Human Understanding*, 1 (Princeton: Princeton University Press, 1972), pp. 98-130.

9. Polanyi, *Personal Knowledge*, pp. 292–294; K. E. Boulding, "The Verifiability of Economic Images," *The Structure of Economic Science*, ed. S. R. Krupp (Englewood Cliffs, N.J.: Prentice-Hall, 1966).

10. W. Dray, *Laws and Explanation in History* (London: Oxford University Press, 1957).

11. For a fuller exposition, see my doctoral thesis, "Models and Theories in Economics," Harvard University, 1969.

12. E. Rotwein, "On the Methodology of Positive Economics," *Quarterly Journal of Economics*, 73, No. 4 (November 1959), pp. 554; D. V. T. Bear and D. Orr, "Logic and Expediency in Economics Theorizing," *Journal of Political Economy* (April 1967), pp. 188-196.

13. An especially vivid exposition of this viewpoint is given by M. Friedman, "The Methodology of Positive Economics," *Essays on Positive Economics* (Chicago: University of Chicago Press, 1953).

14. H. Simon, "Comment," *American Economic Review* (May 1963), pp. 227-236.

15. N. Georgescu-Roegen, *The Entropy Law and the Economic Process* (Cambridge, Mass.: Harvard University Press, 1971).

16. M. Weber, "Objectivity in Social Science and Social Policy," *The Methodology of the Social Sciences*, ed. E. A. Shils and H. A. Finch (Glencoe, Ill.: Free Press, 1949).

17. Polanyi, *Personal Knowledge*; Georgescu-Roegen, *Entropy Law*; S. Wolin, "Political Theory"; D. B. Truman, "Disillusion and Regeneration: A Quest for a Discipline," *American Political Science Review*, 59, No. 4 (December 1965), pp. 865-873.

18. C. Hemple, *Aspects of Scientific Explanation* (New York: Basic Books, 1965).

19. For a somewhat different view, see Lakatos, "Falsification and the Methodology of Scientific Research Programs," Lakatos and Musgrave, *Criticism*, pp. 91-195.

20. J. S. Mill, *System of Logic* (New York, 1858), Ch. 7 and 8.

21. The view that mathematics "fits" physical events better than social ones is not new. See F. Y. Edgeworth, "On the Application of Mathematics to Political Economy," Presidential address to Section F of the British Association, 1889, reprinted in his *Papers Relating to Political Economy*, 2 (London: Macmillan, 1925).

22. J. M. Keynes, *The General Theory of Employment, Interest and Money* (London: Macmillan, 1936), pp. 297-298.

23. Georgescu-Roegen, *Entropy Law*; W. J. Baumal, "Economic Models and Mathematics," *Structure of Economic Science*, ed. Krupp, pp. 88-101.

24. P. Bridgman, *The Nature of Physical Theory* (Princeton, N.J.: Princeton University Press, 1936).

25. P. A. Samuelson, "Economic Theory and Mathematics: An Appraisal," *American Economic Review*, 42 (May 1952), pp. 56-66.

26. M. Friedman, "Lange on Price Flexibility and Employment: A Methodological Criticism," *American Economic Review*, 36 (September 1946), pp. 613-631. The view in this paper is not fully consistent with some in Friedman's later essay, cited previously.

27. This is one point on which I would dissent from Toulmin's analysis in *Human Understanding*, esp. pp. 484-503. M. Weber, "Science as a Vocation," *From Max Weber: Essays in Sociology*, ed. Gerth and Mills (London: Oxford University Press, 1946), esp. pp. 143-144.

28. What is involved is some way of assigning weights to different characteristics of the model, or to accuracy of "fit" with respect to different dimensions of experience.

29. In economics, such implicit valuations are called "shadow prices," and they emerge directly from the mathematics of maximization. See T. Koopmans, *Three Essays on the State of Economic Science* (New York: McGraw Hill, 1957).

30. For an exposition of this approach, see F. M. Scherer, *Industrial Market Structure and Economic Performance* (Chicago: Rand McNally, 1970).

31. J. W. Pratt, H. Raiffa, and R. Schlaifer, "The Foundations of Decision Under Uncertainty: An Elementary Exposition," *Journal of the American Statistical Association*, 59 (1969), pp. 353-375.

32. Not all statisticians would be willing to accept this viewpoint, but I do not believe that the problem can be avoided. Hemple, *Scientific Explanation*, has suggested the notion of "pure epistemetic utility," i.e., the value of the knowledge in and of itself. To me this seems confused since there can be no nonevaluative utility—utility apart from purposes. Further, what is at stake is not the value of answering a question, but the costs of answering it incorrectly in various ways.

33. Weber, "Objectivity," is willing to accept the role of values in choosing concepts, but appears to see "facts" as less problematical. A similar position is taken by R. Dahrendorf, "Values in Social Science," *Essays in the Theory of Society* (Stanford, Ca.: Stanford University Press, 1968), pp. 1-19. For a view similar to that here, see Friedrichs, *Sociology*, pp. 135-165; G. Myrdal, *Objectivity in Social Research* (New York; Random House, 1969); and A. Gouldner, "Anti-Minotaur: The Myth of a Value-Free Sociology," *Social Problems* (Winter 1962), pp. 199-213.

34. One of the major efforts to develop Freudian concepts into a general psychology is to be found in the work of David Rapaport. See particularly his two papers, "The Theory of Attention Cathexis: An Economic and Structural Attempt at the Explanation of Cognitive Processes" and "On the Psychoanalytic Theory of Motivation," *The Collected Papers of David Rapaport*, ed. M. Gill, Chs. 61 and 65 (New York: Basic Books, 1967).

35. K. Lancaster, "Welfare Propositions in Terms of Consistency and Expanded Choice," *Economic Journal*, 68, No. 3 (September 1958), pp. 464-470.

36. J. S. Bruner, J. J. Goodnow, and G. A. Austin, *A Study of Thinking* (New York: Wiley, 1956); H. A. Simon, *The Sciences of the Artificial* (Cambridge, Mass.: M.I.T. Press, 1963).

37. Daniel Moynihan, *Maximum Feasible Misunderstanding* (New York: The Free Press, 1969).

38. Toulmin, *Human Understanding*, Ch. 4 and 5.

39. P. A. Samuelson, "The Evaluation of Real National Income," *Oxford Economic Papers*, New Series, 2, No. 1 (June 1950), pp. 1-29; *The Foundations of Economic Analysis* (Cambridge, Mass.: Harvard University Press, 1948), Ch. 8; R. G. Lipsey and K. Lancaster, "The General Theory of the Second Best," *Review of Economic Studies*, 24, No. 1 (1956), pp. 11-32.

40. Samuelson, "Evaluation of Real Income."

41. A classic in this field is G. Debreu, *Theory of Value* (New York: Wiley, 1959).

42. A good example of this is the work on price cost margins and the effects of industrial structure; see Scherer, *Industrial Market Structure*.

43. Friedrichs, *Sociology*.

GERALD HOLTON

On Being Caught Between Dionysians and Apollonians

How do research scientists go about obtaining knowledge? How *should* they? Today's scientists tend not to be introspective about these questions. During their apprenticeship, they somehow absorb the necessary pragmatic basis and then go about their business, content to leave it to a very few among them to interest themselves in epistemology when some obstinate difficulty blocks advance.

Outside the walls of the laboratory, however, interest in the theory of scientific knowledge runs high among three groups: a small but vigorous set of professional philosophers; students and other laymen who understand that few questions are more practical or urgent today than how knowledge may be reliably gained and where the limits of certainty lie; and critics of culture, including the new Romantics, the remnants of the counterculture movement, and a tiny band of "outsider" scientists and former science students who are disenchanted with the politics of the scientific establishment and are interested in ideological links among knowledge, power, and values.

With all their differences—some of which, as we shall see, are total, unresolvable, and of great consequence—the persons in these three groups have some properties in common. Certain people in each group have a high commitment, eloquence and visibility, which command the attention of a public wider than that of all scientists put together. Nevertheless, they are given, at most, only passing attention by the scientists themselves. This is not surprising: life is short and research is long. Within the laboratory, the accelerating pace and complexity of a scientist's work make heavier demands on him every year. The external world keeps pressing him for additional and ever-more-urgent involvement. In any case, he sees little reason to volunteer for a debate for which he feels he has no particular expertise. Even so, it may be argued that this lack of attention has become a costly luxury, that a whole range of problems—from questions about the validity of the scientific components of public policy decisions to the need for a better understanding of the roots of public conceptions and misconceptions about science—would be clarified by a more widespread apprehension of how scientific ideas are in fact obtained and tested.

In selecting for analysis two of the more popular current views on epistemology, I have little illusion that scientists are, on the whole, likely to see much connection between such a study and the problems which trouble many of them these days. They are preoccupied with such large practical problems as the recent cutback of funding for scientific research, for training, and for jobs.[1] Nevertheless, while the effects of that policy have been tragic, they are fairly rapidly reversible, at least in principle. In a democracy one can hope to replace foolish policies and policy-makers with better ones every few years. World history too, for better or worse, occasionally mandates policy changes in the support of science. H. Guyford Stever, the Director of the National Science Foundation who currently acts also as the President's Science Adviser, noted in a recent talk, entitled "Is It a Science World Again?" that the period of increasing shortages which we have entered is likely to put the recognition of the importance of science on a new basis: "While the atom and space may have made science spectacular, scarcity is making it real."[2]

Science policy in a democracy, however, is dependent not only on the fluctuations of fortune, but also on more long-range factors, one of which is the state of popular understanding of science as a cognitive activity. The time scale of changes in this attitude is far longer than the terms of office of Congressmen, Presidents, or bureaucrats. Don K. Price, in his prophetic address as retiring president of the AAAS in 1968,[3] during the height of concerns over the degradation of science in the scheme of national priorities, warned that the short-range difficulties so evident to scientists should not blind them to the long-range ones. He asked that they look beyond the discomforting and possibly transient "political reaction" of economy-minded politicians and face the "fundamental challenge," which he described as "a rebellion . . . a cosmopolitan, almost worldwide, movement."[4]

Its mood and temper reflect the ideas of many middle-aged intellectuals who are anything but violent revolutionaries. From the point of view of scientists, the most important theme in the rebellion is its hatred of what it sees as an impersonal technological society that dominates the individual and reduces his sense of freedom. In this complex system, science and technology, far from being considered beneficent instruments of progress, are identified as the intellectual processes that are at the roots of the blind forces of oppression.[5]

Don Price is no Pollyanna; he agrees that "we have not learned how to make our technological skills serve the purposes of humanity, or how to free men from servitude to the purposes of technological bureaucracies" (though he adds at once, "But we would do well to think twice before agreeing that these symptoms are caused by reductionism in modern science, or that they would be cured by violence in the name of brotherhood or love"). He calls the rebels pessimists, but says,

I do not think they are even pessimistic enough. To me it seems possible that the new amount of technological power let loose in an overcrowded world may overload any system we might devise for its control; the possibility of a complete and apocalyptic end of civilization cannot be dismissed as a morbid fantasy.[6]

Yet he notes that the intellectual core of the rebellion is not a disagreement over practical proposals to avert catastrophe, but a philosophical aversion to the historic establishment of scientific reductionism—"the change from systems of thought that were concrete but complex and disorderly, and that often confused what is with what ought to be, to a system of more simple and general and provable concepts." The relationship the public perceives between science and politics therefore springs out of the popular theory of knowledge: "The way people think about politics is surely influenced by what they implicitly believe about what they know and how they know it—that is, how they acquire knowledge, and why they believe it."

The issue is even more complex, and therefore more interesting, because of another set of forces. If the scientist—whether he takes notice or not—is confronted by a rebellion based on popular beliefs concerning scientific reductionism, he is also subject to a barrage from exactly the opposite direction, from a group of philosophers who wish to redefine the allowable limits of scientific rationality. Thus the scientist is caught between a large anvil and a fearful hammer. The one is provided by what I might call "the new Dionysians"—by authors like Theodore Roszak, Charles Reich, R. D. Laing, N. O. Brown and Kurt Vonnegut. With all the differences among them, they are agreed in their suspicion or contempt of conventional rationality, and in their conviction that the consequences flowing from science and technology are preponderantly evil. Methodology is not their first concern; they think of themselves primarily as social and cultural critics. But they would "widen the spectrum" of what is considered useful knowledge as a precondition of other changes they desire. They tend to celebrate the private, personal, and, in some cases, even the mystical. Their skill is high and the appeal of their lively prose is large.

If these new Dionysians constitute the anvil, the hammer is wielded by the group I shall call "new Apollonians."[7] They advise us to take precisely the opposite path—to confine ourselves to the logical and mathematical side of science, to concentrate on the final fruits of memorable successes instead of on the turmoil by which they are achieved, to restrict the meaning of rationality so that it deals chiefly with statements whose objectivity seems guaranteed by the consensus in public science. They would "shrink the window" emphatically, discarding precisely the elements which the other group takes most seriously.

Both groups present their cases with the apocalyptic urgency of rival world views. As in most polarized situations, they inflict most damage not on each other, but on anyone caught in middle ground. Indeed, they seem to reinforce each other's position, as cold war antagonists tend to do. In the face of its enemy, each limits the circle of allowable thought and action: one ritualistically heaping scorn on a caricature which it calls rationality, the other on a caricature which it calls irrationality. Each is dissatisfied with how science is done, and does not hide its distaste.

The New Dionysians

Evidence for the existence of the new Dionysians is not hard to find.[8] Although their views may be fashionable, it would be a mistake to think of them as transient fads. The twenty-first century is unlikely to discover among the present new Dionysian writers a new voice, as our century found Nietzsche rather hidden among the nineteenth-century Dionysians; the high level of today's sales of these wares does not seem to be grounded in their lasting literary quality or in the new depths of insight. Yet, even if each of these writers separately lasts only a season, the fact that their message falls on so many believing ears shows that the succession is not likely to wither away soon.

One of their more measured and thoughtful proponents of today, Theodore Roszak, sets the tone of the attack with statements such as these:

What *is* to blame is the root assumption . . . that culture—if it is to be cleansed of superstition and reclaimed for humanitarian values—must be wholly entrusted to the mindscape of scientific rationality.[9]

I have insisted that there is something radically and systematically wrong with our culture, a flaw that lies deeper than any class or race analysis probes, and which frustrates our best efforts to achieve wholeness. I am convinced that it is our ingrained commitment to the scientific picture of nature that hangs us up.[10]

When he proposes to redefine true knowledge as "gnosis," within which traditional science is only a small part of a larger spectrum (that part which seeks merely to gather "candles of information"), one recalls that it is almost exactly a hundred years ago that DuBois-Reymond's essay on "Die Grenzen des Naturerkennens" led to the controversy that culminated in the slogan "the bankruptcy of science," and that George Santayana wrote in *Reason and Science:*

Science is a halfway house between private sensation and universal vision . . . a sort of telegraphic wire through which a meager report reaches us of things we would fain observe and live through in their full reality. This report may suffice for approximately fit action; it does not suffice for ideal knowledge of the truth, nor for adequate sympathy with the reality.

Indeed, today's critics of what they take to be the method and pervasiveness of science fit into a long and often brilliant tradition—Thoreau, Shelley, Coleridge, Wordsworth, Blake ("I come 'in the grandeur of inspiration to abolish ratiocination"), Goethe, Rousseau, Vico, Montaigne, and back to the ancient Greeks: Epicurus is reported to have said, in a letter to Menecaeus,

In fact, it would be better to follow the myths about the gods than be a slave of the physicists' destiny; myths allude to the hope of softening the gods' hearts by honoring them, while destiny implies an inflexible necessity.

As a contemporary version of that tradition, I choose a book that has had a wide popular impact precisely because it did not pretend to have anything so sophisticated as an explicit epistemological message. Long selections from it first appeared in the fall of 1970 in *The New Yorker.* An unprecedented storm of publications followed: it was simultaneously available under six different imprints, one of them going through twelve printings in six months, another through twelve more printings in eight months. It seemed permanently installed on the best-seller list. Seven articles in rapid succession discussed the phenomenon in *The New York Times.* Its meaning was so widely analyzed that another widely circulated book sprang up devoted entirely to reprints of the reviews. Its message was castigated by worthies from Spiro Agnew, to whom it seemed permissive and immoral, to radical activists, who considered it counter-revolutionary; its popular success has never been fully explained. That book is, of course, Charles Reich's *The Greening of America.*

Despite the fact that, less than four years later, the book has faded from view and some details have become out of date, Reich's basic attitude toward nature, science, and rationality is still quite representative of the new Dionysians. In fact, I still find the book more revealing as a study of that world view than more recent ones—somewhat as an art historian interested in the popular understanding of the arts would do well to attend not only to what is hanging in museums, but also to some samples of widely liked *Kitsch.*

Reich's is, on the whole, an optimistic book that promises a kind of paradise or utopia for the United States. But it says little about the problems which the majority of the world's people face. This relatively parochial platform is a hint of the fundamental solipsism that pervades the book. Indeed, the first rule of the new at-

titude Reich espouses, which he calls Consciousness III, is that it "starts with self.
. . . The individual self is the only true reality. Thus, it returns to the earlier
America: 'Myself I sing.' "[11]

To this inward turning, Reich juxtaposes no antithetical command that would
allow the self to be transcended. The direct result is a Ptolemaic, homocentric conception of the world order that may allow many beautiful and satisfying experiences, but doing or understanding science is not among them, nor is any field
of scholarship where the warrant of validity stems not from private enthusiasm but
from some community consensus. For those activities require the recognition that
the individual self is not the "only true reality."

This is a point on which almost all scientists will agree, from beginners to sages.
In its extreme form, this view has been stated perhaps most eloquently by Albert
Einstein in his essay "Motiv des Forschens" (1918). "To begin with," Einstein
said there,

I believe with Schopenhauer that one of the strongest motives that lead persons to art and
science is flight from the everyday life with its painful harshness and wretched dreariness,
and from the fetters of one's own shifting desires. One who is more finely tempered is
driven to escape from personal existence into the world of objective observing and
understanding. . . .

With this negative motive there goes a positive one. A person seeks to form for himself,
in whatever manner is suitable for him, a simplified and lucid image of the world, so to over-
come the world of experience by striving to replace it to some extent by this image. . . . Into
this image and its formation he places the center of gravity of his emotional life, in order to
attain the peace and serenity that he cannot find within the narrow confines of swirling, personal experience.[12]

Later, in the essay "Religion and Science" (1930), Einstein reiterated the point
in these words: "The individual feels the futility of human desires and aims, and
the sublimity and marvelous order which reveal themselves both in nature and in
the world of thought." Einstein thinks of this sympathetically as "the beginnings of
cosmic religious feeling," which, together with the "deep conviction of the
rationality of the universe," he recognizes as "the strongest and noblest motive for
scientific research."

In the constant struggle to go beyond what he called the "merely personal,"
Einstein came, in the end, to agree fully with Max Planck that the final aim of
science is the very opposite of its necessary initial stage of private, even heroic,
struggle. That final aim is "the complete liberation of the physical world picture
from the individuality of separate intellects." In other words, science searches for a
world picture that is "real" insofar as it is covariant with respect to differences in
individual observers.

The starkness of that vision—and not all scientists would follow Einstein and
Planck so far—may be one reason why the new Dionysians seem inevitably
tempted to reach for the word "dehumanizing" when discussing the methods of
science. Yet these methods, to yield testable truths, must go beyond Private
Science, even though they cannot get started without going through that stage first.
Nor do they contradict the fact that human concerns remain central in those activities that have direct societal impact. Thus Einstein said, "Concern for the person must always constitute the chief objective of all technological effort."

Moreover, the path from the merely personal through the projection of a rational world order does, after all, eventually lead back to the solution of complex and pressing human problems—physical, biomedical, psychological, social. Indeed, as I shall note below, it is the only known method for finding such solutions.

But to return to Reich, and through him to the whole movement of which we take him to be an indicator: Another "Commandment" of Consciousness III, Reich tells us, is that it is open "to any and all experience. [Elsewhere he calls experience "the most precious of commodities."] It is always in a state of becoming. It is just the opposite of Consciousness II, which tries to force all new experience into a pre-existing system, and to assimilate all new knowledge to principles already established."[13] With this premise, Reich announces an important theme, one that characterizes the movement: the primacy of *direct* experience—nonreductionistic, unanalyzed, unreconstructed, unordered. This is the guiding attitude, on the one hand, toward music—"the older music was essentially intellectual; it was located in the mind . . . ; the new music rocks the whole body, and penetrates the soul,"[14]—and on the other, toward nature itself.

The new Dionysians are, of course, all *for* nature and the experience of nature, but in a specific way. In one of his most revealing passages, Reich explains that the Consciousness III person

takes "trips" out into nature; he might lie for two hours and simply stare up at the arching branches of a tree. . . . He might cultivate visual sensitivity, and the ability to meditate, by staring for hours at a globe lamp.[15]

(He might also find at that point that "one of the most important means for restoring dulled consciousness is psychedelic drugs." Although Reich does not stridently advocate the use of drugs, he holds that "they make possible a higher range of experience, extending outward toward self-knowledge, to the religious.")

Nature, thus, is what one takes "trips" out into. By nature Reich means "the beach, the woods, and the mountains,"[16] which he claims are "perhaps the deepest source of consciousness. . . . Nature is not some foreign element that requires equipment. Nature is them."[17]

This homocentric view, in which man and nature overlap in a total experience of natural phenomena—an act of imagination without criticism—avoids the very possibility of rational understanding of natural phenomena. And it is meant to do so: "Consciousness III . . . does not try to reduce or simplify man's complexity, or the complexity of nature. . . . It says that what is meaningful, what endures, is no more nor less than the total experience of life."[18]

Even the scientists who are farthest from the usual rationalistic stereotype would have to disagree vigorously. To a mystic like Kepler, experience had a very different function. It triggered a puzzle in the mind, and it was through the working out of such puzzles that, in the view of Kepler and other neo-Platonists, persons could feel that they were communicating directly with the Deity. Newton, at the end of the *Opticks*, expressed an analogous hope for moral benefits to be derived from the study of nature:

If Natural Philosophy in all its Parts, by pursuing this Method, shall at length be perfected, the Bounds of Moral Philosophy will also be enlarged. For so far as we can know by Natural Philosophy what is the first Cause, what Powers he has over us, and what Benefits we receive from him, so far our Duty towards him, as well as towards one another, will appear to us in the Light of Nature.

Even Goethe, though more secular in his expectation, intended that his holistic, noninstrumental approach to nature study would improve the state of science, by opening up new subjects for study—"Optical illusion is optical truth"—and by bringing new types of contributors to science. He writes on the last page of the *Farbenlehre*: "All who are endowed only with habits of attention—women, children—are capable of communicating striking and true observations. . . . *Multi pertransibunt et augebitur scientia.*"

None of these or similar ambitions are, however, reflected in the holism of the neo-Dionysians. What counts there is experience freed from analysis, from questions, even from the perception of complexity itself, as if a short cut through complexity were possible. The "Method" which Newton had in mind in the quotation above, however, consisted of two steps: "As in Mathematics, so in Natural Philosophy, the Investigation of difficult Things by the Method of Analysis ought ever to precede the Method of Composition." It is so to this day in scientific work: first reduction, then synthesis. Einstein, too, in the essay "Motiv des Forschens," wrote with some regret that we have to be satisfied with first "portraying the simplest occurrences which can be made accessible to our experience." More complex occurrences cannot be constructed with the necessary degree of accuracy and logical perfection. He acknowledged that one has to choose "supreme purity, clarity, and certainty, *at the cost of completeness.*" But, he thought, this is only the half-way house: after a world image has been constructed by the method of reduction and simplification, one can hope that, as science matures, it will turn out to apply to natural phenomena as they offer themselves to us, in all their complexity and completeness. The history of science provides a wealth of examples that attest to this truth. So the effort to encompass the totality of experience is in principle possible in science—not at the beginning, but at the end of the two-step process.

Precisely at this point, Einstein introduced a warning that he was to repeat frequently, one which, for quite different reasons, must be as surprising to the new Romantics as to the new positivists. He noted that the reality of human limitations restricts the efficacy of logic; it would be foolish to hide it or deny it, or to restrict the permissible use of reason in science to such narrow ground. From the general laws on which the structure of theoretical physics rests, he feels

it should be possible to obtain by pure deduction the description, that is to say the theory, of natural processes, including those of life—if such a process of deduction were not far beyond the capacity of human thinking. To these elementary laws there leads no logical path, but only intuition supported by being sympathetically in touch with experience [Einfühlung in die Erfahrung]. . . . There is no logical bridge from experience to the basic principles of theory. . . . Physicists accuse many an epistemologist of not giving sufficient weight to this circumstance.[19]

Obviously, I have chosen Einstein because of the clarity, honesty, and independence of his methodological remarks. The process he describes is one most scientists will recognize as applicable to really fundamental work (although the use of the word "intuition" is bound to embarrass some of them). Moreover, almost by definition, the methods an Einstein used cannot reasonably be denied the label "rational," no matter how different they are from the models for rationality set up as strawmen by the new Dionysians or icons by the new Apollonians.

But if Reich has noticed the failure of scientists like Einstein to conform to such models, he does not let on. He feels that nature should be studied neither by induction nor by the analytical-synthetic method, not even if it allows a speculative leap where human limitation makes it necessary and human ingenuity makes it possible. Rather, Reich advocates that one coast through total, unselected experience with one's hands off the wheel and one's rational gearbox in neutral.

Throughout books such as *The Greening of America*, the true enemy is, in fact, not science, not the Corporate State, not the Department of Defense, not even the regrettable failures of science—the cases where scientists or technologists allowed themselves to be used for destructive purposes. The real enemy is rationality itself. Thus we read that the Corporate State has "only one value, the value of technology-organization-efficiency-growth-progress. The state is perfectly rational and logical. It is based upon principle."[20] It would appear that the vision of Saint-Simon had really triumphed in our day.

What, then, is wrong with rationality? Reich gives the answer on the second page of his book, where we read that the rationality of the modern state must be "measured against the insanity of existing 'reason'—reason that makes impoverishment, dehumanization, and even war appear to be logical and necessary." Among the evils of rational thought, discussed at greater length later, are not merely its failures to prevent the recent wars, but the intellectual justifications that were given for those wars.[21] Thus we arrive at the remedy—a recipe for escape from rationality:

One of the most important means employed by the new generation in seeking to transcend technology is . . . to pay heed to the instincts, to obey the rhythms and music of nature, to be guided by the irrational, by folklore and the spiritual, and by the imagination.[22]

Accepted patterns of thought must be broken; what is considered "rational thought" must be opposed by "non-rational thought"—drug-thought, mysticism, impulses.[23]

Technically, one could analyze Reich's many conceptual difficulties in more detail. Thus, as Charles Frankel has accurately noted in a recent article,[24]

The Irrationalist's theory of human nature is steeped in the tradition of the dualistic psychology it condemns. It talks about "reason" as though it were a department of human nature in conflict with "emotions." But "reason," considered as a psychological process, is not a special faculty, and it is not separate from the emotions; it is simply the process of reorganizing the emotions.

But precisely because such flaws are simple to expose, the chief puzzle about the new Dionysians is, and will remain, the large extent of their popular appeal. And here it may be significant to notice an ironic asymmetry. As we have seen, scientists have written about their motivation for turning to their work as if it were an intellectual and emotional turning away from the turmoil within and all around. Reich also wrote at a turbulent time, in the war year of 1970—at the height of another tragic and stupid reign, one which *his* audience, at any rate, seemed to recognize as such even without the subsequent evidence of the secret bombing of Cambodia, the Pentagon papers, the sale of the public trust, the conspiracy to abridge civil rights, the arrogance that led to the Watergate crises, not to speak of the continuation of a senseless arms race and the widening of the world's poverty. Reich, however, charges the horrors of his time to the sovereign rule of reason, and urges his readers to turn inward, thereby abandoning their chief weapon for organizing and validating any realistic attack on the ills he deplores. Yet one key to Reich's wide appeal may be just that he releases his readers from all responsibility for effective action. Furthermore, at a time when so many feel they can only sit by in helpless disbelief to watch the unrolling of an absurd tragedy, he furnishes them with a convenient, safely passive target for their intellectual distaste.

The New Apollonians

Now to the hammer. The school of philosophers who have taken it on themselves to protect rationality in the narrowest sense of the word are also members of a long tradition. Some of their genes can be traced back to the logical positivists of the pre-World War II period, who are themselves descended from a long line of warriors against the blatant obscurantism and metaphysical fantasies that haunted and thwarted science in the nineteenth and early twentieth centuries. Rereading today Otto Neurath's influential essay "Sociology and Physicalism" (1931–1932), one can glimpse the fierce doctrine that helped this school to achieve its victories:

The Vienna Circle . . . seeks to create a climate which will be free from metaphysics in order to promote scientific studies in all fields by means of logical analysis. . . . All the representatives of the Circle are in agreement that "philosophy" does not exist as a discipline, along side of science, with propositions of its own. *The body of scientific propositions exhausts the sum of all meaningful statements.* . . . They wish to construct a "science which is free from any world view."

But the line of descent goes much further back, all the way to Lucretius, to Democritus, to all who undertook the antimetaphysical mission of liberating mankind from the enchantment and terror of superstition. Thus a modern Lucretius, Bertrand Russell, proclaimed that "all these things, if not quite beyond dispute, are yet so nearly certain that no philosophy which rejects them can hope to stand":[25]

That Man is the product of causes which had no prevision of the end they were achieving; that his origin, his growth, his hopes and fears, his loves and beliefs, are but the outcome of accidental collocations of atoms; that no fire, no heroism, no intensity of thought and feeling, can preserve an individual life beyond the grave; that all the labours of the ages, all the devotion, all the inspiration, all the noonday brightness of human genius, are destined to extinction in the vast death of the solar system, and that the whole temple of Man's achievement must inevitably be buried beneath the debris of a universe in ruins.

Although it is no longer fashionable to force the rationalists' message upon a fearful populace with quite so much glee, the ancient division between thematically incompatible world views continues to exist, and is not likely to disappear.[26]

Some of today's most eloquent defenders of rationality are associated with the school of Sir Karl Popper, who himself was influenced, at an early point, by the prewar positivist movement. Out of Popper's own immense and lasting contributions over the decades I shall refer here only to one small portion that happens to have relevance to this particular study. He considers that the rationality of science presupposes a common language and a common set of assumptions which themselves are subject to conventional rational criticism. The contrary opinion is that there may exist cases of individual scientific work that have not been and perhaps never can be subjected fully to such a critique. Popper writes that this so-called "Myth of the Framework" is, "in our time, the essential bulwark of irrationalism."[27]

The progress from one valid stage of scientific theory to another cannot, in his view, break the thread of continuous, rational, progressive development. "In science, and only in science, can we say that we have made genuine progress: that we know more than we did before."[28] To be sure, "an intellectual revolution often looks like a religious conversion." But a critical and rational evaluation of our former views must remain possible in the light of the new ones. If it were not possible, what guarantee would we have that science was indeed accruing a content of truth? What guarantee that the changes in science are indeed a progressive sequence of steps toward objective knowledge, and not merely a sequence of conversion experiences from one unfounded set of beliefs to another?

A critical discussion of this position is, however, made difficult by a set of self-inflicted taboos. Popper writes:

I cannot conclude without pointing out that to me the idea of turning for enlightenment concerning the aims of science, and its possible progress, to sociology or to psychology, or . . . to the history of science, is surprising and disappointing. In fact, compared with physics, sociology and psychology are riddled with fashions and with uncontrolled dogma. The suggestion that we can find anything here like "objective, pure description" is clearly mistaken. Besides, how can the regress to these often spurious sciences help us in this particular difficulty? . . . . No, this is not the way, as mere logic can show.[29]

What, exactly, is at stake here? On one level, it is the definition of where the philosopher of science should look for valid problems and tools. Popper rules out, as of no interest, the context of discovery, and hence the actual working out of a problem by an actual person.

The initial stage, the act of conceiving or inventing a theory seems to me neither to call for logical analysis, nor to be susceptible to it. The question of how it happens . . . may be of great interest to empirical psychology; but it is irrelevant to the logical analysis of scientific knowledge. The latter is concerned only . . . with questions of justification or validity.[30]

Fair enough—though one may not subscribe to Popper's preference, particularly if one's own fascination is with a historical study of the "personal struggle." One may even regret that Popper shares with the large majority of scientists—and, for that matter, with Reich and the new Dionysians—a complete lack of interest in studying the creative act of scientists, thereby denying the possibility of a critique of the imagination.

But where is one to turn for data to examine Popper's logic of discovery and to test out his hypotheses? It is at this point that some modern philosophers of science have recently evolved a technique of criticism that tries to force the understanding of scientific work as far to the right as the new Dionysians wish to force it to the left. Instead of looking at actual case studies in their historic setting—a technique of what they call the "spurious sciences"—they look at a "rational reconstruction" of the event.

Popper himself proposed the technique in a rather gentle way:

Admittedly, no creative action can ever be fully explained. Nevertheless, we can try, conjec-
turally, to give an idealized reconstruction of the problem situation in which the agent found
himself, and to that extent make the action "understandable" or "rationally understandable," that is to say, adequate to the situation as he saw it. This method of situational
analysis may be described as an application of the rationality principle.[31]

This proposal was taken up by others and clothed more dogmatically, by none
more vigorously than by Popper's former student and successor to his chair at the
London School of Economics, Imre Lakatos. In the influential work of this brilliant
author, the opinion of what constitutes a valid study of a historical case is laid
down in such words as these:

In writing a historical case study, one should, I think, adopt the following procedure: (1) one
gives a rational reconstruction; (2) one tries to compare this rational reconstruction with actual history and to criticize both one's rational reconstruction for lack of historicity and the
actual history for lack of rationality.[32]

Lakatos then gives examples of what happens to an historical case study when done
in this style, including his own reconstruction[33] of "Bohr's plan . . . to work out first
the theory of the hydrogen atom [1912-1913]."

His first model was to be based on a fixed proton-nucleus with an electron in a circular orbit
. . . ; after this he thought of taking the possible spin of the electron into account.[34] . . . All
this was planned right at the start.

As it happens, Bohr's early work has been very carefully studied by historians of
science, and this version produced by "rational reconstruction" is an ahistorical
parody that makes one's hair stand on end.[35] Otto Neurath's dictum that
" 'Philosophy' does not exist as a discipline, along side of science, with propositions
of its own" has been stood on its head: the study of the actual work of scientists
does not exist as a discipline, along side of philosophy, with propositions of its own.

The resulting rationalization of actual historic cases, while not without technical
interest in philosophy itself, is so risky an idea and so unacceptable to most
historians of science[36] that one is forced to speculate it may be motivated by higher
stakes than appear on the surface. In the writings of the more extreme members of
the new Apollonians, one senses that their philosophical position is not being
developed simply for its own sake, or for the sake of its potential evaluation in the
crucible of rational critique, but that their ambitions are much larger. They seem to
hope to save scientists from the threat of the irrational, suspecting that scientists
will be unable to do a good job without expert help in deciding which theories are
truly scientific and which are merely pseudoscientific. Lakatos acknowledges
sadly,[37]

If we look at the history of science, if we try to see how some of the most celebrated
falsifications [of hypotheses] happened, we have to come to the conclusion that either some
of them are plainly irrational, or that they rest on rationality principles radically different
from the ones we just discussed.

Hence rational reconstruction; hence the effort to replace the "naive" version of
methodological falsification with a "sophisticated version . . . and thereby rescue

methodology and the idea of scientific *progress*. This is Popper's way," Lakatos tells us, "and the one I intend to follow."[38]

Hanging over the whole stage is the shadow of David Hume, with his message, as Popper puts it, that "not only is man an irrational animal, but that part of us which we thought rational—*human knowledge*, including practical knowledge—is utterly irrational."[39] The new Apollonians dedicate a major effort to the disproof of this specter, with particular attention to scientific reasoning. But their ambitions are even larger than that: to save mankind from obscurantism, astrology, and revolution. Thus Lakatos writes that a recent theory of scientific progress—which allows the role of exemplars rather than of logical proof alone—makes "scientific change a kind of religious change."[40] Such a theory, he says, not only poses a threat to technical epistemology, but "concerns our central intellectual values," hence affecting "social sciences . . . moral and political philosophy." Moreover, it "would vindicate, no doubt unintentionally, the basic political credo of contemporary religious maniacs ('student revolutionaries')." Elsewhere,[41] Lakatos is led so far as to speculate on the possibly sinister personal influence of the author of such a theory: "I am afraid this might be one clue to the unintended popularity of his theory among the New Left busily preparing the 1984 'revolution'."

Now we recognize what is really at stake: civilization itself. These philosophers of rationalism see themselves as the soldiers at the gates, fending off a horde of barbarians. Popper himself has, of course, made no secret of his mission. Long before the new Dionysians were as prominent as they are now, he said that the conflict with advocates of irrationalism "has become the most important intellectual, and perhaps even moral, issue of our time."[42] Irrational attitudes and the flagging of the critical habit, he warns, could well open the way for demagogues who promise political miracles. One must preserve what has been gained, with all its short-comings, for "our present free world, our Atlantic Community . . . ruled by the interplay of our individual consciences . . . is the best society that has ever existed."[43] Lakatos, for his part, warns that a work on the nature of scientific programs with which he disagrees is "a matter for mob psychology," "vulgar Marxism," and "psychologism," and has even triggered "the new wave of sceptical irrationalism and anarchism."[44]

Thus each of the two opposing groups is imbued with a sense of urgency to save the Republic from the hands of the other. Each thinks following a proper process for gaining valid knowledge is a key for salvation, and proposes to clarify the understanding of that process. One side condemns the scientists for being too rational; the other chides them for being too irrational. Caught in between, scientists, virtually without exception, pay no attention to either side, not even to defend themselves against grotesque distortions of what it is that they really do.[45] Without a challenge, they hand over the public platform to the propagation of two sets of quite different but equally erroneous answers to such questions as those posed at the beginning of this article: how do scientists actually go about gaining knowledge, and how *should* they?

Postscript

This is not the place, nor is it my intention, to build, on the analysis of the symptoms, a prescription for a cure. A deeper involvement of research scientists in discussions concerning their methods would surely improve the understanding of science—including their own. The four phases of scientific work which obviously rest on rationality, by any definition of the word, could well benefit from more modern analysis—namely, rationality in the deductive portions of private theorizing; rationality in the structure of a theory once it has been worked out moderately well; rationality in the process of communication and validation among scientists operating in the area of public science; and the perception, at least among our more exalted spirits, of an underlying rationality and uniqueness in the world order seen through science—perhaps the only order open to human perception which is not a Rashomon story, different for each observer.

In addition, sound pedagogic materials are needed to show that there are processes at work in science-making which, while they are acts of reason, cannot be forced into the logical-analytical framework. Entering into such processes are the ways by which new ideas arise and are handled during the nascent moment; the sources of individual thematic choices, and the reasons for cleaving to them; the connection between the elementary concepts, both of science and of everyday thinking, and the complexes of sense experience; and the eternally surprising fact that we so often find the logically simple suitable for building a theory of nature's phenomena.

As Peter Medawar has courageously observed, the hypothesis of the interaction of essential dual components may still be the most fruitful one.

Scientific reasoning is an exploratory dialogue that can always be resolved into two voices or episodes of thought, imaginative and critical, which alternate and interact. . . . The process by which we come to form a hypothesis is not illogical but non-logical, i.e., outside logic. But once we have formed an opinion we can expose it to criticism, usually by experimentation.[46]

This is not compromising between rationality and irrationality. On the contrary, it is widening the claim of rationality, as well as the scope of much-needed research on the nature of scientific rationality in practice. In direct opposition to both groups I have analyzed, Medawar holds that

the analysis of creativity in all its forms is beyond the competence of any one accepted discipline. It requires a consortium of the talents: psychologists, biologists, philosophers, computer scientists, artists and poets would all expect to have their say. That "creativity" is beyond analysis is a romantic illusion we must now outgrow.[47]

Whether for pedagogic purposes or as a field of research, whether as a part of philosophical analysis or as a key to a study of politically significant intellectual rebellions and reactions, the methods by which humans gain scientific knowledge are themselves much in need of more thorough scientific study. Possibly the worst service the new Dionysians and the new Apollonians render the cause of understanding ourselves is that their antithetical attacks continue to discredit the accommodation of the classically rationalistic with the sensualistic components of knowledge. We should, rather, strive to acquire a clearer notion of how actual mortal beings, with all their frailties, have managed to use both these faculties to grasp the outlines of a unique and fundamentally simple universe, characterized by necessity and harmony. Such knowledge, one hopes, may even be of practical use at a time when the whole species seems to depend, for its very survival, on tapping all the resources of reason for the generation of new ideas that are both imaginative and effective.

REFERENCES

1. The statistics are striking in terms of funds, but even more so in terms of lost opportunities for talented young people; in physics alone, about 40% fewer graduate students in the U.S. will be getting a Ph.D. degree 4 years hence than did 4 years ago, using projections of the number of students now in the pipeline. *Physics Manpower 1973* (New York: American Institute of Physics, 1973), pp. 45-47.

2. Advance Release of Remarks at 140th Annual Meeting of American Association for the Advancement of Science, San Francisco, February 27, 1974, p. 1

3. Don K. Price, "Purists and Politicians," *Science*, 163 (3862), 1969, pp. 25-31; the address was delivered on December 28, 1968. It is surely significant that Price was speaking as the first social scientist in many decades to be an AAAS president.

4. One aspect of the cosmopolitan nature of the undercurrent of reaction against science, at least in the developed countries, was symbolized for me by a "Questionnaire on Science and Society," distributed by the Editorial Board of the Soviet journal *Literaturnya Gazetta* to Soviet as well as foreign scholars at an international congress in Moscow in the summer of 1971. Among the questions, in Russian and English, were these:

    • Could, in your opinion, rapid development of science lead to some undesirable consequences?
    • Do you think that some fields of scientific research may be "taboo" from the moral point of view? If so, which? why?
    • May there be any reasons for stopping a successful line of research? If so, what are they?
    • Does scientific research by itself foster high moral qualities in men?
    • Does it not seem to you that after a period of extreme popularity of the exact sciences among youth, a cooling off may set in?

5. Price, "Purists and Politicians," p. 25.

6. Price, "Purists and Politicians," p. 31.

7. The terms "new Dionysians" and "new Apollonians" are based not directly on the characteristic identifications with Greek mythological figures, but on the related usage given in Nietzsche's *Birth of Tragedy*. A. Szent-Gyorgyi (in *Science*, 176 [1973], p. 966) has recently proposed the revival of the Nietzschean usage of Dionysian and Apollonian characteristics within science; he identifies the Dionysian elements with intuition, and the Apollonian element with the logical-rational. (A useful discussion is in Y. Elkana, "The Problem of Knowledge in Historical Perspective," *Ellenike Anthropistike Etaireia* 24 (Athens, 1973), pp. 200-201.) Both these characteristic opposites have generally been credited with doing useful work within science; however, this does not apply to the *new* Dionysians and the *new* Apollonians as here defined, since both are active outside science.

8. *Time* magazine, in a series called "Second Thoughts about Man," had an installment not long ago, entitled "Reaching beyond the Rational." In it, *Time* announced that it had "been examining America's rising discontent with entrenched intellectual ideas: liberalism, rationalism and scientism. . . . This week, the Science section considers the repercussions for science and technology. It finds a deepening disillusionment with both." (April 23, 1973)

    As Don K. Price points out elsewhere in this issue, even the Congressional Research Service, which is not noted for extreme ideas, not long ago reported that the new trend in American culture "implies throwing out the scientific method, the definition of effects, and the search for cause . . . through the process of rational analysis."

9. *Where the Wasteland Ends* (New York: Doubleday, 1972), p. xxx.

10. "Some Thoughts on the Other Side of This Life," *New York Times*, April 12, 1973, p. 45.

11. Charles Reich, *The Greening of America* (New York: Bantam Books, 1970), pp. 241-242.

12. The essay "Motiv des Forschens" has been republished in an English translation in *Ideas and Opinions by Albert Einstein* (New York, 1954), pp. 224-227. The difference between Reich's rather comfortable self to which he seems to repair gladly and Einstein's more pessimistic perception of the self is striking.

    The remarks quoted above are closely related to the famous passage in Einstein's later "Autobiographical Notes" of 1946, in which he noted that when leaving "the religious paradise of youth," as a result of reading popular scientific books as a child of twelve, he had made "a first attempt to free myself from the chains of the 'merely personal,' from an existence which is dominated by wishes, hopes, and primitive feelings. Out yonder there was this huge world, which exists independently of us human beings and which stands before us like a great, eternal riddle. . . . The contemplation of this world beckoned like a liberation." In P.A. Schilpp, ed., *Albert Einstein, Philosopher-Scientist* (Evanston, Ill., 1949), p. 5.

13. Reich, *The Greening of America*, p. 251.

14. *Ibid.*, p. 266.

15. *Ibid.*, pp. 279-280.

16. *Ibid.*, p. 284.

17. *Ibid.*, p. 285.

18. *Ibid.*, p. 426.

19. For a further analysis of the Einsteinian methodology, see my article, "The Mainsprings of Discovery," *Encounter* (April 1974), pp. 85ff.

20. Reich, *The Greening of America*, p. 95.

21. As Reich puts it, "Consciousness III is deeply suspicious of logic, rationality, analysis, and of principle. Nothing is so outrageous to the Consciousness II intellectual as the seeming rejection of reason itself. But Consciousness III has been exposed to some rather bad examples of reason, including the intellectual justification of the Cold War and the Vietnam war. At any rate, Consciousness III believes it is essential to get free of what is now accepted as rational thought. It believes that 'reason' tends to leave out too many factors and values." *Ibid.*, p. 278.

22. *Ibid.*, p. 414.

23. *Ibid.*, p. 394.

24. C. Frankel, "The Nature and Sources of Irrationalism," *Science*, 180 (1973), p. 930.

25. Bertrand Russell, "A Free Man's Worship" (1903), *Mysticism and Logic* (London: Longmans, Green, 1919), pp. 47-48.

26. The basic difference has been put well in an article, entitled "On Reading Einstein," by Charles Mauron, which T. S. Eliot himself took the trouble to translate in 1930 for publication in his journal, *Criterion*. Mauron compares the two basically antithetical epistemological approaches:

    The first holds that any profound knowledge of any reality implies an intimate fusion of the mind with that reality: we only understand a thing in becoming it, in living it. In this way St. Theresa believed that she knew God; in this same way the Bergsonian believes that he knows at the same time his self and his world. The second type of opinion, on the contrary, holds that this mystical knowledge is meaningless, that to try to reach a reality in itself is vain, inasmuch as our mind can conceive clearly nothing but relations and systems of relations (*The Criterion*, 10 [1930], p. 24).

27. K. R. Popper, "Normal Science and Its Dangers," *Criticism and the Growth of Knowledge*, I. Lakatos and A. Musgrave, eds. (Cambridge: Cambridge University Press, 1970), p. 56.

28. *Ibid.*, p. 57.

29. *Ibid.*, pp. 57-58. For another discussion of what he dismissed as the "subjectivist" approach see K. R. Popper, *Objective Knowledge* (Oxford: Clarendon Press, 1974), p. 114.

30. K. R. Popper, *The Logic of Scientific Discovery* (1934; New York: Harper, 1965), p. 31.

31. Popper; *Objective Knowledge*, p. 179. However, in the last sentence of the same chapter a certain ambiguity emerges as to how seriously Popper took his proposal; he writes that his aim was to delineate "a theory of understanding which aims at combining an intuitive understanding of reality with the objectivity of rational criticism" (p. 190).

32. I. Lakatos, *Criticism and the Growth of Knowledge*, p. 138. Italics in original. See also I. Lakatos, "History of Science and Its Rational Reconstruction," *Boston Studies in the Philosophy of Science*, 8, ed. R. C. Buck and R. S. Cohen (Boston: D. Reidel, 1971), pp. 91-146, 174-182.

33. *Ibid.*, p. 146.

34. A footnote at this point laconically adds: "This is a rational reconstruction. As a matter of fact, Bohr accepted this idea only in his [paper of] 1926."

35. The example has been analyzed by T. S. Kuhn, *ibid.*, pp. 256-259, and by Y. Elkana, "Boltzmann's Scientific Research Programme and Its Alternatives," *Some Aspects of the Interaction between Science and Philosophy*, ed. Y. Elkana (New York: Humanities Press, forthcoming). See also Brian Easlea, *Liberation and the Aims of Science* (London: Chatto & Windus, 1973).
There is no space here to go further into the distorting effects of the rational reconstructionist's view of the progress of science. One or two short examples must suffice. Lakatos assures us that a theory has to undergo "progressive problem shifts" to remain scientific. If the advance is made with the use of ad hoc proposals, the "progressiveness" is spoiled and such programs become "degenerating," so that one has to "reject" them as "pseudoscientific." Since, however, ad hoc proposals frequently do figure in actual cases of what is widely acknowledged to be successful scientific work, he and his followers are forced into making the attempt to rescue Lorentz' ad-hoc-prone work from any possible charge that it might not be a single theory constantly undergoing "progressive problem shifts." To do this requires, however, new definitions of "ad hoc," "novel fact," etc., which are patently ad hoc themselves; moreover, they, in turn, entail a number of distortions of well-known historical facts. For an analysis, see A. I. Miller, "On Lorentz's Methodology," *British Journal of the Philosophy of Science* (April 1974).

36. For example, see I B. Cohen, "History and the Philosophy of Science," *The Structure of Scientific Theory*, ed. F. Suppe (Urbana, Ill.: University of Illinois Press, 1974), pp. 308-349, and his comments, pp. 351 ff.

37. Lakatos, "Methodology," p. 114.

38. *Ibid.*, p. 116.

39. K. R. Popper, *Objective Knowledge*, p. 90.
A strenuous attempt to "rescue" Einstein has recently prompted Gerig Gutting (in *Philosophy of Science*, 39 [1972], pp. 51-68) to an analysis which solemnly concludes that "any intuitions Einstein had . . . took their place in a logically coherent argument." The evidence from Einstein's own testimony that there were occasionally elements that did in fact not yield to conventional, logical analysis is dismissed by overamplification: to allow that would amount to making a case that the discovery of relativity theory as a whole "was derived essentially from a private intuition."

40. Lakatos, "Methodology," p. 93.

41. Lakatos, "History of Science and Its Rational Reconstructions," p. 133.

42. I. Lakatos, *Open Society and Its Enemies*, 2 (London: Routledge, 5th edition, 1966), p. 244.

43. K. R. Popper, *Conjectures and Refutations* (London: Routledge, 1963, 2nd ed. 1965), p. 375.

44. Lakatos, "Methodology," p. 278.

45. In truth, those scientists who write textbooks all too often make matters worse by providing distortions of their own, namely by presenting severely rationalized versions of the scientific process—just as there implicitly hangs over the desk of all scientists, while they write up their research results for publication, the admonition attributed to Louis Pasteur: "Make it seem inevitable."

There are good reasons why that is the case. But, in this respect, the hammer of the new Apollonians finds its most malleable materials in those who prepare didactic presentations. Usually, they provide indeed only "candles of information"; at worst, their products appear to have the character of unchallengeable, closed (and therefore essentially totalitarian) tracts. It may well be that the anarchic reconstructions of science on the part of the new Dionysians are largely reactions to the view of science that emanates from the rationalistic reconstructions of the new Apollonians and of text writers—for *that* is perhaps all they (like most students) ever get to see of "science."

I have discussed the stages of progressive rationalization in the development of didactic materials in an early draft of this paper, entitled "Science, Science Teaching, and Rationality," delivered in September 1973 at the meeting of the University Committee on Rational Alternatives (Proceedings in press).

46. P. B. Medawar, *Induction and Intuition* (Philadelphia: American Philosophical Society, 1969), p. 46.

47. *Ibid.*, p. 57.

ANDRÉ MAYER AND JEAN MAYER

Agriculture, The Island Empire

FEW SCIENTISTS think of agriculture as the chief, or the model science. Many, indeed, do not consider it a science at all. Yet it was the first science—the mother of sciences; it remains the science which make human life possible; and it may well be that, before the century is over, the success or failure of Science as a whole will be judged by the success or failure of agriculture.

The present isolation of agriculture in American academic life is a tragedy. Not only does it deprive us of the most useful models of the systems approach to human affairs, but it puts us—and the world—in mortal peril. It is, furthermore, a relatively recent aberration in our intellectual history. American academicians tend to be curiously limited in their view of science. Discussing the history of science or its social role, they seem, over and over, to fall back on the same examples. They see mathematics, astronomy, and physics as representing the leading edge of scientific progress and the normative pattern of development throughout history, and consider physics, in its organization and methodology, the paradigmatic modern science. The life sciences, perennial laggards, are often subsumed under the heading "biomedical"—as if they could be dealt with in the context of the professional treatment of human illness. Yet through most of history, agriculture has been the most advanced science, and today it remains in some respects the "biggest" science of all.

When human beings first learned the cycle of plants and seeds, they were scientists. As they learned when and how to plant, in what soil, and how much water each crop needed, they were extending their understanding of nature. This knowledge was not less scientific for having been discovered and transmitted by people who could not read or write. No scientist performs a greater act of faith in the predictability of the operation of natural laws than the farmer who plows a part of this year's harvest back into the earth.

The description and domestication of tens of thousands of varieties of the main cereals was started at least six thousand years before Darwin and Mendel, and the slow work of improving yields through genetic selection begun. The domestication and selection of farm animals is at least as ancient. Agriculture stimulated man's interest in measuring time and area—in climate, season, and the calendar—and thus gave rise to astronomy itself. More recently, modern biology, physiology, biochemistry, and soil chemistry benefited from the interest of their founders—such scientists as Justus von Liebig and Jean Baptiste Boussingault—in practical agricultural problems. Louis Pasteur's work on silkworms, animal diseases, and beer-making led to the bacteriological revolution, and hence to much of modern medicine.

Perhaps agriculture's strongest claim to consideration as a scientific undertaking has come with the rise of cybernetics, or systems theory—the theory of how the diverse elements of systems interact through time to produce change. Even in oral traditions, agriculture has always been perceived as a system; hundreds of generations of farmers have understood the interaction of stock, soil, water supply, climate, supply and demand. Today these elements are understood better than ever; they can be quantified; they can even be controlled. More than ever, the science of agriculture stands at the center of a broader system integrating human society and its physical environment. The further study of this system demands the coordination of all the sciences from physics to sociology, but we find ourselves ill prepared, individually and institutionally, to meet the challenge.

II

The failure of our secondary schools and liberal arts colleges to teach even rudimentary courses on agriculture means that an enormous majority, even among well-educated Americans, are totally ignorant of an area of knowledge basic to their daily style of life, to their family economics, and indeed to their survival. It also means that our policies of agricultural trade and technical assistance, as important to our foreign relations as food production is to our domestic economy, are discussed in the absence of sound information, if indeed they are discussed at all.[1]

The world today is the scene of a somber race between two contrary agricultural phenomena. On the one hand, after three decades of slow improvement, serious food shortages have developed on a global basis, as a result of population increases in developing areas and of changing patterns of consumption in industrialized countries. On the other hand, agricultural practices are still undergoing the process of rapid change known as "the Green Revolution." The future of civilization depends on which trend prevails. It is essential that the public (particularly in the United States, the leading agricultural nation) understand the situation and what is at stake.

The American people are well aware that food prices are rising. But if they rely on the explanations voiced by the government and the media, they cannot understand why. They learn that outmoded price-support policies, localized drought, and the "wheat deal" with the Soviet Union have caused temporary dislocations, which can be relieved by bringing land back into cultivation, by normal rainfalls, and by a wiser foreign trade policy. The one factor which is (rightly) presented as a long-term problem is the population explosion in developing countries, and even here "miracle grains" are promised to give us hope. The world population continues to increase at a rate slightly above 2 percent per year, or, more graphically, to double with each generation. The fact that world food production in the past decade has risen at a rate of 2.5 percent per year tempts us to conclude that, however slowly, the world is gaining on its food requirement, and that, while it is obviously desirable to slow down population growth for other reasons, the food situation, at any rate, is not critical.

Alas! A closer analysis shows that these conclusions are unjustified.[2] Over four-fifths (55 million people) of the annual increment in world population occurs in poor countries. Most of the increase in food production has occurred in rich countries. As a direct result of this growing imbalance, the poorer nations are increasingly dependent on the richer ones for food. In the thirties, Latin America was the largest exporter of grain in the world: between 1934 and 1938, it exported an average of 9 million tons a year. In 1972, it imported 4 million tons, and its deficit is growing. Between 1934 and 1938, Asia exported 2 million tons per year; today it is grossly deficient and had to import 35 million tons in 1972. China, once the sole exporter of soybeans, now imports soybeans from the United States. At present, practically all the world's exports of cereals come from North America (84 million tons in 1972) and Australia (8 million tons).

Developed countries also exert pressure on the world's food supply, due to the increase in their consumption of animal products, which in turn utilize, indirectly, an ever-growing amount of grain, soybeans, and even fishmeal. In the United States, for example, our direct consumption of cereal in bread, breakfast cereals, pastry, and "convenience foods" is 150 pounds per person per year, as compared with 400 pounds per person per year in a poor but adequately fed country. Our total cereal utilization, however, including feed grain which reaches our tables in the form of animal products, is 2200 pounds per person per year. Between 1952 and 1972 our beef consumption increased from 62 to 116 pounds per person per year, our chicken and turkey consumption from 27 to 52 pounds. Meat, of course, is enormously expensive to produce; poultry yields only about 15 percent of the calories of the grain used to produce it, beef only 7 percent. Protein yields are of the same order. Interestingly, consumption of milk and eggs, which are produced more efficiently than other animal products, has decreased slightly during the same period.

Americans presently consume far more meat and poultry per capita than most of the rest of the developed world: 50 percent more than the West Germans, more than twice as much as the Russians, seven times as much as the Japanese. But the gap is closing. Even where, as in Japan, fish is a major element in the diet, affluence is bringing a marked shift in favor of red meat. This insidious tendency, occurring in all developed countries, absorbs the bulk of the increase in agricultural production. The wealthy nations include not only the exporters of grain, but also their leading customers (in 1973, the Soviet Union with 28 million tons, and Japan with 17 million). The result has been a rapid decrease in cereal reserves available to feed developing countries in the event of an unfavorable climatic change or a biological catastrophe such as a food plant disease of epidemic proportions.

Facing the pressures of these twin increases—in population and in consumption of animal products—what can we do? An immediate answer would be to increase our grain and bean production, and our fish catch. This involves formidable difficulties, however, as yet little understood by the American public.

Just as, day after day, the U.S. press reports new "medical miracles" which, despite the ballyhoo, have had little or no cumulative effect on our adult life expectancy, so at least once a month, a new agricultural miracle is reported—"miracle rice," "miracle wheat," "miracle corn," or "miracle sorghum." It may seem that these should solve the food problems of the world. For several reasons, however, they do not. In some cases, uninformed reporters assume that because a new high-protein corn or sorghum has been bred, it can and should be cultivated immediately on the surface of whole continents. But sometimes the yield of these new varieties is less than that of the old ones, and the world is short of food in general as well as of protein. Furthermore, disease resistance has to be bred into the new varieties, since the risk of destruction by a fungus, a virus, or an insect is increased when thousands of square miles are planted in a few varieties where before there were literally thousands.

In fact, "miracle grains" are not simply new seeds which can replace the older ones and automatically give much larger crops. They are fertilizer-responding varieties which can indeed considerably increase yields but only if cultivated under optimum conditions. In general, increasing the use of fertilizer increases the need for water, and hence creates the need for more comprehensive irrigation—a sophisticated undertaking in itself. Irrigation systems must be designed to prevent the erosion of the topsoil, as in the Dust Bowl, and its "poisoning" by excess salinity, as in Mesopotamia, North Africa and the Indus Valley.

Since increasing irrigation and the use of fertilizers encourage the growth of weeds as well as of crops, they create an imperative need for herbicides. Conservation of the increased crop requires pesticides, as well as additional storage capacity (silos). Increased yields may make mechanization necessary and certainly increases its usefulness. Inasmuch as all these needs must be provided before the crop can be planted, development agencies must arrange for agricultural credit so that the new tools can be bought, and provide agricultural extension services to teach their use.

A country hoping to take full advantage of the "miracle" grains must thus mobilize a great deal of trained manpower, and support its agricultural effort with rapid industrial development or massive imports to provide the agricultural chemicals, the irrigation and farm machinery, and the concrete for storage facilities.

The less obvious consequences of agricultural modernization may sometimes be among the most serious. In an agricultural system which depends on intensive labor, small farms are more efficient than latifundia, and distribution of land to small farmers may increase production. With the introduction of crops that require large amounts of capital to obtain large yields, and of machinery that operates most effectively on large units, land reform becomes counterproductive unless cooperatives are created and credit made available to them. Furthermore, large landholders who were ready to cede their unproductive land for a modest compensation become reluctant to do so when the land's potential profitability increases. Thus, the social implications of the Green Revolution, while certainly profound, are not always politically progressive.

Clearly, what one is dealing with is not simply the replacement of one variety of seed by another, but a technological and social revolution. Failure of a single factor in this complex system is enough to compromise the success of agricultural development. And the rate of change required is unprecedented in history.

The advent of the "energy crisis" and the enormous increase in the price of petroleum products endangers agricultural development in the Third World to a degree which most people have yet to understand. Increased food production in poor countries is entirely dependent on petroleum products. They are used to produce fertilizers, herbicides, and pesticides; to build and run irrigation systems; to fuel farm machinery; to transport agricultural products; and even to dry certain crops (new varieties often have a higher water content than older strains). The new difficulties in obtaining energy and raw materials, added to the long-term economic, social, and technological demands of modernization, suggest that the progress of the "Green Revolution" is going to be slow, at best.

Other possible solutions to the world food problems, like alternative sources of energy, have been the subject of massive publicity but scanty research. The world fish catch increased from 21 million tons in 1950 to 63 million tons in 1968; since then the total catch has declined. Experts agree that most of the thirty or so main species of commercial fish are overfished. It is likely that this decline in resources will continue. While the slack can be taken up, and probably made up for many times, by aquaculture of fish, mollusks, and perhaps crustaceans, there has been little discussion of the feasibility of large-scale production capable of yielding the millions of tons of protein the world needs. The possibility of producing food from unicellular organisms, algae in particular, enjoyed a flurry of attention for a time as an adjunct to the space program, as a potential component of an oxygen and food regenerating system for interplanetary travel by a few individuals. When planning for long manned flights was abandoned, however, development of that idea was virtually stopped—irrespective of the massive needs here on earth.

Quite apart from the immediate nutritional crisis, the long-term progress of the Third World depends on agricultural development. Under a variety of economic systems—in Uruguay, the Ivory Coast, India, Taiwan, and the People's Republic of China—the most successful development plans have been those which focus on agriculture, and organize industry and distribution to serve it. The current revolution in the world market suggests that, in the future, comparative economic advantage may lie increasingly in agricultural rather than industrial production. Thus, the same increase in agricultural prices that makes dependence on food imports an ever greater threat to the survival of poor nations offers them the hope of prosperity if they can become exporters of food.

III

Intellectually and institutionally, agriculture has been and remains an island—a vast, wealthy, powerful island, an island empire if you will, but an island nevertheless. As it developed into an intellectual discipline in the nineteenth century, it did so in academic divisions which were isolated from the liberal arts center of the university, and which have grown no less isolated as they developed into massive schools, experiment stations, and far-flung extension services. Instead of drawing on the resources of other schools, they produced ancillary disciplines parallel to those in the arts and sciences: agricultural chemistry, agricultural economics, and rural sociology. Agriculture also developed its own scientific organizations; its own professional, trade, and social organizations; its own technical and popular magazines; and its own public. It even has a separate political system—executive departments at the state and federal levels, and legislative committees (in Congress, the House and Senate Agriculture Committees and a House appropriations subcommittee)—which operates with remarkable independence. The strength of this complex is formidable. No President of the United States can be elected against the farmers, in spite of their declining number. Congress and the state legislatures have tended to give agriculture what it wanted. Support for research in agriculture has been subject to fewer vagaries than that in other fields, even medicine.

For most of its history, the United States was a predominantly agricultural country. Agriculture has been an area of particular emphasis and success for American science. Thus it has played a central role in the formation of American scientific institutions and American attitudes toward science. At the same time, in large part *because* of its early success and broad clientele, agriculture has become separated from the mainstream of American scientific thought.

In the early years of American independence, agriculture was the nation's most important scientific interest. Agricultural improvements concerned the two "national" scientific academies as well as specialized groups. The Founding Fathers, many of them innovative planters, placed agriculture at the center of an Enlightenment concept of science broad enough to include society, politics, and sometimes even theology. George Washington encouraged nascent agricultural societies, and sought federal support for the diffusion of agricultural knowledge. Congress agreed that the cause demanded the "patronage of the General Government," but appropriated no funds. Thomas Jefferson, who presented one of the most important scientific statements of the age in the guise of an article on a new plow moldboard, attempted to use the American Philosophical Society, of which he was president, to find a cure for the Hessian fly grain blight.[3]

Because small-scale farmers—the bulk of the populace—could not support private research, Northern farmers and their representatives pressed for the establishment of a federal agricultural bureau. Before 1860, their success was limited, though a statistical program and distribution of seeds and information was begun under the auspices of the Patent Office, and later of the Interior Department. Further federal action was blocked by strict constructionists, mostly Southerners who saw in the efforts to expand the role of the central government in agriculture or science a threat to states' rights. Even James Smithson's half-million-dollar bequest for a national scientific institution set off a protracted constitutional debate. Thus, up to 1861, federal scientific research was restricted to exploration, navigation, geodetics, and various activities of military and naval personnel. But with the outbreak of the Civil War, proponents of federal agricultural research suddenly found that most of their opposition had seceded from the Union.[4]

Because of this radical shift in the balance of political power, 1862 was the *annus mirabilis* of American agricultural science. On May 15, the Department of Agriculture was established as an independent (though not yet cabinet-level) agency, with a mandate far broader and clearer than that of its predecessor in the Interior Department. Five days later, the Homestead Act provided for free distribution of federal land to farmers who occupied and worked it. On July 2, the Morrill Land Grant Act appropriated funds from the sale of public land to support agricultural and mechanical colleges in every state. The scientific import of all this activity was not evident at first; the legislation was framed and pushed through, not by the scientists who established the National Academy of Sciences a year later, but by farmers, politicians and journalists. The department's role evolved slowly and the nature intended for the land-grant institutions was unclear. The acts of 1862 had, however, given scientific institutions a secure constitutional position for the first time; they had erected the scaffolding for a coordinated yet decentralized research and educational system; and they had assured agricultural science of substantial public support for the future.

During the first twenty years of the Department of Agriculture, internal political and organizational problems hamstrung its scientists and alienated their peers outside the government. The research divisions, organized according to university-style disciplines, proved unsuited to their task. Attempts to transfer the department's research functions to established universities failed, however, because such universities tended to be private eastern institutions which could neither meet the needs of the farming public nor win its support. Science developed slowly, too, in the land-grant colleges, where early professors of agriculture often lacked practical understanding of their subject.

In a triumph of political and scientific pragmatism, science did eventually gain a good deal of power in the Department of Agriculture. Beginning in 1881, the entire program was rebuilt. The department as a whole was upgraded to the point that its head won cabinet rank in 1889. A solid body of support for agricultural science developed in Congress. The research divisions were restructured around problems rather than disciplines. The Hatch Act of 1887 established the Agricultural Experiment Stations which, under W. O. Atwater, articulated agricultural research at the federal and state levels. Simultaneously, the land-grant colleges began to come into their own as research institutions, and as dispensers of scientific knowledge.[5]

The reorganization was a scientific as well as a political success. Pasteur had already proven that the study of practical agricultural problems could lead to basic discoveries, and in 1889-1893 a team under Theobald Smith in the Bureau of Animal Industry of the Department of Agriculture, studying the Texas fever that plagued the western cattle industry, extended Pasteur's work by demonstrating the role of secondary hosts in disease transmission. This success, although particularly brilliant, was only one of many that the new system's increased specialization and higher standards made possible.

The department's scientific development continued after the turn of the century. The Pure Food and Drug Act gave the department regulatory functions that involved prescriptive powers and an enlarged scientific staff. New research bureaus expanded the department's interest.[6] Extension services were broadened, and brought down to the school level through the 4-H movement. Well before World War I, a remarkably well-integrated, self-contained system had emerged—a system which proved to be not only the glory of American agricultural science but its tragedy as well.

For two generations, it was this system that shaped most Americans' attitudes toward science. In the 4-H clubs, the colleges, the experiment stations, and the extension programs, people came in contact with a science that was benevolent, useful . . . and limited. Rural America was astonished that professors would presume to tell farmers how to farm, and at first their doubts were justified. By 1890, however, the fledgling agricultural institutions were deluged with requests for soil analysis and for advice on fertilizers, seeds and even machinery.[7]

On the other hand, if farmers came to see science as essential to their well-being and to press for government support earlier than other Americans, this did not mean that they understood the nature of scientific research. They perceived land-grant colleges and experiment stations as service centers where experts dealt with practical problems and gave immediate answers. The leaders of the institutions and of the Department of Agriculture often encouraged this view of science in an effort to develop an ever larger and more vociferous farming constituency which would, in turn, press the state legislatures and Congress for more funds and broader programs. They played down the more abstract values and more distant applications of science. These men, many of them scientists, did not see any conflict between their public relations campaigns and first-class science, but by appealing for votes rather than for understanding they let slip the opportunity to mold a new constituency for science as a whole. Thus, that portion of the American public which was most supportive of research—government-financed research at that—was also the group which was least prepared to support science in its intellectual aspects.

The agricultural scientists' failure to popularize academic ideals of research may be as much an effect as a cause of the isolation of agriculture within our universities. When the modern American university began to emerge after the Civil War, agriculture seemed destined to take an honored place. Higher education from Charles William Eliot down was pervaded with a utilitarian ideal of the university as a social service institution—the ancestor of the "Wisconsin Idea" and the "multiversity." Stanford operated an extension service; Harvard offered a few courses in agriculture, and even an aesthete like Cornell's Andrew D. White took pride in his university's leadership in agriculture. Public and private schools, though increasingly distinct in finance and governance, were scarcely distinguishable by their size or curriculum.[8]

The utilitarian idea, however, was giving way before the combined force of the ideal of the liberal arts college—a modernized version of the pre-university philosophy—and the research ideal, modeled on the German university with its emphasis on theoretical science. New distinctions were made; in 1890, Columbia flew in the face of Pasteur's dictum—"there is no pure science and applied science, only science and the application of science"—creating separate schools for pure and applied science. The mood of the country with regard to utilitarian learning was changing, and although it was not the sole victim, agriculture suffered especially badly. In several senses, Baltimore, Cambridge, Chicago, New Haven and New York did not offer fertile soil for agricultural undertakings, and agriculture was relegated, for the most part, to less distinguished institutions. Social as well as academic snobbery no doubt played a part: it was one thing to train the sons of merchants for the elite Forest Rangers (as at Yale), and another to teach farmers' sons to be farmers. Finally there were pedagogical considerations. Despite the objections of faculty and alumni, President A. Lawrence Lowell was glad to let Harvard Business School students learn to sell hypothetical bananas. But he knew that the study of animal husbandry would entail a more rigorous application of the "case method" and was unprepared to have real "hogs at Harvard."[9]

The general move toward separating the theoretical and the practical would have been serious enough if it had affected only the liberal-arts-oriented private institutions. In fact, however, it also alienated agriculture from those divisions of the university concerned with social policy—an alienation which persists, even in institutions where the study of agriculture still flourishes. That colleges of agriculture seem anomalous in great universities such as the Universities of Wisconsin or California is more than a betrayal of history; it is an intellectual disaster.

IV

Although the independence of agriculture has ensured the power and prosperity of its large-scale practitioners and clients, it has been tremendously costly. For lack of effective outside criticism, a great deal of agricultural research has proceeded on assumptions which are very much open to question. Thus, much of the genetic research carried out in our agricultural schools and experiment stations has been pursued without attention to nutritional values. New varieties of tomatoes, for example, have been selected for hardness to help them survive mechanical picking and processing; the fact that they might be lower in vitamins and minerals has been judged of little or no importance. The significance of the protein content of cereals versus their yield in terms of human as well as animal nutrition—incidentally, a very complex problem, the answer to which may vary from one area to another—has not been the object of well-publicized debates, as it should be. There should also be more discussion of the risks involved in replacing an enormous multiplicity of strains by a few varieties—a result of the Green Revolution—and of the urgent need to create entirely new varieties, such as fertilizer-responding beans and soybeans, or even, through genetic engineering, entirely new species. The desirability of breeding and feeding cattle so as to lower the saturated fat content of beef is a goal on which medical schools could be unanimous, but one which they have not really communicated to schools of agriculture. There is no lack of nonphysicists criticizing technological developments based on physics, or of nonphysicians criticizing recent trends in medical research, development, or applications, but there is a serious lack of scientific critics from outside looking at agriculture in an informed and constructive way.

The lack of criticism of agricultural policy has been no less serious in the political and economic spheres. In Congress, only politicans identified with a farming interest have been willing to serve on the Agriculture Committees and subcommittees. This self-selection has tended to foster large-scale government programs designed to benefit narrow classes of producers without regard for consumers or even an overall production policy. Because we lack viable alternatives to price-support programs tailor-made for special interests, we have continued to institutionalize an economy of plenty long after the world picture had become one of scarcity rather than surpluses. In particular, tens of millions of acres which could have produced crops for needy populations were taken out of cultivation. It is only in 1973 that the last fifty million acres hitherto left fallow were put back into service in an effort to stabilize consumer prices at home, to improve the balance of payments and to replenish grain reserves. Because programs to feed the poor—commodity distribution, food stamps, free school lunches—are the responsibility of the Agriculture committees, they have been alternately neglected and bartered for votes in favor of price supports for commodities like cotton. That the isolation of agriculture committees had become self-perpetuating was exemplified in 1969 by Congresswoman Shirley Chisholm's well-publicized rejection of an appointment to the House Agriculture Committee, which she considered irrelevant to the interests of her urban constituents.[10] At the same time, she was deploring the inadequacy of programs to feed the poor, apparently not realizing that hers could have become a key appointment in changing the attitude of the committee which controls those programs.

The situation in the Executive Branch is just as bad. Essentially, the Department of Agriculture functions on its own, with its own organizations, its own research, its own press, its own domestic policies, and its own foreign policy. Traditionally, it has thought of itself as having five functions, which, in decreasing order of importance, are as follows: it is the electoral agent of the President in the farm states; it represents farmers in the government, especially those who operate medium-sized and large farms; it tries to facilitate the sale of our agricultural commodities abroad; and—lagging far behind—it makes an effort to protect the consumers of agriculture; and it feeds the poor. Until the late sixties, it fulfilled these roles on its own—some well, some poorly—with relatively little criticism or interference from outside the agricultural community of interest. More recently, the failure of poverty and consumer programs, and the massive sale of cereals to the Soviet Union which has caused increases in U.S. food prices and impaired our ability to meet nutritional commitments to Third World nations, has awakened public concern.[11] The formulation of a domestic nutrition policy began with the White House Conference on Food, Nutrition and Health in 1969; this year's Senate Study on National Nutrition Policy, with the cooperation of the State and Agriculture Departments, extends the reappraisal to the foreign sphere.

If the President's Government Reorganization Plan, delayed by Watergate and the subsequent diversion of Congressional interest, is finally carried out, programs to feed the poor will be taken over by HEW, and some of the consumer-protection programs, such as meat and poultry inspection by regulatory agencies. The Department of State would like to cooperate with the Department of Agriculture in creating a unified foreign agricultural trade policy with a view to insuring three objectives: the improvement of our balance of payments (agriculture now provides our most profitable exports); the use of our surpluses as a constructive tool of foreign policy (the wheat sale to Russia which permitted Near East countries to constitute war stocks needed for the 1972 adventure is a negative example); and the preservation of the United States' traditional role as a bulwark for poor countries against famine—a role now threatened in India and Bangladesh by the depletion of our reserves. The possibility of at last creating world famine food banks, the need to integrate agriculture into our national system of priorities in allocating energy resources, and a number of other developments all indicate that the administrative isolation of agriculture cannot continue indefinitely.

V

The reintegration of the resources and traditions of the agricultural science sector into the nation's intellectual life will have to be a major goal of both academic and political leaders.

For many years prior to 1969, social thinkers in the United States believed that they could take American agriculture for granted. Ours is, after all, an urban industrial nation. Our amazing productivity, built up by government-backed applied research, provided an embarrassment of riches; the main problem was inventing devices to prevent production from flooding domestic and world markets with unwanted commodities. Land was retired from cultivation; farmers were paid not to produce; and the government had to buy produce and dispose of it through Food for Peace programs, "sales" to poor countries for useless soft currencies under the P.L. 480 program, the School Lunch program, and commodity distribution to the poor. Self-selected agriculture committees in Congress threw together jerry-built programs for particular crops or particular years, with reasonable confidence that no one except their farm constituents would show any interest in their proceedings. To middle-class wage earners, food was merely one component of the cost of living, and a declining one at that. There seemed to be no need for a national food and agriculture policy.

The radical changes that have taken place since 1969 in the domestic agricultural and nutritional situation and since 1972 in the world food market have transformed both the political outlook of agriculture and the demands on agricultural technology and management. In a world of shortages, nutrition promises to become the greatest international problem, and the ability of the United States to export food emerges as her best guarantor of continued economic viability. Those who would plan, predict, or even understand the future can no longer afford to ignore agriculture. And, in considering it, they cannot afford to overlook the vast body of expertise that already exists, often within their own institutions.

One of the fields of study in which the separation of agriculture from other disciplines is most clearly dysfunctional is that of nutrition—an area in which American achievements have been outstanding, but curiously skewed. In Europe, bioenergetics—studied in human subjects by people like Antoine Lavoisier, Max von Pettenkofer, Karl von Voit, Max Rubner, and E. P. Cathcart—was carried on in a variety of academic institutions, including medical schools. In contrast, in the United States, at least until the late 1940's, the men who led the breakthroughs in nutrition—W. O. Atwater, Herbert Osborn and Lafayette B. Mendel, Elmer V. McCollum, and Conrad Elvehjem—began or conducted their research in agricultural experiment stations or colleges of agriculture. Interestingly, the two major figures in antibiotics, the other basic medical development in America in the same era—René Dubos and Selman Waksman—also started their work in agricultural institutions. It is revealing that, in this case, the obvious applications to medical care for the acutely ill and the fact that in Britain research on antibiotics was conducted by well-known physicians, led medical schools to adopt the field of antibiotics at once.

The field of nutrition, however, has never been adopted by medicine. Public education in nutrition has been the province of the Department of Agriculture, home economics teachers, and football coaches. In the universities, research and teaching in nutrition was moved from colleges of agriculture to schools of home economics and eventually to schools of public health. To this day, some of the best medical schools in the country teach virtually no nutrition, even when, as at Harvard, there is a thriving department of nutrition next door. Nutrition suffers from its isolation, for a discipline whose current chief effort is to fatten animals for slaughter is ill adapted to meet the needs of a human society, least of all in advanced countries where the main nutritional problem is obesity. Ironically, however, human nutrition has now become alienated from agriculture as well as from medicine, to the extent that much research in animal nutrition and genetics is conducted without regard to the food value of the animal.

Another field of great current interest in which agricultural science has an important part to play is that of ecology. At some universities—Berkeley, for example—environmental studies are taught in the college of agricultural sciences; even so, however, the subject is often discussed (even at Berkeley) without direct reference to agriculture. That this is foolhardy should be obvious to anyone who has ever glanced out the window of a transcontinental jet. Moreover, closer attention usually reveals that agriculture *is* being referred to, but indirectly and in bits and pieces; pesticides, for example, are discussed at length with very little understanding that these poisons may be part of a system that keeps people alive. Without agriculture, we cannot develop a humanistic ecology.[12]

Rising food prices and more widespread shortages may mean that the intellectual isolation of agriculture will end. However, the development of intelligent and informed criticism will not come easily. In the past, criticism has come mainly from nonacademic and, however effective otherwise, not particularly vocal bodies such as the Rockefeller Foundation. Universities have largely failed at their task of encouraging dialogue among disciplines. The great universities with large agricultural schools have failed most strikingly. Nevertheless, theirs remains the greatest opportunity. Even urban institutions are not without some resources for providing informed criticism—demographers, botanists, hydrological engineers, economists, agribusiness experts, and nutritionists. These peripheral fields, however, are no substitute for the competence in agriculture itself required of a university whose many schools are all interested in the well-being of American society and the future of poorer countries. We need a change, both in states of mind and in institutions, if agriculture is to benefit from the intellectual evaluation it deserves and needs. If, as a result of continued insularity, agricultural, natural, and social scientists allow nutritional disaster to overtake large segments of the human race, they will deserve the criticism they receive.

REFERENCES

1. For a brief overview of the government role in domestic nutrition matters, with particular attention to the problem of feeding the poor, see Jean Mayer, "Toward a National Nutrition Policy," *U.S. Nutrition Policies in the Seventies*, ed. Jean Mayer (San Francisco: W. H. Freeman, 1973), pp. 1-10. This book contains discussions, by leading authorities, of a wide range of current issues in the field of domestic nutrition policy.

2. A fine, up-to-date analysis of the world food situation, and of the perils of American "callousness" in this area, is Lester R. Brown, "The Next Crisis? Food," *Foreign Policy*, 13 (1973-74), pp. 3-33. The statistics on international trade are cited on pp. 6, 21.

3. Daniel J. Boorstin, *The Lost World of Thomas Jefferson* (New York: Henry Holt, 1948); Jefferson's "description of a Mould-board of the least resistance, and of the easiest and most certain construction" is in *Transactions of the American Philosophical Society*, IV (1799), pp. 313-322; Brooke Hindle, *The Pursuit of Science in Revolutionary America 1735-1789* (Chapel Hill: University of North Carolina Press, 1956), pp. 355-367 (on Agricultural Science in the early Republic) and *passim;* A. Hunter Dupree, *Science in the Federal Government: A History of Policies and Activities to 1940* (Cambridge, Mass.: Harvard University Press, 1957), pp. 3-19. In our historical discussion of federal policies we draw primarily on Dupree's book, still the most comprehensive study of the evolution of American scientific institutions. Of particular value for the history of agricultural science is a series of works by A. C. True: *A History of Agricultural Extension Work in the United States 1785-1923; A*

History of Agricultural Education in the United States 1785-1925; and A *History of Agricultural Experimentation and Research in the United States 1607-1925* (U.S. Department of Agriculture *Miscellaneous Publications,* Nos. 15, 36, 251; 1928, 1929, 1937).

4. Dupree, *Science in the Federal Government,* Ch. II-VI, pp. 110-114; True, *History of Agricultural Experimentation and Research,* pp. 18-40.

5. Dupree, *Science in the Federal Government,* Ch. VIII ("The Evolution of Research in Agriculture 1862-1916"); True, *History of Agricultural Education,* pp. 95-191; T. Swann Harding, *Two Blades of Grass: A History of Scientific Development in the U.S. Department of Agriculture* (Norman, Okla.: University of Oklahoma Press, 1947), pp. 6-44.

6. Dupree, *Science in the Federal Government,* pp. 172-183.

7. Margaret W. Rossiter, "Justus Liebig and the Americans: A Study in the Transit of Science, 1840-1880" (Ph.D. thesis, Yale University, 1971), describes the initial failure of soil chemistry to gain acceptance among farmers, and its eventual success—a key victory for Agricultural Science in the United States. See also Roy V. Scott, *The Reluctant Farmer: The Rise of Agricultural Extension to 1914* (Urbana, Ill.: University of Illinois Press, 1970).

8. On American conceptions of higher education in the second half of the nineteenth century, see Laurence R. Veysey, *The Emergence of the American University* (Chicago: University of Chicago Press, 1965), Part One. Veysey writes from the perspective of the liberal arts and sciences; unfortunately, there exists no comparable interpretation that deals with agriculture in higher education. Besides True's work, there are general studies of the land-grant colleges—Earle D. Ross, *Democracy's College: The Land-Grant Movement in the Formative Stage* (Ames, Iowa: Iowa State College Press, 1942); Edward D. Eddy, Jr., *Colleges for Our Land and Time: The Land-Grant Idea in American Education* (New York: Harper, 1957)—and histories of individual institutions. Charles E. Kellogg and David C. Knapp, *The College of Agriculture: Science in the Public Service* (New York: McGraw-Hill, 1966) is a recent survey of the current status of these schools.

9. Melvin T. Copeland, *And Mark an Era: The Story of the Harvard Business School* (Boston: Little Brown, 1958), Ch. II, *passim.* Veysey points out that, in terms of educational philosophy, business administration was simply the urban equivalent of agriculture (*The Emergence of the American University,* p. 112).

10. *New York Times,* Jan. 30, 1969, p. 16.

11. Jean Mayer, "USDA: Built-in Conflicts," *U.S. Nutrition Policies in the Seventies,* pp. 205-207.

12. Perhaps this knowledge of their ultimate importance explains why agricultural scientists, though long aware that misapplication of their work could bring catastrophe, have not suffered from the Faustian *Angst* that afflicts their counterparts in the pure sciences. Application is an integral part of their science. Even the most enlightened utilitarianism has its drawbacks, but academics who have been deprived of support, or who have awakened to find that their basic research was funded with very pragmatic ends in view—that, in spite of their attempts to free themselves from political pressures, they are servants to the military—may no longer feel that they have nothing to learn from their colleagues in agriculture.

DON K. PRICE

Money and Influence: The Links of Science to Public Policy

PEOPLE IN THE social and physical sciences are not immune to the characteristic American tendency, in matters of politics, to be mildly manic-depressive, or at least subject to sudden swings from supreme self-confidence to self-pitying gloom. A decade ago their leaders were charting the increase in federal funds for science with the same zest that Wall Street shows during a bull market, and glorying in the influence of scientific advisers throughout the federal government. But now the money they get has been declining, and so has their freedom to spend it as they wish. Worse still, they have been expelled from the inner sanctum of political influence, their official advisory position within the executive office of the President. And so they (or in this collection of essays, we) are engaged in diagnosing what is wrong with science in its own opinion and in that of the general public. It is hard to tell whether the trouble is deeply rooted in a changing conception of science and a crisis in the relation of civilization to technological power, or whether it is a more superficial ailment, which may cure itself or be cured by modest remedies.

The argument that we are faced with a fundamental challenge is a tempting one. When politicians began to vote public funds to support science after World War II, none of them indulged publicly in epistemological theorizing. But it is plausible to suppose that they accepted rather uncritically the assumptions that underlay scientists' requests for funds—namely, that science is the reliable way to gain knowledge, that the knowledge gained by basic science will be translated automatically into useful and practical ends, and that this process will contribute to material progress, which is naturally a good thing.

Obviously some critics now attack these assumptions, on moral as well as material grounds. On the material side, we—at least, some of us—now turn some of the cherished aims of a generation ago upside-down. We see our national military strength as an invitation to war rather than as a safeguard; we see the conquest of disease as creating a problem of overpopulation; we see industrial growth as signaling the imminent destruction of the environment. On the moral side, science is challenged both by those who seek to restore traditional values and by those who prefer to explore new modes of consciousness.[1]

There is no doubt about the politicians' loss of faith in the automatic beneficence of technology; the political clout of the environmentalists is testimony to that. But is there a loss of faith in science in a more fundamental and philosophical sense?

It is hard to know whether the disillusion of the academic critics has influenced the actions of politicians. There are certainly straws in the wind which cannot be ignored. A recent public opinion poll shows a sharp drop in popular respect for "the people running the scientific community" (I wonder who *they* are!) and a belief that "science is making people so dependent on gadgets and machines, people don't know what nature is anymore."[2] The recently re-elected head of the Southern Baptist Church, himself the retiring president of a chemical corporation, proclaimed that "Christian ethics and virtue died as our scientific and technological age was born."[3] And a general survey of U.S. national goals prepared for the Congress by the Congressional Research Service, which is not addicted to radical and esoteric ideas, quoted such fundamental critics of science and technology as Jacques Ellul, E. J. Mishan, and Theodore Roszak and noted that the new trend in American culture "implies throwing out the scientific method, the definition of effects, and the search for cause . . . through the processes of rational analysis."[4]

It is quite likely, I suppose, that such general ideas have some practical effect on political action. Laymen and politicians are more likely to be affected by observing the cracks in the collective morale of the scientific community than to be persuaded by esoteric theories. Nevertheless, I doubt that the practical problems of money and influence—which I take to be at the heart of the "public policy" issue—are for the time being very deeply affected by philosophic qualms. It seems more probable that the loss of support and status have resulted from a combination of events quite outside the control of scientists, together with the difficulties they have had defining for themselves (and for politicians) the role that they should play in relation to political authority. These difficulties—problems of academic self-governance, of professional ethics, and of political obligation—are indeed related to fundamental questions regarding the nature of science. But they can be stated in simpler and more practical terms—terms which pose the problem less as one of irreconcilable ideological conflict, and more as a manageable and miscellaneous set of difficulties. This is an emotionally disappointing diagnosis. It is much more inspiring to be told that one's ailments are caused by a great crisis in the moral order. Nevertheless, it is better not to preach to the patient, or to psychoanalyze him, if the prescription he needs is diet and exercise.[5]

Let us look first at the federal financial support for science—the research and development programs—with respect to their terms and conditions as well as the amount of support they provide, and second at the arrangements by which scientists influence policy and programs.

Federal Funds for Science

President Nixon's 1973-1974 budget and the program decisions that accompanied it have posed difficulties for science, and are often described by scientists as "an unprecedented reaction." There is no doubt about the practical difficulties of many institutions. The slight increase in the dollar amounts allotted for research does not keep up with inflation. Furthermore, money has been shifted away from general categories in which the purpose and direction of research is left to grant recipients, and into categories of applied research toward politically popular—though perhaps unattainable—ends. Efforts have been pushed to distribute research funds more widely, reducing their concentration in a small number of institutions. And funds for the training of graduate students have been drastically curtailed.

Now it may be fair to call this general movement a reaction, but not an unprecedented reaction: indeed it goes back, at least to some extent, to the way the U.S. government traditionally supported research before World War II. It does not cast doubt on the fundamental utility of science as a means of acquiring knowledge, or even on the expectation that basic knowledge will be translated into useful applications that will further human progress. On the contrary, it shows an impatient confidence in these traditional ideas, and a determination to take political shortcuts to make them effective. The fundamental philosophical basis for this reaction is not any new alternative cognitive system; it is the old-fashioned Jeffersonian confidence that science will be furthered better by very pragmatic and applied approaches than by scholastic theorizing.

The system that embodied this old approach was the agricultural research program, the largest scientific research program in the federal government before World War II. It had three key characteristics: it (1) related research directly to the mission of the Department of Agriculture to help the farmers, (2) supported research exclusively in public institutions, and (3) allocated funds for research mainly to state colleges and experiment stations by automatic formulas related to population and other factors.

The sales pitch for a new approach after World War II—its tone set by the Vannevar Bush report—politely ignored the old approach, but systematically departed from it. It proposed to support basic research with no immediate connection to any applied purpose, and to support it in private as well as public institutions; and it proposed to allocate funds not by formula or geographic area, but on the basis of the scientific merit of specific projects, as judged by eminent scientists in private life.

This new sales pitch was presented, of course, by the leaders of the wartime Office of Scientific Research and Development. Their successful experience lent plausibility to their argument that basic science would be translated automatically into desirable applications. Had not theoretical physicists, for example, proceeded directly from their mathematical abstractions to the design—indeed to the operational development and strategic planning—of radar, proximity fuses, atomic bombs, and other weapons? Within the administration of wartime research programs, the unity of will to defeat Hitler broke down the boundary between theoretical science and national purpose, and seemed to justify the demand of the scientific community for a new approach to the support of basic science, in which the key decisions would be entrusted to the leaders of scientific institutions.

The new approach, however, was only partially instituted. The delay in enacting the National Science Foundation legislation, and the controversy over the degree to which it should be exempted from the conventional type of accountability to political authorities, gave other agencies time to move into the picture. Within the second decade after World War II, the pattern was set: nine-tenths of the federal funds for research and development went into applied research and development, and of the fraction devoted to basic research, nine-tenths was controlled by departments and agencies which granted funds in the belief that their support of some basic research would be justified by the indirect and long-run contribution it would make to their practical missions. Even the National Science Foundation, the one agency dedicated to the support of basic science for its own sake, gained its modest appropriations on the plea that, by improving the fundamental quality of American science, the new approach would, in the long run, bring indirect and unpredictable practical benefits.

Seen from today's perspective, the new approach—which dominated the National Science Foundation and the National Institutes of Health and had some influence in most other departments and agencies—was fantastically successful. It raised American theoretical science from second-class status to preëminence in the world. Yet the new approach was not based on a clear-cut decision of national policy, but rather on a series of compromises that came to seem less satisfactory as the national focus of political interest changed. Even though this dissatisfaction shows up most clearly in the recent Nixon policies, it was becoming apparent during earlier administrations in several ways.

The most general difficulty appeared within the range of research programs that the Office of Scientific Research and Development supported during the war, and that continued to receive most support in the post-war years—those concerned with military problems, atomic energy, aeronautics and space programs, and medicine. The leaders in these fields helped bring about a gradual transition from wartime to peacetime research, maintaining the links between government on the one hand, and universities and industry on the other. But strains developed within the system. They were based in part, of course, on differences of opinion between scientists and politicians. Perhaps even more important, however, were the differences between those scientists whose main interest remained with departmental missions, and those whose primary concern was basic science in relation to higher education.

The main difficulty came, perhaps inevitably, from the central compromise of the system. Most of the money came from the budgets of departments that had to pretend to Congress that the grants would contribute to the success of their missions. Many of the grants, however, were awarded to universities on terms that required no specific acomplishments; thus they came to be looked on as contributions to the ordinary educational and research purposes of the university, within the broad limits of some program or discipline.

While budgets were generally increasing, this central ambiguity caused no great conflict. When priorities shifted and money became tight, a conflict arose between the priorities dictated by a department's view of its mission, and those which would advance science generally in alliance with higher education. On this issue, university scientists found themselves in opposition not mainly to politicians who disapproved of the support of science, but to a more difficult combination: scientists committed to the mission of a given department, lobbyists advocating it to Congress, and politicians seeking ways to justify spending money for a particular type of scientific activity by connecting it with a particular purpose.

Even within the National Science Foundation, which was established with no mission other than the advancement of science, the pressure to establish some obvious connection with practical applications has been too great to resist. The RANN (Research Applied to National Needs) program, and others with specific applied purposes, are growing, and less than half of NSF expenditures now go to Science Research Project Support. Furthermore, the pressure in the direction of practical application has been coming, over at least the past decade, from the most sympathetic Congressional supporters of the Foundation, such as Congressman Daddario and Senator Kennedy.

The reaction within the universities against the Vietnam war has been accompanied by an increase of interest—especially among younger scientists—in dedicating their research to social policies that they consider liberal and humane. The fears of atomic warfare, of irreversible damage to the environment, of tyrannical use of techniques of computation and communication, and of the immoral use of genetic engineering have all contributed to a demand not only that the relation of science to society change, but also that science develop a new way of knowing based on an acceptance of human values.[6] The determination to serve only purposes of a socially oriented nature has certainly involved many scientists and organizations of scientists in programs of social action.

The main result of this has been to give new emphasis and status to the social sciences. In such fields as health, housing, and welfare, the types of questions at issue may indeed involve the biological and physical sciences, but usually they also involve broader questions of economic cost, administrative feasibility, and social and psychological relationships. To the extent that such issues are subject to disagreement among political parties or factions, it is difficult to elicit extensive financial support to do research connected with them.

Scientists, both within the government and outside of it, took action to prevent business and government from doing what they would otherwise have done. The most notable example is the movement for protection of the environment, which has had a conspicuous impact on public policy by pursuing a wide spectrum of activities, from efforts within the Congress and the administration to develop a new system of technology assessment, to the promotion of propaganda and lawsuits to stop particular actions or classes of actions.

To the extent that such activities have given publicity and prominence to a field of policy, they may have contributed to the willingness of the government to support research in relevant disciplines, and they have certainly encouraged a new type of critical and systematic analysis of the effects of technology on society. But to the extent that they have put scientists into organized opposition to economic or political interests, they have probably made politicians more wary about supporting research without constraining its purpose and scope.

As a federal republic, the United States has never operated on a unitary basis, making administrative decisions on the basis of uniform standards. It has usually tempered the justice of a merit system with the mercy of patronage of one kind or another. By patronage I mean nothing necessarily illegal or improper—only the use of a governmental program not merely to accomplish its ostensible salutary purpose most efficiently, but also to give incidental material benefits to certain persons or groups, or to make sure that they are widely distributed.

Science in particular and the academic world in general have now become so important as channels for federal spending that new kinds of patronage have developed. One kind of patronage is distributed on a geographic basis. For example, the funds for early research on public works or agriculture could not have been distributed to the satisfaction of Congressmen on a purely rational basis, so a combination of political pressures and automatic formulas was developed. In contrast, the new approach for the support of science initiated by the Bush report led to a concentration of project grants in a small number of universities. The beneficiaries of this concentration, of course, considered it the due reward of their merit. Those who were left out, however, considered it favoritism; they were scientists too, and their views and those of their regional educational associations carried weight with their local Congressmen. To accommodate the various states and regions which had no universities qualified to compete for project grants, programs of financial support had to be invented to create new centers of excellence, able to compete on an equal basis. Well before the election of President Nixon, special institutional grants were devised for this purpose, and the formulas for distributing graduate fellowships were altered to make sure that recipients were distributed broadly among universities, rather than allowed to concentrate in the most attractive centers of research.

Another kind of patronage, of course, is designed to favor organized groups of one kind or another, as when the civil service system is adjusted to give preference to veterans of the military services. After the initial successes of the civil rights movement, women and various minority groups, especially blacks, began to turn their organized efforts to demanding more affirmative action with regard to equal employment opportunities. As one would have expected, this movement was neither initiated nor supported by the reactionary wing of politics, or by a conservative administration dedicated to protecting private institutions from bureaucratic interference. It was, on the other hand, welcomed most enthusiastically in liberal academic circles, at least at first. Now, however, the affirmative action program poses difficult problems for universities, especially in the natural sciences, for the requirements of the Department of Health, Education, and Welfare are forcing universities, on pain of losing scientific research grants, to expand the employment of women and minority groups, and most of the money involved in the prospective penalties has been going to the natural sciences, the very disciplines in which women and blacks are conspicuously scarce. It remains to be seen whether the affirmative action program can accomplish its constructive purpose without in effect converting its hopeful and idealistic goals into patronage quotas.

It is tempting, but not very realistic, to lump together all the reductions in support for science and all the new bureaucratic constraints on the terms and conditions of such support, and attribute them to a coherent political ideology of the current Presidential administration. It is not only unrealistic, but tactically misleading, however, because it suggests that a change in administration will eliminate the political troubles of science. The fact is that most of the trends that many scientists now consider troublesome began before the Nixon administration took office, are supported by some of the most enduring habits in American political life, and were initiated or supported by substantial segments of the scientific community.

The principal financial worry of the scientific community is not likely to be that the aggregate amount of federal expenditures for research and development is insufficient. Science (which for political purposes includes technology) has become a critical mass in politics, and politicians are unlikely to stop supporting it or to lose faith in its practical benefits. The problem is likely to be too much faith, or too much uncritical faith, rather than too little. For uncritical faith in specific results leads Congress to channel support into applied technology rather than basic science, and to impose constraints and conditions that may benefit particular regions, institutions, or groups at the expense of the highest quality research and education.

Heavy reliance on the project grant system controlled by panels of outside advisers may have led certain programs too far in the direction of supporting science for its own sake—too far, at least, in terms of the maintenance of the necessary political support. On the other hand, the swing away from that system and the imposition by political authority of applied goals for research threaten to overbalance the system in the opposite direction.

What approach would make it possible for basic science to get federal funds with fewer strings attached? Without pretending to outline a political strategy for science, let me suggest two points. First, those universities and research institutions with the most independence and the least reliance on federal funds obviously have the most bargaining strength. Thus the central core of research as well as teaching programs might benefit from greater efforts to build support based on private financial sources or on state appropriations. Accordingly, I suspect that the more eminent basic scientists, who traditionally stand near the liberal or radical end of the American political spectrum, will be tempted to move in a conservative direction on issues involving the provision of private funds for university science. It is the liberal or populist wing of politics that seeks to repeal the tax incentives encouraging large gifts for educational and scientific purposes, and to reduce the privileges of private foundations. But within the American political spectrum there is nothing inherently liberal or conservative about the idea of protecting the autonomy or immunity from political interference of certain types of private (or even public) institutions for reasons of public interest, even when they are granted governmental support. This is a point on which scientists and university leaders need to rethink their basic strategy.

To the extent that federal funds must be relied on—and the amounts of money involved make heavy reliance necessary—it may be better to face frankly the fact that higher education generally needs federal support, and to avoid the temptation to skew university budgets by heavy reliance on specific research and training grants. General support grants from the federal government would, of course, carry the dangers of public control. But if the rich American experience with the university support by state legislatures is any indication, political control is most likely to deal with the incidental aspects of education (such as employment practices) rather than with its central intellectual content, which the academic community seems reasonably able to protect.

My second point is that scientists will need to reassess their traditional attitude toward administrative authority. They have feared bureaucratic interference with scientific freedom, and consequently distrusted bureaucrats. This attitude is especially strong among those who seek to put science at the service of more liberal and humane policies, and to cut its connections with the military-industrial complex. Yet the federal operating departments that have granted research funds with the fewest restrictions and the highest degree of freedom are probably the military research programs, and of course the military services are the most highly organized career services with the greatest bureaucratic power of all the career systems within the federal government.

If this seems a paradox, it is not hard to explain. An administrative agency cannot be understood in isolation: it is a part of a symbiotic system with the Congressional committees that control its powers and programs and money. An agency can delegate only as much discretion to its grantees as is delegated to it by the Congressional committees. When constraints and requirements are attached to a grant of funds, it is always difficult to tell how much they reflect the personal opinion of the administrator, and how much his reaction to or anticipation of Congressional demands.

Military services, especially since World War II, have been given more money and more discretion than civilian departments. Congress disagrees far less over the purposes they serve, and consequently over the political implications of their research, than over those of novel and controversial social programs. In considering this contrast between the relative bureaucratic autonomy of the military service and the typical vulnerability to political interference of the civilian agency, it is important to note that political interference with scientific research programs is poorly described by such negative words as "constraints." More often than not, politicians try to interfere with research not in order to stop it or hamper it, but to apply it prematurely. Their emphasis on "applied" rather than basic research is this kind of interference: they say, in effect, "Stop wasting time on esoteric theory, and put your science to practical use." This kind of political interference grows not from any dislike of science, and certainly not from any idealistic or philosophical distrust in its cognitive capacity or material purposes, but from an excessively naive faith in it which amounts almost to an intoxication with the belief that science can save us.

The bureaucratic authority and independence of the military services make it possible for them to distinguish, in their organization and procedures, among basic research, applied research, development, and operational testing and evaluation. This system has the disadvantage of conservatism; if it had not been shortcut by the OSRD, we would not have had some of the spectacular weapons invented and used during World War II. But its conservatism may protect against the opposite danger: generals and admirals often have the motive, and sometimes the ability, to resist pressure from their political superiors to proceed too hastily to put into operation a weapon which is still in the stage of research, a gleam in an inventor's eye.[7]

By contrast, when an important program of sound reform is in the experimental and research stage, it may easily be corrupted by pressure to put it into operation generally and immediately. A notable example is the poverty program, in which the mistakes of community action agencies might have been avoided if more time had been taken in a limited number of communities to conduct the proposed experiments and to observe their results.[8] The pressure to expand what should be experimental research into an immediate and widespread action program may seem to come from the Congress or the President, or from an administrator seeking to make the program more attractive to politicians and to anticipate their reactions; in either case, the cause is the same—inadequate expertise and strength within the administrative agency—and the effects are the same—to prostitute scientific research and to destroy its capacity to be of practical service in the long run.

The civilian program most like the military in this respect is that of medical research. This is partly because its scientific content is more impressive and incomprehensible to laymen than that of the social sciences, and partly because the organized strength of the medical profession takes the place of the organized strength of the military hierarchy. But it has to resist the same pressure toward a premature or unwarranted transfer from one stage to another along the spectrum from basic research, to applied research, to practical development, and on to operational or commercial use.[9] These stages correspond to functional roles within the political system—those of the scientists, the professionals such as engineers or physicians, the administrators, and the politicians (corresponding, in the business world, to top management and directors).[10] In a complex society, unless these roles are kept sufficiently distinct and institutionalized with appropriate degrees of autonomy, the independence of basic research is especially vulnerable.

Scientists have long understood one half of this problem, and sought to protect their freedom of inquiry from political interference. But they may not yet realize how great an impact science is having on public policies, and consequently how necessary it is, if science is to be protected from politics, that politics be protected from science. Or more accurately, perhaps, that politicians be protected from the temptation to escape their responsibility for decisions by asking science to give immediate answers to questions that are inherently unscientific.

The Political Status and Influence of Science

The problem of sorting out the functional roles of the scientist or professional from those of the administrator or politician becomes especially tricky in the higher levels of the governmental hierarchy. Here we have seen what many scientists consider a disaster: in 1973 the President abolished the President's Science Advisory Committee and transferred out of his Executive Office the Office of Science and Technology, and its head, who had served also as Chairman of PSAC. What does this mean for the status of science in terms of the second aspect of this paper: its influence in the major policy decisions of government?

At the outset, while considering financial support for science, I suggested that the willingness of politicians to vote funds had not, as a rule, been much affected by the way some scholars and intellectuals now distrust science as a way of knowing. On the other hand, in a much more specific and professional sense, those politicians directly responsible for dealing with issues with a high scientific content have probably become far more sophisticated about the limitations of pure science as a source of answers to complex policy questions. And their new sophistication is analogous to—if not learned from—what scientists themselves have been saying about the limitations of science.

Scientists as a group now need to acquire a similar sophistication (as many of their leaders have already done) about the strengths and limitations of their role in public policy—a sophistication that would lay the basis for something like a professional ethic for policy advisors.

The first step is to appreciate what science as such cannot do, which is very different from saying what scientists cannot do. Some leading scientists, troubled by some of the academic disillusion with science as a method of knowledge, have discussed the way the abstraction and specialization of science limits its application to complex concrete problems. Victor F. Weisskopf, for example, writes that a complete description in scientific terms of any phenomenon (he uses the example of a Beethoven sonata) may "not contain the elements of the phenomenon that we consider most relevant." He goes on to speculate that, just as medieval Europe suffered from the predominance of religious over scientific thought, we may now be suffering from the reverse. Practical decisions, he argues, must be "based on judgments that are outside the realm of science."[11]

Practical politicians need not go so far as to accept the idea that there are aspects of knowledge which no science can touch. All they need to understand, in order to appreciate the limitations of science as a guide to policy, is that each complex problem involves aspects to which several scientific disciplines are relevant, and that a master of one discipline is not likely to be reliable as a synthesizer of all of them.

Some of the best scientists understand this most clearly. When members of various Congressional committees, perhaps overwhelmed by the difficulty of the problem, appealed to physicists to tell them what to do about the nuclear test ban treaty, the physicists had to explain that the crucial aspects of the problem were not those of physics. More recently, the same stand-off has occurred in debates over environmental pollution and the energy crisis. The scientists have been teaching the politicians that pure science cannot be translated simply and immediately into practical decisions.

As the physical and biological scientists point out their own limitations to the politicians, they sometimes appeal to the social sciences to cover the gaps in their competence. Accordingly, the social status of social scientists has risen: in recent years a few have been admitted to the National Academy of Sciences and appointed to the President's Science Advisory Committee. Some social sciences, furthermore, have contributed to efforts to synthesize various disciplines in a broader systems approach, which can sometimes transcend the limits of specialization and deal with problems involving high degrees of uncertainty. This effort is laying the basis for a more professional application of science to policy questions.

The politician, however, will still have reason to be skeptical about the complete objectivity of scientific advice. Two scientists may agree completely on the scientific aspects of a policy issue, yet disagree violently on its other aspects, and themselves sometimes be unaware of the difference. This is lack of objectivity at its most elementary level.

On a more complex and difficult level, there is the recent tendency of some scientists, especially social scientists, not only to reject the classic distinction between facts and values, but to argue that the moral commitment of the scientist requires him to put his science at the service of a political cause. In rebellion against colleagues who themselves were rebelling against social conservatism and scientific orthodoxy, Alvin Gouldner argued in 1968 that "the myth of a value-free social science is about to be supplanted by still another myth, and that the once glib acceptance of the value-free doctrine is about to be superseded by a new but no less glib rejection of it."[12]

A bias even more difficult than that of political prejudice is the one that comes from the fundamental method of science. The radical philosopher Herbert Marcuse has charged that "the mathematical character of modern science determines the range and direction of its creativity, and leaves the nonquantifiable qualities of *humanitas* outside. . . ."[13] His argument is given some weight by the fact that even physical scientists point out the difficulty of systems analysis: if in order to seem objective one takes into account only things that are easy to quantify, one leaves out the most important human and political factors.[14]

This bias is aggravated by politics—less by partisan prejudice than by the desire of politicians and administrators to find a noncontroversial basis for action. The simplest example is the use of objective tests, rather than those permitting discretion in grading, in a civil service examination; at a more complex level, consider the preference of the politician for financing hard sciences more generously than soft sciences, or the tendency of judges to cite objective statistical data whenever they serve their purposes. This effort to avoid bias by relying on the most objective data may itself build a tremendous bias into our social choices.

Clearly, however, some procedural or institutional checks, some special ethical or professional restraints, are needed to help political authorities know when scientific advice is valid, and when the limitations of science make it misleading. And politicians have learned from scientists themselves that they need help in making such a distinction. It is possible, of course, for such checks to be developed within either scientific or governmental institutions.

One of the things that scientific institutions (including scientific societies and universities) are set up to do is to determine the objectivity and quality of scientific work within particular disciplines. The specialization of each discipline, its abstraction from practical purposes, and the publication of its findings enable its practitioners to protect it from economic or political pressures, and its referees to audit the validity of new contributions. In subjects which cut across disciplines, however, especially ones in which intense political and moral controversy affects immediate political decisions, scientific institutions are made impotent by their very virtues. A scientist does not like to rule against another scientist except within a common discipline, on the basis of standards or tests on which they both agree. This limitation has made scientific institutions vulnerable when their members undertook to debate current political issues, or the scientific aspects of such issues. The most obvious recent example concerns genetic and racial problems. The unwillingness of the National Academy of Sciences and the American Association for the Advancement of Science to conduct formal studies of inherited racial differences in intelligence, the violent student protests at various universities against the lectures of scientists who have published works on the subject, and the careful avoidance of the topic by governmental research agencies suggest how hard it is to find procedures within scientific institutions to determine what is scientifically valid and what is not when the problem has complex moral and political ramifications.

It is hard to devise such checks or procedures at high political levels for a different set of reasons. A political authority is compelled to make decisions whether or not it knows the relevant facts, and sometimes before it has time to gather or consider them. If it is to associate scientists in this process, it must know the terms of the bargain.

It is one thing, of course, for a scientist to advise Congress or the President in the way any newspaper pundit does—publicly, usually without access to the politician's secrets or his personal opinions, and always without responsibility for the consequences of a decision. A more intimate and effective participation in the politician's deliberation is another matter, and it is by no means easy to set up the ground rules for such a relationship.

Let us consider the case of the Presidency. The Executive Office of the President was created as recently as 1939; before that, there were only rudimentary arrangements for providing a President with staff advice or assistance of any kind—administrative or professional, to say nothing of scientific. The assumption on which the Executive Office was established is analogous to the one about the political status of scientific knowledge. It was assumed that facts were politically neutral, and the administrative career staff and scientific advisers should be too, while policy should be determined by political values, and elected politicians held responsible for value judgments. Everyone with any political sophistication knew that this assumption was a myth. But a useful myth, for on no other basis would a President desire an institutional staff, or the Congress let him have one.

The natural scientists were latecomers within this staff system. The New Deal of Franklin D. Roosevelt tried to involve them through the work of the National Resources Planning Board,[15] but the political gulf between FDR and the conservative leaders of the scientific community was much too wide, even before Congress decided to put an end to such nonsense as planning by abolishing the board. Although the wartime OSRD was in the Executive Office of the President, its leaders did not recommend that the peacetime National Science Foundation which they proposed should remain close to the seat of political power; on the contrary, they urged that it be given a status of detachment and isolation, as much like that of a private foundation as possible.

A decade or so later (Sputnik provided the final push) it became more clear that scientific factors had to be taken into account at the highest policy levels. In order to integrate national policy across departmental lines, some central point of view, informed by scientific expertise, was obviously necessary. The natural scientists expected and received a status within the Executive Office at the center of political power that raised new issues, or old issues in a new degree, concerning their status, the basis for their selection, and the extent of their latitude in policy and public politics.

It was assumed from the beginning that the scientists could retain their status in private scientific institutions while in official positions. The members of the President's Scientific Advisory Committee served part time in the White House, while continuing their work in universities (or occasionally in business or government). They were even permitted to keep secret the names of those scientists whom they brought in as members of subordinate advisory panels.

They were of course, formally appointed by the President. Nevertheless, the President usually selected them with little or no personal acquaintance, and almost no advice from partisan political sources; he ratified the co-option of scientists who were eminent academically, concerned with public issues, and reasonably compatible with each other and the President on policy issues. The PSAC maintained the formality that its chairman was elected by its members, a unique and awkward Constitutional anomaly that could be overlooked since they always elected the man the President chose as his Special Assistant for Science and Technology. Many scientists liked to think of PSAC as the ambassadors of the scientific community within the constitutional system, not as merely one set of Presidential staff advisers, among many others.

For a time, these peculiarities in their status and selection caused no great difficulties with respect to their method of operation or the latitude they were granted in policy and politics. Under Presidents Eisenhower and Kennedy, their most important role was to serve as an independent check on the technological experts from military departments; their agreement on basic values with the President made it possible for them to work as a part of the Executive Office system, in effective cooperation with the economic, budgetary, and political staff. Nevertheless, they came under frequent criticism from scientific leaders for not defending the point of view of the scientific community vigorously enough, and for being too docile as subordinates of political authority.

Their docility, however, probably did not contribute as much as their independence to their abolition. Their advice to the President did not always prevail; their outlook was academic and intellectual, and often conflicted with the industrial or operational interests of executive departments, or with the political showmanship of the President himself. It was impossible to keep it a secret that the PSAC advised against the NASA plans for going to the moon, and were unsympathetic with such technological programs as the supersonic transport.

Even so, executive staff work presented too many official constraints for many scientists. It seemed in the interest of the leaders of science—especially of the National Academy of Sciences—not to confine themselves so narrowly to the White House. Some of them cultivated instead a relationship with the committees in the Congress most sympathetic to science. Some, through the Committee on Science and Public Policy within the Academy, built up a more active agent for expressing the policy goals of leading scientists. Others, in the American Association for the Advancement of Science and many of the specialized scientific societies, involved themselves more and more publicly in discussions of policy and political issues. This diversity of policy roles would have been acceptable under normal conditions. Or at least it might have evolved into an acceptable arrangement, with some scientists accepting the constraints involved in serving as official advisers, and others (or the same ones at different times) choosing to act publicly as Congressional witnesses or public spokesmen for their scholarly societies. But the political strains of the Vietnam war period were too great, and the formal abolition of the PSAC was accepted by Congress with hardly a word of protest. Obviously the rift between the political views of the Nixon (and for that matter of the Johnson) administration and the typical attitude in the community of basic science was too great to permit the continuation of an intimate and confidential relationship. Which side one chooses to blame is a partisan matter and need not be argued here.

The long-term prospects for the status of scientific advisers to political authority, however, is not so clear-cut. First of all, the conflict between radical or liberal scientists and conservative politics is not necessarily a permanent one. It is true that the more theoretical scientists, among them the most respected leaders of scientific institutions, tend to be less conservative than their colleagues in more applied science, or than engineers or medical doctors.[16] On the other hand, exceptions can be found to this tendency, just as Republicans find labor leaders, and Democrats business leaders, for Presidential Cabinets. The nature of issues and political attitudes toward them change. It is not hard to imagine a situation in which the main political issue for scientists became the degree of freedom of universities and research institutions from government control, and on this issue the theoretical scientists might seem very conservative indeed, arguing for a nineteenth-century, if not a medieval, conception of the freedom of the intellectual estate from secular politics.

Second, scientists are not unique in finding it uncomfortable to advise politicians with whom they are out of sympathy. The Joint Chiefs of Staff and the Council of Economic Advisers have similar problems. But PSAC was different in two ways: first, its advice put it more directly in conflict with the operational decisions of a wide range of departments and agencies, in something like the way any budget office is in conflict, and second, the professional connections and part-time status of its members made them feel more obliged than other advisers (especially budget officers) to assert their independence from political authority.

Nevertheless, in a period involving less political strain, it should be possible to work out new arrangements and a tolerable code of professional ethics for Presidential science advisers. The guiding principle of such a code must be that the more intimately one shares in the secrets and influence of political authority, the less one can be guided either by the abstractions of a scientific discipline or by the sentiments of a private constituency.

The desire of scientists for an independent status may even lead them to prefer an advisory status outside the White House. A formal location so close to the center of political power is uncomfortable for any group with professional scruples. Economists can stand it because in this context no one expects economics to be an exact science, and because the Council of Economic Advisers has nothing direct to do with advocating grants to support research in the economics discipline. Generals and lawyers, on the other hand—all of them with policy interests cutting across broad political concerns—may likewise find it useful to be established in departments outside the Executive Office. The President, for example, has attorneys on the White House staff, but gets his official advice on the relation of law to his program from the Department of Justice, which must bridge the gap somehow between the politics of Presidential policy and the professional standards of the courts. The National Science Foundation (the director of which, Dr. H. Guyford Stever, was designated as the President's Science Adviser in July of 1973) may seek to develop an analogous role, and its friends may renew the effort to strengthen its status by naming it an executive department.[17]

The President is Constitutionally entitled to get his scientific advice from such an arrangement if he sees fit, and almost any advisory pattern he wished to use would be preferable to one he ignored. Under the new arrangement, however, the foundation will be handicapped by its obligation to cultivate the favor of the Congressional committees that vote its money and authority, an obligation which might—on such difficult issues as military weapons policy—weaken its position as a critic of other agencies on behalf of the President.

For the time being, and until a President feels comfortable in a close personal relationship to his science advisers, I am inclined to think that a poor second-best arrangement is better than none, and that policy advisory relationships would evolve more satisfactorily if diffused than if concentrated. The White House can easily have access informally to a great wealth and variety of scientific advice, managed discreetly by inconspicuous staff members, just as it gets political advice from a variety of sources. No one expects a President to pay unique attention to the political advice of the chairman of his national party committee. In the immediate future, both the President and the scientists might be more comfortable if he got his scientific advice through a variety of sources—such as the National Academy and National Research Council, the various nonprofit research corporations, and appropriate scientific societies—and if he contracted for such advice through a variety of channels, including the National Science Foundation, various parts of the Executive Office (such as the National Security Council), and appropriate Executive departments.

Such an arrangement would protect the Presidency, and until a better and more sophisticated set-up can be restored, it might also protect the scientific community. For, at any given time, each scientist would have various options. He could, for example, offer his advice from a posture of complete personal and scientific independence (as an individual professor or a member of a purely scientific organization); by choosing this option he would insure that no private club gained a monopoly on scientific advice to politicians. At the other extreme, he could accept the political obligation that goes with complete commitment as a staff member in a fully official and hierarchial relationship to the President. Fortunately, the conventions of the American political system do not require that a scientist choose either extreme, or any role, once and for all. They do make it desirable, however, that he be aware of the inevitable trade-off between influence and independence, and that his idea of the bargain be the same as that of the official he is advising. Once the fundamental nature of the relationship is understood all around, it seems clear to me that the Executive Office of the President should contain a small but strong unit for scientific advice on all aspects of major policy to which it is relevant.

Conclusion

The freedom of scientific institutions depends on a tacit constitutional bargain, in which it is understood that decisions on public policy which involve political interests and power are not to be made by science. Neither are they to be made by scientists unless they subject themselves to the usual tests of political responsibility. That bargain depends on the fundamental assumptions that the scientific mode of knowledge is inherently limited in its ability to deal with major questions of public policy, and that it would destroy the freedom of both science and politics to institutionalize a single ideology or body of knowledge for purposes of controlling our thoughts and actions.

Within these limits, it does not seem to me that the legitimacy of science as a guide to public policy is under serious challenge. The current difficulties in the relationship of science to politics are mainly the result of the traditional tendency of politicians to demand too much too soon in return for the support they give to education and research in the sciences. They are partly the result, on the other hand, of the normal tendency of scientists to try to have their cake and eat it too—to influence policy decisions as political advisers without losing their status in privileged academic sanctuaries.

I see no immediate danger of a crisis as long as each of the two groups—politicians and scientists—has the status and public support to defend its essential position. The inevitable friction may be reduced and the issues mediated not only by personal understanding on both sides, but also by the existence of two buffer groups between raw political power and pure scientific knowledge. These are the scientific professions such as engineering and medicine which apply the abstractions of science to practical purposes, and the professional managers and administrators who less systematically help politicians understand ways to make use of science and the professions within programs of action. As the impact of science and technology on policy continues to increase, the importance of these two kinds of vocations will increase, and their competence will be crucial in defending the legitimacy of science to politicians and the public.

REFERENCES

1. Charles Frankel, "The Nature and Sources of Irrationalism," *Science*, 180 (June 1973), pp. 927-931; and Edward Shils, "The Attack on Rationality and the Future of Science," unpublished paper.

2. Harris poll, *Boston Globe*, February 17, 1972.

3. *The New York Times*, June 13, 1973.

4. The Evolution and Dynamics of National Goals in the United States, prepared by Dr. Franklin P. Huddle at the request of Senator Jackson pursuant to Senate Resolution 45, Serial No. 92-2 (Washington, D.C.: U.S. Government Printing Office, 1971).

5. I have no doubt, however, that the current political problems of science are related in a long-range and fundamental way to changes in moral and political ideas. See my "Purists and Politicians," *Science*, 163 (January 1969), pp. 25-31.

6. Everett Mendelsohn, "A Human Reconstruction of Science," *Boston University Journal* (Spring 1973).

7. This is quite different from the question whether or not to deploy a weapon that has been fully developed, or to use it in combat.

8. D. P. Moynihan, *Maximum Feasible Misunderstanding* (New York: The Free Press, 1969); and James L. Sundquist, *Politics and Policy* (Washington, D.C.: The Brookings Institution, 1968).

9. For case illustrations, see the Appendix by Joseph D. Cooper to *The Quality of Advice*, Vol. 2 of *Philosophy and Technology of Drug Assessment* (Washington, D.C.: the Smithsonian Institution, 1971).

10. These distinct roles are considered at greater length in my *The Scientific Estate* (Cambridge, Mass.: Harvard University Press, 1965).

11. V. F. Weisskopf, "The Significance of Science," *Science*, 176 (Spring 1972), p. 138.

12. Alvin Gouldner, "The Sociologist as Partisan: Sociology and the Welfare State," *The American Sociologist*, 3, No. 2 (May 1968), p. 103.

13. H. Marcuse, "The Individual in the Great Society," *A Great Society?* ed. B. M. Gross (New York: Basic Books, 1968), p. 74.

14. Murray Gell-Mann, "How Scientists Can Really Help," *Physics Today*, 24, No. 5 (May 1971), pp. 23-25.

15. The NRPB, earlier known as the National Resources Committee, had a subordinate Science Com-

mittee which had unsatisfactory relations with the abortive Science Advisory Board of 1933-1935, established by President Roosevelt under the National Academy of Sciences and the National Research Council. See Lewis E. Auerbach, "Scientists in the New Deal," *Minerva* (Summer 1965).

16. E. C. Ladd, Jr., and S. M. Lipset, "Politics of Academic Natural Scientists and Engineers," *Science*, 176 (June 1972), pp. 1091-1100.

17. For Dr. Stever's own assessment of his role, see his statement before the House Committee on Science and Astronautics, July 17, 1973.

DAVID Z. BECKLER

The Precarious Life of Science in the White House

As EXECUTIVE OFFICER of the President's Science Advisory Committee and assistant to all six Presidential Science Advisers, I was involved, for some twenty years, with science and technology policy development in the White House and Executive Office of the President. From this perspective, it is clear that the science and technology advisory function must be carried out in close interaction with the Presidential decision-making process. Such interaction has resulted in many important benefits that could not otherwise have been realized. Yet, in the course of successive administrations and changing national concerns, experience has also shown that these benefits are dependent on the environment in which the science and technology function is exercised. To assess the implications of recent changes in the White House science and technology functions and their future evolution, one must view them in a historical perspective as part of the changing overall environment of policy and program formulation at the Presidential level.

Early in 1973, President Nixon sent a Reorganization Plan to the Congress to abolish the Office of Science and Technology in the Executive Office of the President. About the same time, he terminated the White House post of Science Adviser and accepted the *pro forma* resignations of the members of the President's Science Advisory Committee (PSAC). The civilian functions of the Office of Science and Technology (OST) were transferred to the Director of the National Science Foundation (NSF), and the security functions to the National Security Council (NSC).

Thus, in one fell swoop, the President eliminated the entire White House science and technology mechanism that had been painstakingly erected in the years following the Soviet Sputnik in 1957. Unfortunately, the President's action did not reflect a careful assessment of the strengths and weaknesses, past accomplishments and future potential of the science and technology mechanism in the White House. Rather, it appeared to be the result of a hasty decision taken on the basis of general considerations.

Although the action stimulated little reaction at the time, a tide of questions has arisen in recent months as to the rationale underlying it. There have been hearings by the House Science and Astronautics Committee, and various bills introduced in Congress which would establish special organizations in the White House to deal with science and technology—a Science and Technology Resources Council, for example, a Solar Energy Research Council, and a Biomedical Research Panel in the Office of the President.[1]

The energy crisis has fueled these second thoughts as to the wisdom of the "science on tap but not on top" attitude of the Administration. The creation of an

Energy Research Advisory Committee to the Federal Energy Office (FEO) and the appointment of Dr. Alvin Weinberg as the FEO Director of Research and Development in effect restored a White House science and technology mechanism, but one directed to a specific problem. Every time there is a crisis, will the nation have to respond with a multibillion dollar crash research and development program? Should we have a White House science and technology lighthouse for our ship of state?

Ever since the White House science and technology mechanism was set up, certain forces have limited the influence of scientific and technical judgments on decision-making at the White House level. Basically, these are forces within the White House structure which tend to reject the presence of specialized points of view. The analogy of the heart transplant suggests itself. The "alien" science and technology mechanism functions effectively only as long as the forces of rejection are continually suppressed. In the end, if the environment does not become more favorable to the acceptance of specialized advice, the transplant fails. Clearly, an understanding of the forces of rejection is essential if a viable White House science and technology mechanism is to be designed.

Major forces that affect the stability and vitality of the White House science and technology mechanism include:

(1) the President's perception of science and technology as an instrument of national policy contributing to the solution of critical national problems, particularly politically visible ones;

(2) the effectiveness of channels of communication between the President and the science and technology mechanism (which tend to close unless positive pressure is exerted to keep them open);

(3) the potential threat posed by the science and technology mechanism to other sources of advice and decision-making in the Executive Office of the President, such as the NSC and the Office of Management and Budget (OMB), and the heads of the departments and agencies;

(4) the credibility and confidentiality of the science and technology mechanism in terms of its members and its institutional loyalties, as perceived by the President and his staff;

(5) the relationships of the Presidential Science Adviser and his professional staff to key personnel in other White House units and in the federal departments and agencies;

(6) the level of the White House "generalists' " distrust for specialists on the President's staff, and fear of being pressured from within by narrow expertise and advocacy;

(7) budgeting as it is affected by political demands for short-term decisions that tend to drive out long-term thinking and planning, and by the disposition of budgetary authority over program and policy decisions.

The Post-Korean Period

The White House science and technology mechanism was born of the Korean War; in April, 1951, the question of how to mobilize science and technology in a national emergency situation resulted in the creation of a Science Advisory Committee to the Office of Defense Mobilization (ODM). President Truman stated that the new committee was being established within the ODM so that it would be "in a direct position to participate in the mobilization program as it affects scientific research and development."

The new Science Advisory Committee coasted downhill without influence and without attention until 1953, when a debate in the National Security Council involving the human and technological feasibility of the proposed Distant Early Line to detect Soviet bombers over the far Northern reaches of this continent led to its first major task. President Eisenhower kept the committee alive by asking it to assess the capabilities of science and technology for minimizing the hazard of surprise attack against the United States, in spite of the rapid advance of Soviet military technology.

During the mid-1950's, some suggested that the President should appoint a full-time Science Adviser. Others, however, felt that the advice of the Science Advisory Committee could be channeled to the President through the Director of the ODM, who, as a statutory member of the NSC, could frequently see the President at NSC meetings and during individual appointments. A full-time Science Adviser might not have equal real clout. As a senior member of the President's staff observed at the time, a Presidential Science Adviser would have power and influence in proportion to the number of direct contacts he had with the President; it might well be that the President would rarely find reason to see his Science Adviser. Such a situation, widely recognized at top bureaucratic levels, would greatly weaken the influence of a Science Adviser. This line of argument, sound in the circumstances of the time, prevailed, and the ODM continued to provide a sheltered home for the Science Advisory Committee, from which it could emerge at the onset of each successive crisis.

Response to Sputnik

In the fall of 1957, the ODM Science Advisory Committee had just completed its second major study for the NSC: *Deterrence and Survival in the Nuclear Age*. It dealt with the relative value of active and passive measures to protect the civilian population in case of nuclear attack. There were indications that this report would have considerably less impact than the committee's earlier study on surprise attack. The report's recommendations for a nationwide fallout shelter program were heading into stiff resistance due to the prevailing atmosphere of stringent economy.

What was scheduled to be a routine meeting of the Science Advisory Committee with the President, on October 15, 1957, happened to occur eleven days after the launching of the Soviet Sputnik. In the wake of a traumatic national reaction, this turned out to be a prophetic meeting. According to my informal notes on the occasion:

Dr. I. I. Rabi, Chairman of the ODM Science Advisory Committee, remarked spontaneously that from the committee's point of view, most matters of policy coming before the President have a very strong scientific component. Not only a technical but a scientific point of view plays a role. He did not see around the President any person who would help keep the President aware of the scientific considerations, as in the economic field. He did not see the scientific point of view put forward in a way to give daily opportunities to influence attitudes. He observed that science was, in a sense, being called in after the fact. There was no continuous involvement. The President said that he agreed with this, and that more than once he had felt this need. But the lines of organization were frozen. The Office of the President was crammed and inadequate. Congress has traditionally been jealous with respect to this Office. However, something could probably be worked out. It is not the entire answer, but it would help to have someone who could see the scientific problems and bring in more specific ideas—a special assistant trained as a scientist. Dr. Rabi felt the President should have a person with whom he could live easily. The President asked General Goodpaster to mull this over. Dr. Killian pointed out that a committee such as the Science Advisory Committee could provide proper backup for such an individual. The committee could be given recognition and status so the individual would not be isolated from the scientific community.

The President said that he had felt a need for such assistance time and again. An office with such a person would develop a chronological record of discussions that have been made involving scientific matters. He cited that early in the scientific satellite program, he emphasized the importance of the psychological aspects. Yet secondary matters entered into the question, and the primary need was forgotten. He noted that events have proved that he was right. There was a need for a person to keep needling in order to see that the original purpose is carried out.

On November 7, 1957, President Eisenhower addressed the nation on TV and radio, announcing that he was creating the Office of Special Assistant to the President for Science and Technology and appointing James R. Killian, Jr. to fill it with the aim of assuring that the nation's entire science and technology program for responding to the Soviet challenge would be carried forward in a closely integrated fashion. The President noted that one of the greatest and most glaring deficiencies of the citizenry was their failure to give high enough priority to scientific education and to the place of science in national life. Two weeks later, the President reconstituted the ODM Science Advisory Committee as the President's Science Advisory Committee (in the White House).

What followed during the remainder of President Eisenhower's term was a remarkable series of actions directly traceable to the new science and technology mechanism, including:

(1) the creation of NASA as a civilian space effort to build on the foundation of the National Advisory Committee for Aeronautics;

(2) the establishment of the Director of Defense Research and Engineering as the number three man in the Department of Defense (DOD);

(3) major improvements in the long-range ballistic missile program, including new emphasis on solid propellant engines;

(4) the acceleration of ballistic missile early warning capabilities;

(5) major advances in our technical capabilities for antisubmarine warfare and photographic intelligence gathering;

(6) recommendations that led directly to the establishment of the Arms Control and Development Agency;

(7) assessments of the desirability and technical feasibility of a nuclear test ban which led, in the Kennedy Administration, to the successful consummation of the atmospheric test ban treaty;

(8) the establishment of the Federal Council for Science and Technology to coordinate federal research and development.

During this period, a number of positive factors tended to neutralize the forces of rejection I alluded to earlier, and to contribute to the effectiveness of the science and technology mechanism. In the first place, since the President had established the mechanism, he was anxious to use it in a way that would establish public confidence in it, and provide direction and momentum to our military and space research and development efforts. Second, the Special Assistant for Science and Technology had easy access to the President and regularly attended meetings of the NSC. This was greatly facilitated by the fact that the Special Assistant for National Security Affairs, General Robert Cutler, felt that his function was to assist direct access to the President. There was also the fact that military and space issues were so pressing they provided an important bridge between the President and his Science Adviser—one which could carry other issues involving science and technology as well.

Third, the military and space problems faced by the White House had such a large science and technology component that they called for the special expertise and experience of the Science Adviser and members of the PSAC. Since most of those involved had been associated in important posts in military research and development efforts during World War II, they functioned as an unusually effective team. Fourth, this was a time when scientists and engineers had great credibility and respect in the public eye. They were treated with unusual deference by Congressional committees. They were regarded with awe by public officials who professed ignorance in dealing with the technologically sophisticated problems. Finally, the White House/Executive Office organization was relatively small, and the President's reliance on his Special Assistant for Science and Technology gave the science and technology mechanism considerable impact. Since, by contrast, science and technology management in other federal agencies was relatively weak, the heads of those organizations did not regard the White House science and technology mechanism as a threat or a competitor.

This was the honeymoon period for the White House science and technology mechanism. Almost all the forces at work during this period reinforced its position in the White House structure. Although George Kistiakowsky's succession to Killian's post brought a different style and emphasis, he too enjoyed a close relationship with the President and was able to sustain the momentum generated by Killian and the scientific consultants to PSAC.

The Post-Eisenhower Era

Considering the normal upheaval occasioned by a change in the Presidency, particularly when the political party changes too, it is remarkable that the full-time staff of the Office of the Special Assistant for Science and Technology survived the inauguration day in 1961. President Kennedy's appointment of Jerome Wiesner as his Science Adviser was widely anticipated. Because of his prior and close association with the President, particularly during the campaign, Wiesner enjoyed a uniquely close and informal relationship with his boss.

At the same time, however, there were new forces at work that tended to move the science and technology mechanism in increasingly uncertain directions through the later Kennedy years and the Johnson and Nixon Administrations. The very style of President Kennedy generated a diversity of ideas and idea initiators at the top of government. His philosophy was that imaginative new programs are usually generated outside the departments and agencies. In this context, the White House science and technology mechanism lost its uniqueness as an intellectual foothold at the pinnacle of government.

In the post-Eisenhower era, communication between the President and his Science Adviser has consistently been at its best in the early part of each Administration. Organizational relationships among members of new Presidential teams are fluid at first, but gradually crystallize as jurisdictional lines harden and the staff power structure becomes more firmly entrenched.

After President Kennedy appointed McGeorge Bundy as Special Assistant for National Security, the Science Adviser found it more difficult—although still possible—to reach him directly on national security affairs. This situation was intensified in the Johnson and Nixon Administrations to the extent that the Science Adviser's lack of direct access to the President on national security matters prevented him from bringing his special viewpoint to bear directly on major policy issues in the prosecution of the Vietnam war. This weakening of direct communication with the President was partly due to the emergence of a sizable NSC staff concerned with detailed national security policy formulation as contrasted with the very small NSC coordinating staff of the Eisenhower Administration.

Similarly, the President's Science Advisory Committee was restricted in its ability to deal with the full range of national security issues. After initial efforts which led to the establishment of the Arms Control and Disarmament Agency and the consummation of a nuclear test ban, the committee no longer had active panels dealing with arms control issues (with the exception of biological warfare). By the end of 1972, most of the PSAC panels on military technology had been transferred to the Office of Science and Technology.

Since the President's direct involvement with science and technology issues related mostly to national security matters, the Science Adviser's lack of direct communication with the President in these areas greatly weakened his ability to bring other issues to the President. As a consequence of this lack of direct contact with their Science Advisers, Presidents Johnson and Nixon were largely unaware of the activities and effective performance of the OST and PSAC.

Nevertheless, despite their limited access to the President, Science Advisers continued, throughout the entire span of the White House science and technology mechanism, to exert substantial influence on the military and space programs. Across the Potomac, a succession of Secretaries of Defense saw the advantages of having another "window on the world," and often regarded the White House science and technology mechanism as a source of thoughtful, highly professional, and seasoned judgments on military technology and its tactical and strategic implications, unencumbered and undistorted by jurisdictional lines of thought. They recognized the dangers of the parochial in-house thinking that tends to capture the heads of large bureaucracies. The Bureau of the Budget (BOB) also welcomed the Science Adviser as an independent authoritative source of analysis and options that sharpened BOB's own assessments of military programs. This was also true for outer-space programs. The reports of PSAC and an OST-directed interagency task group led directly to Presidential approval in 1969 of a balanced program of scientific and manned space exploration and application.[2] A PSAC study in 1971 (unpublished) scaled down NASA proposals for a fully recoverable space shuttle booster and orbiter to a partially recoverable shuttle system, a change which reduced estimated costs by 50 percent.

Through the intervention of the Science Adviser, an additional launch was approved for the Apollo series that greatly contributed to our knowledge of the moon but added relatively little to our basic investment in the manned lunar operation.

The White House science and technology mechanism was given a statutory base by Reorganization Plan Number 2 of 1962, which established the Office of Science and Technology in the Executive Office of the President, and transferred to OST certain statutory responsibilities from the National Science Foundation. At a meeting with PSAC near the end of his term, President Eisenhower expressed concern lest the office of his Special Assistant for Science and Technology and the PSAC be lost in a change of political leadership. He hoped that a way could be found to make them permanent staff mechanisms, like the BOB, with a professional staff essentially nonpolitical in character. Senator Henry Jackson's Senate Subcommittee on National Security Staffing and Operations (of the Senate Government Operations Committee) expressed a similar concern. Under President Kennedy, Richard Neustadt conducted a study of White House organization. He concluded that a statutory base was needed for the White House science and technology staff, but noted that the staff's professional, nonpolitical character and size were inappropriate to members of the President's official family. It was finally decided to propose the establishment of the OST, but that the separate title of Special Assistant to the President for Science and Technology should be retained, and that the same individual who filled that office should also be appointed Director of the OST. This change was an effort to change the administrative status of science and technology in the White House without downgrading its importance or role. This conversion to OST had, in fact, no effect on Wiesner's White House status; his close relationship with President Kennedy was overriding.

As recent events have shown, however, the institutionalization of OST within the Executive Office of the President did not assure its continuing existence. In later years, its apolitical character in a politically sensitive environment gave it an ivory tower "outside" image, and it was often accorded arms-length treatment by key members of the Presidential staff. Unlike the BOB, OST had no Presidentially delegated authority to make its voice heard in the highest councils. Attendance by the Science Adviser at NSC meetings became rare. Although one of the main reasons for the organizational change had been to permit the OST Director to serve as an Administration spokesman before the Congress on scientific and technical matters, the office, since it was established by Presidential Reorganization Plan rather than Congressionally initiated legislation, was given little support on the Hill. The Congress considered the office a part of the President's "staff," and showed little interest in its functions and responsibilities. At this point, the White House science and technology mechanism fell between two stools and lacked a strong constituency either in the White House or in the Congress. Even the scientific and engineering community did not feel that the Science Adviser represented their interests, although, paradoxically, the White House staff based some of their opposition on their feeling that he did. Thus, it was not surprising that the Congressional hearings on Reorganization Plan Number 1 of 1973, dealing with the dismantling of the mechanism, evoked little debate either within or outside of government.

The range of the problems faced by the White House science and technology mechanism and the diversity of its membership began to change after the Eisenhower years, in ways that weakened it. The PSAC did a study in 1961 which resulted in the creation of a functional research and development unit in the U.S. Development Assistance Program. At President Kennedy's request, it undertook a broad review of the National Institutes of Health biomedical research, and, later, a study of the nation's long-term needs for scientists and engineers. Other special PSAC studies were initiated as the result of discussions between President Kennedy and the heads of foreign governments. A PSAC report in 1963 on the use of pesticides recommended the exploration of biological alternatives to the use of persistent pesticides.

Due to the expanding breadth of the matters put before PSAC, its work was carried out increasingly by specialist panels, a practice which diluted the corporate identity of the committee as a whole. The psychological effects of this change were intensified in post-Kennedy years as PSAC's direct contacts with the President diminished. The loss of coherence in PSAC was further aggravated by the nature of its membership, which rotated on a four-year basis and became increasingly diverse as more people with biomedical, social, and industrial backgrounds were included.

Although PSAC came to lack the spirit and sense of challenge it had had in earlier years, it continued to play an essential role in reviewing panel reports, in serving as a sounding board for the Science Adviser, and in lending its professional authority and stature to the White House science and technology mechanism.

During the Johnson Administration, there was a gradual shift in the responsibilities of the staff of the OST. Although they had served primarily as staff members for specialized study panels, they now acted in an individual professional capacity, providing expert advice and a coupling between the work of the consultant panels and the policy and programming mechanisms of the White House. In the face of obstacles to contacts between the Science Adviser and the President, the staff moved to maintain effectiveness by strengthening lateral communications with White House and agency staff members. At times, the desires of individual staff members for independence of action operated to curb the use of outside consultants. As a result, during and subsequent to the Johnson Administration, the OST was perceived increasingly as a source of individual experts rendering specialized advice, rather than as a policy entity represented by the Science Adviser. Although this change inevitably detracted from the strength of the OST as a corporate whole, the new mode of operation was generally effective, considering the practically nonexistent communication channels to the President.

The shift in national priorities in the late 1960's toward increased emphasis on domestic problems diminished the authoritativeness of the White House science and technology mechanism. Its views continued to carry special weight in assessments of military and space technologies, for in this area the government is the sole source of funds, the regulator of performance requirements, and the principal customer for the resulting products. The development of products and services for use by the public, however, is determined by an entirely different set of relationships between federal agencies and outside organizations and institutions—one which neither the White House/Executive Office organization nor the science and technology mechanism was equipped to deal with. The scientific and technological opportunities available for solving domestic problems must be assessed in relation to economic, social, financial, institutional and political considerations, including nontechnical as well as technological options. In its energy and environmental analyses, the OST/PSAC attempted to bring to bear such broader considerations, but it lacked the staff resources and credibility to delineate the pros and cons of a full range of alternatives for the Presidential decision. Moreover, other powerful groups, including the Office of Management and Budget, Domestic Council, Council of Economic Advisers, Environmental Quality Council, and White House staff members, assumed jurisdiction over the broader aspects of domestic problems with which the science and technology mechanism was grappling. As a consequence, OST and PSAC reports and memoranda addressed to the President tended to get sidetracked in White House staff coordination procedures.

The inability of the White House science and technology mechanism to deal with the broader aspects of technology policy is reflected in the creation of other specialized agencies in the Executive Office of the President (the National Aeronautics and Space Council, the National Council on Marine Resources and Engineering Development, the Council on Environmental Quality and, most recently, the Federal Energy Office). The Council on Environmental Quality, for example, was established by the Congress in the face of initial opposition by the Nixon Administration, on the grounds that the existing Cabinet Committee on the Environment, supported by the OST staff mechanism, was sufficient, and that specialized councils in the White House impede balanced consideration of national problems by the President. The Congressional committees, however, felt that the White House science and technology mechanism was much too narrowly based to deal effectively with the broad economic and political aspects of environmental protection. Similarly, when the three-man Energy Policy Office in the OST, dating back to 1967, ceased to be a realistic response to the energy crisis, a separate energy policy office was created to report directly to the President. The Space and Marine Councils no longer exist. And, at the moment, the FEO overshadows the Council on Environmental Quality. It is not unthinkable that they, too, will disappear in time, however, as the Environmental Protection Agency and the proposed Federal Energy Administration (or an equivalent agency) take over the main burden of policy and program development in these areas.

There is admittedly some logic behind the establishment of such *ad hoc* mechanisms to deal with urgent national needs. Yet a problem persists: we need a strong and continuing science and technology function in the White House to suggest the tools and options needed to mitigate the next "crisis," and to provide the President with a balanced view of science and technology as a whole and of the trade-offs among competing problems. The answer undoubtedly lies in maintaining a strong central science and technology mechanism to serve as a common resource for the ad hoc arrangements.

Throughout its entire life, no leak of privileged or classified information was ever traced to the White House science and technology mechanism, to its full-time staff, or to outside consultants on the PSAC and its panels. Nonetheless, the presence of outside scientific advisers within its ranks has been a matter of concern to the White House and Executive Office staffs. Irrespective of administration, the politically charged environment of the White House fosters an acute sensitivity to any possibility of opposition to Presidential policies or premature disclosure of information bearing Presidential decisions.

The White House science and technology mechanism was damaged by two events involving scientific consultants. Both involved highly controversial issues in the Congress on which President Nixon placed the full weight of his office. The first concerned the deployment of an antiballistic missile system whose military effectiveness was seriously questioned in open Congressional testimony by former Presidential Science Advisers in 1969. Their public stand against the President was associated with a similar position taken by the PSAC within the tight wraps of national security. Although they were not PSAC members at the time, their close identification with the committee cast its shadow on PSAC relationships with the President.

The second and more damaging incident involved the SST program. The chairman of the SST panel of PSAC testified and campaigned publicly against the SST, dissociating himself from his PSAC connection by emphasizing that he was speaking in a personal capacity entirely on the basis of publicly available information. A subsequent legal action, taken by an environmental organization under the Freedom of Information Act, forced White House release of the PSAC panel report which questioned the economic and environmental feasibility of the proposed SST. Although the actual release occurred after Congressional rejection of the SST by a cliff-hanging vote in the Senate, the substance of the report was generally known beforehand. This rejection permanently scarred PSAC in the minds of the President and his staff.

The question of confidentiality of advice poses increasingly difficult problems for the future of any scientific and technical advisory mechanism in the White House. There is an inherent conflict between, on the one hand, the President's needs for confidential scientific and technical advice given in such a way that he can balance all the relevant factors in arriving at decisions (in both the ABM and SST cases, the President considered certain nontechnical aspects overriding), and, on the other hand, public pressures to make the activities of advisory committees open, according to the Freedom of Information Act which requires public disclosure of committee reports, and the Federal Advisory Committee Act which requires that committee meetings, with limited exceptions, be open to the public.

Although it did not involve questions of confidentiality of advice, the relationship of PSAC to the President and his immediate staff suffered from the increased political activism within the scientific community itself. Former PSAC members were visible on the board of the Federation of American Scientists which launched a potent lobby against controversial Presidential proposals such as the ABM. Ex-PSAC members were also identified with key political competitors of the President, including the various "Scientists and Engineers for Presidential Candidates." And there was the general polarization between the academic community and the Johnson and Nixon Administrations on the Vietnam war. All in all, fine distinctions between PSAC members and ex-members and between Science Advisers and ex-Science Advisers were inevitably blurred by the political sensitivity of administrations which tended to view PSAC and the Science Adviser as representatives of the scientific/academic communities rather than as independent and faithful advisers to the President.

It is a well-known phenomenon in the federal bureaucracy—perhaps in any bureaucracy—that preoccupation with short-term problems and opportunities tends to discourage and drive out longer-term initiatives. When the science and technology mechanism pushed the science and technology accelerator pedal in response to long-term problems, it was, by and large, discounted as an advocate of larger budgets for science and technology—a self-interest group promoting science and technology rather than solutions to national problems. On the other hand, when the White House science and technology mechanism produced analyses that questioned program proposals, it was sought out and listened to (with notable exceptions in the case of the ABM and SST). When it generated politically attractive projects or contributed to a Presidentially initiated study of a recognized problem, its counsel was deemed valuable. But when, in 1965, it urged action to deal with environmental problems before the lakes filled with algae, and to bring about a better balance between nuclear and fossil fuel research before the pressure of Arab intervention, there was little or no interest or response at the top of government.

Other Visible Accomplishments

This account of the science and technology function in the White House may well overemphasize troubled waters. Clearly, the forces of rejection—whether latent or hostile—were always present. It is remarkable that the function survived various vicissitudes; it is even more remarkable that it continued to affect policies and programs to the very end of its active life. Space does not permit more than a skimming of the highlights of its continuing influence. The national security classification of many reports prevents my revealing the enormous impact they had on our military and intelligence capabilities. Here, however, are some brief indications of the range of the grasp and influence of the White House science and technology function. (See list of PSAC reports at the end of the article.)

(1) PSAC studies (internal) on innovation and experiment in education led to the establishment of the National Institute for Education. Earlier PSAC studies resulted in regional educational laboratories and the Model School concept to deal with the educational problems of the culturally deprived and segregated individuals in our society (1964).

(2) A few words inserted into President Johnson's State of the Union Address in 1966 legitimized federal research on human reproduction and methods of fertility control, and resulted in substantially increased research budgets in this field and new action programs in the Agency for International Development.

(3) The Presidential Statement on Government Patent Policy was drafted, providing for the first general guidelines for the disposition of rights to inventions made under government contracts (1963).

(4) A major report on the quality of the environment was the first comprehensive report on the subject prepared within the government that influenced attitudes in the executive and legislative branches and served as an important guide for policy and legislation when the environmental crisis became a political issue. It set forth the first general Statement of Principles for government policy in this area (1965).

(5) Due to President Kennedy's concern that the quality of biomedical research might have suffered through rapid expansion of support, the biomedical grants program of the National Institutes of Health were examined in depth.[3]

(6) A PSAC study on the effective use of the sea provided a head start for the National Commission on Marine Resources and Engineering Development (1966).

(7) OST initiatives led to the President's proposals to the Congress on highway and vehicle safety legislation (1966).

(8) A three-volume report on *The World Food Problem* served as a primer both for developing countries and for American farmers. It emphasized the critical interdependence of increased food production and family planning programs in developing countries, as well as the relationship of agricultural development to overall economic development (1967).

(9) A PSAC study on the effectiveness of biological warfare led directly to President Nixon's decision to propose an international ban on biological warfare.[4]

(10) The first Presidential Message to the Congress on energy was drafted by the OST (largely a research and development message), and was followed by the establishment of an Energy Policy Office within the OST (1971).

(11) A broad-ranging report on energy technologies in 1964[5] was followed by a second, even more comprehensive study in 1971-1973 (under the Federal Council for Science and Technology)—a study which provided the principal building blocks for the Chairman of the AEC's report to the President in December 1973 and subsequent proposals to the Congress in January 1974.

(12) The President gave his first Message to the Congress on Science and Technology, setting forth a coherent framework and strategy for relating federal science and technology programs (1972).

(13) Initiatives in the international sphere included new bilateral arrangements with Eastern European countries and a proposal which led to the creation of the U.S.-U.S.S.R. Commission on Scientific and Technical Cooperation agreed upon at the Summit Conference (1972).

(14) A continuing series of classified studies resulted in major contributions to strategic weapons developments and defenses, to our naval warfare capabilities, and to the expedited introduction of sophisticated weapons systems such as the laser-guided bomb in Vietnam.

Even after the termination of PSAC, postmortem reports continued to flow which will enrich the dialogue and guide constructive action in areas of central long-term concern:

(15) A report on *Chemicals and Health* dealt broadly with needed improvements in the scientific basis for regulating the very large number of chemicals introduced into society (1973).

(16) A study on *Youth: Transition to Adulthood* raised fundamental questions

about the institutions that bring youth into adulthood and proposed a number of changes in them (1973).

Thus, a small band of 25 full-time professionals together with 200 part-time consultants—with a budget that never rose much beyond $2 million—made their presence felt despite an increasingly adverse climate.

Judgment Without a Trial

It remains the task of future historians to unravel the threads that shrouded the White House science and technology mechanism. In Reorganization Plan Number 1 of 1973 and in the related Congressional hearings, the administration claimed that it disbanded the mechanism mostly in order to streamline the White House staff, and to delegate specialist functions to the agencies where there are more adequate resources. In my view, there are *post hoc* judgments which attempt to define the cure without an understanding of the underlying causes of the malaise. Although the demise of the White House science and technology mechanism was due more to lack of strong support than to overt opposition, the following factors loomed in the background of the final judgments underlying the hasty decision in the late fall of 1973. (It is interesting to speculate whether a similar decision would have been reached six months later.)

Perhaps the greatest influence was the cumulative effect of the years of strain between the White House and the academic and intellectual community over the Vietnam war. This strain deeply eroded the position of the White House science and technology mechanism, which was viewed by key White House staff members as an island of academia within the White House—as a part of the scientific community, whose loyalties were profoundly divided. This was, of course, exacerbated by the PSAC's nonconcurrence with Administration positions in the ABM and SST.

Those who determined the final decision to disband the science and technology mechanism lacked knowledge of its contributions. As far as it is known, the Special Assistant for National Security—the most important single "customer" on the White House staff—was not consulted. People at higher levels were unaware of the web of collateral relationships that had developed in the space and domestic program areas among OST and OMB, the Domestic Council and other White House staffs. Since he lacked a close relationship to the President, the Science Adviser himself was not consulted or advised before the decision was taken.

Still other factors lurked in the background affecting the decision. The New Technological Opportunities (NTO) exercise in the fall of 1971 failed to meet White House expectations; despite the fact that the President had by-passed the Science Adviser to give the NTO assignment to William Magruder (after Congressional rejection of the SST program which Magruder headed), this worked to the detriment of the White House science and technology mechanism. In September 1971, President Nixon addressed the Congress, saying that he would in due course present new programs "to ensure the maximum enlistment of America's technology in meeting the challenges of peace." With the help of OST staff members, and the promise of new money, Magruder conducted a dragnet search for new technological opportunities which stimulated intensive efforts in all of the civilian departments and agencies. The resulting proposals disappointed the White House staff. In most instances, they involved substantial increases in budgetary support of selected on-going programs, they did not have the desired political appeal, and they were criticized by OMB and others for not taking nontechnological alternatives into account. Furthermore, they did not consider fully enough the economic, social, and institutional measures and investments needed to take advantage of the programs they suggested. And finally, some of the proposals raised, but failed to answer, basic questions as to the role of the federal government vis-a-vis private enterprise. (Nonetheless, there were some meritorious proposals that have since been funded.)

There was a feeling in the White House and OMB that the OST/PSAC mechanism was "input" rather than "output" oriented, and that this characterized much of the federal research and development effort. Thus, both the 1972 Budget Message and the Science and Technology Message of March 1972 stressed the need for a "strategic approach" to the planning and management of research and development. Certain problem areas were identified for priority treatment, such as clean energy, delivery of health care, transportation, and natural disasters. However, the call for "top-down" management of research and development that reverberated in the White House corridors in the fall of 1971 was soon forgotten, for there was little inclination to give the OST the budgetary clout necessary to forge coherent, focused programs across the jurisdictional lines of independent agencies. The OST was a classical case of responsibility without authority.

Furthermore, the OST and PSAC were perceived within the White House/Executive Office as staunch advocates of basic science, critical of big technological initiatives—a perception reinforced by the academic backgrounds of many PSAC and OST staff members. Thus, even if there had been a desire to exert a top-down research and development management function, the White House might well have looked elsewhere for what it considered a more appropriate mechanism, as it did in the Magruder-NTO exercise. Unaware that its life span would be foreshortened, the OST in the fall of 1972 was deeply engaged in a self-examination aimed at developing the organizational capability to provide executive management oversight of federal technology programs. To do so would have required an accommodation with the OMB that could not easily be achieved without explicit Presidential encouragement and backing.

Functions for the Future

This has been largely a retrospective look at the White House science and technology policy mechanism. What about the future? Events of past years have put the science policy mechanism to a severe test, to the point of failure. The test has revealed basic deficiencies, not so much in the basic concept of the science and technology mechanism, as in the lack of an adequate structure in the White House for analyzing longer-range domestic policy and program alternatives, to which the science and technology mechanism could relate its output. Experience in the national security area where the NSC staff carries out the policy assessment function has demonstrated that there can be an effective coupling with a separate science and technology mechanism.

The predictive capabilities of the White House are deficient outside of the national security area. A mechanism is needed which would enable the President to look ahead, beyond preoccupation with the budget and the short-term projections of federal programs, to the longer-range objectives and needs of the nation—energy, natural resources, the environment, transportation, food production, etc. Such a mechanism requires analytical resources to forecast and define future problems and needs, to assess the long-term consequences of ongoing programs, and to illuminate the advantages and disadvantages of alternative choices. Science and technology must both contribute to and be guided by such forecasts.

Experience with the White House science and technology mechanism reveals how difficult it is to project a long-term point of view in an atmosphere dominated by an annual budget and a four-year Presidential outlook. It is instructive to note that some of its reports had their greatest impact in the national security area where long-range concepts, planning mechanisms, and related methodologies and analytical capabilities—such as those of the Rand Corporation—are considered essential to guard against future threats. Unfortunately, impending crises of *domestic* concern have not been accorded similar attention. The PSAC and OST reports on the environment and energy, for example, resulted in little action until a nationwide crisis became evident. The PSAC report on the *Quality of the Environment* in 1965 included a Statement of Principles, the first of which declared, "The public should come to recognize individual rights to quality of living, as expressed by the absence of pollution, as it has come to recognize rights to education, to economic advance and to public recreation." Another stated that "the special importance of the automobile as a source of pollution problems should be clearly recognized." Clearly this underscores the need topside for broad, longer-range program development backed by willingness to commit research and development resources today to provide options for tomorrow.

With more staff and analytical capabilities, and more persistence, the White House science and technology mechanism might have been able to catalyze Presidential response before the signs of crisis. But it is difficult to get the President, any President, to move out with vigorous programs that are preventive rather than remedial. Resources are always insufficient even in relation to short-term needs, and the Congress and the public have a penchant for dealing with pressing rather than emerging problems. This short-term point of view may be changing, however. Witness the response to the energy crisis and "Project Independence." There is also discussion concerning the long-term outlook for minerals and food production. We shall see how durable this change of attitude is after the present crisis has eased.

The Congress has created the Office of Technology Assessment as an indication of its concern about the need for better maps to guide our future national programs and policies. No corresponding step has yet been taken at the Presidential level to develop a separate professional capability to analyze and assess program and policy options. To a significant degree, the White House science and technology mechanism performed this function, albeit from a relatively narrow vantage point.

The pluralistic organization of the Executive branch has great strength. It permits the delegation of authority and responsibility required for effective direction and management of the complex federal science and technology enterprise. Not only would central planning and management be less effective in this respect, but it would be more susceptible to serious errors of omission or commission in program direction. On the other hand, agency research and development programs do not add up to fill national needs as a whole. Scattered efforts dealing with domestic problems of energy, environment, transportation, health care, or urban development often do not reinforce each other in relation to common program objectives. There are gaps. There are conflicting and overlapping efforts. This fragmentation of research and development is just one aspect of a widespread deficiency in federal organization, for the agency missions and programs as a whole exhibit the same lack of coherence in relation to functional problems. The President recognized this situation in his Department Reorganization Proposals. To knit the pieces together, however, including the science and technology components, would still require a capability at the White House level to analyze national program and policy alternatives and priorities and to formulate coherent research and development strategies.

Pressures for action—in the environmental area, for example—highlight the need for a much closer connection between science and technology and the regulatory functions of government. To regulate intelligently in the public interest, the regulatory agencies themselves must strengthen their scientific and technical capabilities which have not kept pace with the demands on them. But White House leadership is necessary, for individual agencies tend to respond to short-term needs and pressures from the entities subject to regulation, and to lack of assured research and development funding and scientific and technical traditions. Only a White House level function can assess the trade-offs and interrelationships among actions taken by different regulatory agencies. The OST commissioned a study, the first of its kind, on Cumulative Regulatory Effects on the Costs of Automotive Transportation (RECAT) which attempted to establish the cost/benefit relationships of energy consumption versus environmental concerns, of bumper heights versus air bags, among a host of independent, yet interrelated, regulatory decisions affecting automotive transportation. In several important instances, the report cast doubt on the wisdom of actions taken by individual regulatory agencies without assessing their broad implications.

Conclusion

Science and technology are of vital importance to the achievement of the objectives of almost every federal department and agency. It follows that they must have highly developed scientific and technical capabilities as well as the operational capabilities needed to accomplish their missions. This decentralized approach is a great source of strength in the U.S. government in that it stimulates agency initiatives and a diversity of approaches and viewpoints.

At the same time, we have seen ever-increasing overlaps and interdependencies in the programs of federal agencies in relation to problems of national concern, such as those involving energy, the environment, transportation, and urban development. The interrelationships of the research and development programs of domestic agencies must be appraised and guided in order to achieve the most effective use of our scientific, technical and fiscal resources.

Overall priority-setting, program assessment and integration have long been carried out at the Presidential level, principally in the preparation of the

President's annual budget proposals to the Congress. Preoccupation with the budget process has tended to result in a short-term outlook which gives inadequate consideration to longer-range program options, directions, and implications.

The value of a White House science and technology mechanism to support and guide both the short-term budgetary and long-term assessment functions has been amply demonstrated. However, a science and technology mechanism insulated from the President, with unclear relationships to other White House Executive Office agencies, is in an unstable position. What is clearly needed is an institutionalization of that mechanism which makes its relationships within the White House Executive Office and its direct channels to the President (to be used sparingly) entirely clear. Thus, an assessment of the role of a White House science and technology mechanism must begin with an examination of the broader framework of White House-Executive Office organization for assisting the President in formulating national policies and programs.

As has been pointed out, there is no identifiable "customer" for the longer-range scientific and technical assessments of the science and technology mechanism, such as its reports on energy and the environment and its more recent report on youth. This lack of a customer has often resulted in the mere distribution of such longer-range reports without any systematic evaluation, appraisal and follow-up of their conclusions and recommendations.

What is needed then is a White House/Executive Office organization to develop policy objectives and options to assist the President—one which can assess and interrelate scientific, technical, economic, social, political and institutional factors. It must be future oriented, but not entirely long range in its outlook or it will fall into disuse, as have many so-called policy planning staffs. Only if it has its feet firmly planted in the real problems before the President can it survive with its head in the clouds of the future. It must be closely meshed with the budget formulation and management functions, yet able to provide the President with independent and sometimes conflicting views for his consideration.

A report of the President's Advisory Council on Executive Organization under Roy L. Ash provided the conceptual basis for a White House organization that could satisfy these needs. The resulting Presidential Reorganization Plan Number 2 of 1970 established the Office of Management and Budget to prepare the annual budget and oversee its execution, emphasizing program evaluation and coordination and serving as the President's principal arm for the exercise of his management functions. And it established the Domestic Council to assess national policy needs, develop forecasts to help define national goals and objectives, identify alternative ways of achieving objectives, and recommend "consistent, integrated sets of policy choices."

The Domestic Council, consisting of the heads of the departments and agencies with domestic missions, is well suited to coordinate the ideas and talents of government agencies to affect common objectives. However, the roles of OMB and the Domestic Council have not evolved the way they were planned. The OMB, through its responsibility for the preparation of the federal budget, has exercised a *de facto* control over program choices not strongly counterbalanced by the Domestic Council. The Domestic Council, on the other hand, has been heavily preoccupied by the President's assignment to provide "rapid response to Presidential needs for policy advice on pressing domestic issues."

If the intended balance between OMB and the Domestic Council could be established, they would provide a hospitable home for a White House science and technology mechanism which could serve both the short-term needs of OMB and the broader, future-oriented needs of the Domestic Council. To accomplish this would require substantial enlargement and professional upgrading of the Domestic Council staff, together with a Presidential commitment and sufficient authority to put it on an equal footing with OMB in presenting program choices to the President.

The White House science and technology mechanism would also need direct access to the President and authority to intervene in the decision-making process where it involved the research and development programs and plans for federal departments and agencies. The last Presidential Science Adviser, Edward David, proposed that the science and technology mechanism be given "authorization" authority over the federal research and development budget similar to that of authorizing committees of the Congress. In any case, a viable White House science and technology mechanism must have clearly understood prerogatives, or else the forces of rejection will prevail. To survive and serve the President and the nation, a White House science and technology mechanism must be one of several interdependent parts of a reconstituted White House organization. Designing such an organization will require the same multidisciplinary skills and outlook as operating it, but it should be a priority task. For science and technology, the future is now.

Appendix

PUBLIC REPORTS OF THE
PRESIDENT'S SCIENCE ADVISORY COMMITTEE

1957

    Deterrence and Survival in the Nuclear Age—November 7, 1957—Declassified
    January 10, 1973

1958

    Introduction to Outer Space—March 26, 1958
    Improving the Avaliability of Scientific and Technical Information in the United States—December 7, 1958
    Strengthening American Science—December 27, 1958

1959

    Panel on Seismic Improvement—January 5, 1959 (Released June 12, 1959)
    The Argus Experiment—March 26, 1969
    High Energy Accelerator Physics, A Proposed Federal Program in Support of—May 17, 1959
    Education for the Age of Science—May 24, 1959

1960

    Food Additives—May 14, 1960
    Scientific Progress, the Universities, and the Federal Government—November 15, 1960

1961

Research and Development in the New Development Assistant Program—May 24, 1961
Ad hoc Panel on Environmental Health—June 6, 1961
Waterlogging and Salinity in West Pakistan—August 1961
Project West Ford—October 3, 1961
Ad hoc Panel on Non-Numerical Information Processing—December 1961

1962

Science and Technology in the Department of State—February 27, 1962
Science and Agriculture—January 29, 1962
Bioastronautical Panel—February 15, 1962
Foot and Mouth Disease (report of the Scientific Mission to Republic of Argentina on)—January 31, 1962 (Released March 3, 1962)
Strengthening the Behavioral Sciences—April 20, 1962
Research Support Abroad Through Grants and Contracts—September 4, 1962 (Released September 20, 1962)
Meeting Manpower Needs in Science and Technology—December 12, 1962

1963

Science, Government, and Information—January 10, 1963
Some New Technologies and Their Promise for the Life Sciences—January 23, 1963
Large Scale Scientific Actions With Possible Environmental Effects—January 29, 1973
Ad hoc Panel on Radioastronomy Frequencies—March 13, 1963
Report of the Panel on High Energy Accelerator Physics of the General Advisory Committee to the Atomic Energy Commission and the President's Science Advisory Committee—April 26, 1963
Use of Pesticides—May 15, 1963
Engineering and Science Technicians—July 1963
Statement by the President's Science Advisory Committee on the Nuclear Test Ban Treaty—August 24, 1963

1964

Innovation and Experiment in Education—March 1964
Report on Manpower Statistics—July 13, 1964

1965

Federal Organization of Science—April 9, 1965
Cotton Insects—April 23, 1965
Restoring the Quality of Our Environment—November 1965

1966

Handling of Toxicological Information—June 1966
Effective Use of the Sea—June 1966

1967

The Space Program in the Post-Apollo Period—February 1967
Computers in Higher Education—February 1967
Privacy and Behavioral Research—February 1967
The World Food Problem (Vols. I and II)—May 1967
The World Food Problem (Vol. III)—November 1967

1968

    ITCAP (International Technical Cooperation and Assistance Panel)—February 28, 1968

1969

    Ad hoc SST Review Committee—March 30, 1969
    Preschool and School Education (The Importance of Research and Development in)
      —October 15, 1969
    The Biomedical Foundations of Manned Space Flight—November 1969

1970

    The Next Decade in Space—March 1970
    Report on Science—June 1970
    Ad hoc Air Traffic Control—December 1970

1971

    Report on 2, 4, 5-T—March 1971
    *On Increasing National Productivity Through Educational and Technological
      Change*—May 5, 1971
    Panel on Science and Technology Policy—September 1, 1971

1972

    Improving Health Care Through Research and Development—March 1972
    Chemicals and Health—December 1972

1973

    Scientific and Educational Basis for Improving Health—September 1973
    Youth: Transition to Adulthood—June 1973

REFERENCES

1. Technology Resources Survey and Applications Act of 1974 (S. 2495), Solar Energy Research Act of 1973 (S. 2819), and Biomedical Research Act of 1974 (S. 3023).

2. *The Post-Apollo Space Program: Directions for the Future,* Space Task Group Report to the President (September 1969).

3. *Biomedical Science and Its Administration* (The White House, February 1965).

4. Statement by President Nixon on Chemical and Biological Defense Policies and Programs, November 25, 1969.

5. *Energy Research and Development and National Progress,* Office of Science and Technology, June 5, 1964.

EMILIO Q. DADDARIO

Science Policy: Relationships Are the Key

APPEARING BEFORE a Congressional committee in 1970, A. Hunter Dupree, professor of history at Brown University, showed that he was a prophet as well as a historian when he said, "The present organization charts that I have seen put the structure for science and technology, which has been now for two decades building, nowhere near the interior circle where the President must make priority decisions."[1]

Lee DuBridge, then the Director of the Office of Science and Technology in the Executive Office of the President, as well as his Science Adviser, replied to that assertion when he appeared as the next witness.

I was a little amused to note that apparently Dr. Dupree in his testimony . . . said that the Office of Science and Technology had been wiped off the organization charts. I don't know what organization charts he is looking at, because that was certainly news to me! I think it must be news to the President, too, since he gave a special reception in honor of the President's Science Advisory Committee the other evening. I don't think he realized we were off the organization charts.[2]

Dupree knew, of course, what the organization charts looked like, but his observations were a way of saying that the Office of Science and Technology did not really have the ear of the President and that it was not effectively involved in top-level decision making. Neither then nor now can White House receptions and ceremonies make up for this deficiency. That all of the White House structure for science and technology was liquidated in January, 1973 is not in itself important. For some time up to its dismantling it had been unable to carry out the legal responsibilities assigned to it by President John F. Kennedy in 1962.

It is no longer seriously questioned that the federal government must assume the primary responsibility for supporting science and its applications to public programs. Indeed, the network of agencies throughout the federal bureaucracy which deal with one aspect or another of science and technology is astonishingly complex. Yet during the past few years these agencies have frequently pursued their separate tasks with little coordination, suggesting that policy making has been the handmaiden of expediency. The question, then, is whether science and technology are seriously and effectively considered in the highest councils of government.

For there does need to be a specific place in government for the coordination, sponsorship, and advocacy of government science and technology, and a group of scientists at that focal point on whom the President can depend and whom the scientific community respects. If the lower levels of the federal bureaucracy are to be responsive to the science policy of the President, then there must be an organizational structure which will coordinate their efforts. Equally important, there must be evidence that there *is* a coherent science policy, which, in turn, implies the need for a broad understanding of the multiple tasks that need to be done, as well as for sufficient outlays of government funds in their support.

These, then, are the basic ingredients around which a constructive science policy can be formed. Unfortunately, however, science policy making seems to have become synonymous with justifying programs already underway, or with improvising others on the spur of the moment to meet one expedient or another. One need only review the makings of the federal research and development budget for fiscal year 1973 to perceive that expediency, not judicious cerebration, lay behind it. In times of tight budgets and inflationary surges, science has always suffered, but lately even the most trivial budget issues have not been too small to prevail over science.

Although the budget for Fiscal Year 1975 includes substantial increases in dollar amounts for science and technology, we must not lose sight of the facts that these increases are tied to the energy crisis and must be measured against the inflationary spiral. When additional support is so precisely related to special problems, it is almost always accompanied by supporting language which reminds us that we work best under the harshest of conditions. It is indeed true that many good things have resulted whenever we have forced science and technology to meet a crisis of one type or another. In FY 1974, for example, it is estimated that the National Science Foundation will obligate $400 million for basic research, and $89 million for applied research. For FY 1975, the estimates jump to $570 million and $144 million, respectively, with $113.5 million of the basic research increase allocated for research in the energy field. Such quick adjustments make us wonder, however, if we should so often depend on ad hoc relationships and haphazard funding to give sustenance and life to our scientific research. Somehow we must make it a matter of policy to work out a course of action within which the key is continuity and predictability of support in keeping with recognized national goals and objectives. Unfortunately, we must still conclude that science policy makers have had less influence in recent years on funding for basic research and the rate and direction of technological change than have the advertising account executives who have been so prominent in high councils.

The growing and changing role of science and technology in the affairs of American society calls for more, not less, attention to the planning and support of scientific research and development. It continues to be tempting to assume that the democratic process has the inherent capability to take us on the right technical course, or at least to permit us to muddle our way through. This assumption is no longer tenable, however, for our future as a nation may depend on our ability to bring science and technology to bear on national and world problems. We need to strengthen this ability, but the prospects of doing so seem uncertain and remote at this time.

In addition, the shortcomings in our science policy may be due in large part to our having spent so much of our effort on the structure and organization of science and technology within the government, while downgrading and virtually ignoring the social values and choices presented by that same science and technology—those upon which the legitimacy of any government science policy must ultimately rest.

We have come to assume, perhaps, that the effectiveness of science policy makers is assured merely by depositing them within the Executive Office of the President and the various departments and agencies of government. The belief is widely held that scientists need only to be placed in the inner court in order for reason to prevail. But the record shows that in recent years there has been more frustration than satisfaction in the hearts and minds of Presidential science advisors, and that those who have responsibility for giving advice on science to the heads of the departments and agencies of government are regarded more as accountants than as movers.

A science policy framework of sorts has now been set up, and is headed by the Director of the National Science Foundation, although apparently not in his capacity as Director. Even if we assume that the Director of the Foundation has sufficient time to spare to take on an additional role as Science Adviser and Chairman of the Federal Council for Science and Technology, two fundamental questions remain: Does he have the ear of the President? Does he have the ear of the Cabinet? Without the former he is not likely to have the latter.

Consider the three largest R & D items in the federal budget: defense, space, and health. The Department of Defense is specifically excluded from the new science advisory mechanism; and it is no secret in Washington that the Administrator of the National Aeronautics and Space Administration does not look to the Director of the National Science Foundation for guidance on the space program, nor does the Department of Health, Education and Welfare with respect to health research. It is doubtful, moreover, whether any other department head would look for policy guidance to the head of an equal department. Perhaps the Science Adviser's effectiveness could have been enhanced if the President had issued an executive order explicitly and dramatically delegating science policy authority to the Director of the National Science Foundation. Unfortunately, the appointment was announced by nothing more than a vague and ambiguous letter, which seemed an afterthought to the notice evicting the former Science Adviser from the White House in January, 1973, when Reorganization Plan Number 1 was sent to the Congress. Moreover, this plan explicitly excluded military R & D from the scope of the NSF Director's concerns, despite the fact that the late White House science apparatus worked hardest in precisely that area. One would hope that this serious weakness could be cured by a Presidential directive making explicit the authority of the Science Adviser, rather than leaving it to chance and political ability to earn a place at the decision-making table, wherever that may be in Washington these days. It may be, however, that the ambiguity and weakness of the charter given to him—if a mere letter can be taken as such in the federal bureaucracy—bespeaks a disposition to get scientists out of sight and out of mind. The change which has taken place has not yet proven itself. What we must look for are positive results related to the basic arguments in favor of the change. We see nothing yet to support the fundamental premise of its proponents that the various federal agencies have become so sophisticated about science and technology that the coordinating or advisory functions once performed by OST are no longer necessary. It also remains to be determined whether it is in fact more advantageous to use ad hoc consultant groups when they are needed than to maintain a permanent organization—an argument which arose from the contention that the OST became bureaucratic.

II

There is something to be learned from the whimsies and ironies of science policy over the past fourteen years, particularly from the contrasts between the Reorganization Plans of 1962 and 1973. Since 1969, we have come a long way: back to Square One. Let us briefly review the steps up the hill and back down again.

Sputnik was an event compelling enough to usher science policy into the White House. When President Eisenhower appointed James R. Killian, President of the Massachusetts Institute of Technology, as his special assistant for science and technology, he gave him both the responsibility and the authority to oversee the space and missile R & D program. "Through him," the President said,

I intend to be assured that the entire program is carried forward in closely integrated fashion, and that such things as alleged interservice competition . . . shall not be allowed to create even the suspicion of harm to our scientific research and development program. Moreover, Dr. Killian will see to it that those projects which experts judge have the highest potential shall advance with the utmost possible speed. He will make sure that our best talent and the full necessary resources are applied. . . . In looking to Dr. Killian to discharge these responsibilities I expect him to draw upon the full abilities of the scientists and engineers of our country.[3]

From that base, a major and direct relationship between the President and science developed. Since drawing upon scientific advice required advisers, President Eisenhower shifted the Science Advisory Committee of the Office of Defense Mobilization to the White House, renamed it the President's Science Advisory Committee, enlarged its membership, and added to its authority by giving it the power to initiate studies on its own. Then, to coordinate R & D programs within the federal departments and agencies, he established the Federal Council for Science and Technology in April of 1959, making the Science Adviser its chairman.

It turned out, however, that the Science Advisory Committee and the Federal Council were not enough to elevate the voice of science policy above the threshold of noise in the White House. For this reason, the Office of Science and Technology was created under the Reorganization Plan No. 2 of 1962, to take over the science policy responsibilities of the National Science Foundation. The Science Adviser was given a full-time professional staff, and specifically designated to advise and assist the President on the development of important government policies, plans and programs in science and technology. The authority was taken from the National Science Foundation and assigned to him to coordinate and sponsor basic research and education in the sciences and to evaluate scientific research programs in the federal departments and agencies.

Despite the increased resources that this office provided the Science Adviser, what transpired over the next ten years was a struggle for survival which culminated in Reorganization Plan No. 1 of 1973 and the razing of the White House science structure. This move just about brought us back to where we started from. Much of the authority that had passed from the National Science Foundation to OST in 1962 returned to the Director of the National Science Foundation. It is important to note that the recent changes have left certain responsibilities and functions unclear:

(a) Is the Director of NSF really the Science Adviser to the President, or simply on call to the Executive Office?

(b) What is the real channel, if any, by which the Director of NSF can approach the President?

(c) What, exactly, is the function of the Federal Council under the new arrangement?

(d) How does the Science and Technology Policy Office make its work and findings available to the people who need it?

(e) In what ways is the Director of the NSF, as the chief science figure in the Administration, able to affect the Office of Management and Budget in its allocation of funds for natural resources, energy, and basic scientific research?

The rationale given for the 1973 *coup de main* must have been devised by someone so preoccupied with his ax and block that he did not take the time to read the argument for the 1962 reorganization—namely, that

the [National Science] Foundation, being at the same organizational level as other agencies, cannot satisfactorily coordinate Federal science policies or evaluate programs of other agencies. Science policies, transcending agency lines, need to be coordinated and shaped at the level of the Executive Office of the President, drawing upon many resources both within and outside of Government. Similarly, staff efforts at that higher level are required for the evaluation of Government programs in science and technology.

In 1973, however, "certain" undefined responsibilities, formerly vested in the White House Office of Science and Technology, were transferred to the Director of the Foundation on the basis of the argument that "the research and development capability of the various executive departments and agencies, civilian as well as defense, has been upgraded." Translated: "The kids have now grown up to be strapping and aggressive teenagers and, therefore, they need less guidance from higher authority." Clearly, the earlier argument has been declared inoperative.

It serves little purpose to lament this banishment of science from the White House and to urge the President to reconsider. If the President preferred not to have a scientific advisory committee within the perimeter of the Executive Office Building, threatening to offer frank appraisals of the Supersonic Transport, the Anti-Ballistic Missile, and other programs backed by the White House, it was his prerogative to clear the halls. He also cleared away an illusion that advice was being heard in the White House inner circle because advisers were in the vicinity. An adviser without an advisee is the first to go in a housecleaning, and in its last few years the President's Science Advisory Committee, although its membership was as distinguished as ever, was like the proverbial tree crashing in the vast and secluded Siberian forest, unheard because there was nobody there to listen.

III

Let us look beyond the present while trying to make the most of it, and prepare for a time when there is a bona fide auditor. For it is idle to suggest what the science policy structure of this Administration should be when it does not want science in its house. Perhaps it is idle to dwell on organizational structure in any case, for structure is only a means, though one whose importance cannot be ignored. What is important is the future relationship between science and the Presidency.

If we put the question this way, we find that the answer is self-evident. Science policy was at its zenith in the White House when there was a relationship of mutual respect and understanding between the scientists and the President's inner circle. The advisers from the scientific community had a fair hearing. Their advice did not always prevail, nor should it have. But they were heard and, in return, they listened. There is no substitute for such a relationship and no way to achieve it, except to have the right people involved, and the choice of people is up to the President. That is why we must wait.

In the meantime, however, the scientific and technical communities in this country should speak out frankly on the issues, for when they do this they are at their best. Banishment is not all bad. It can increase freedom of expression, and provide an incentive for scientists to call the shots as they see them.

Having had the opportunity, since leaving the House of Representatives, to look first-hand at the uses to which science and technology are put, I am more than ever convinced that the scientific community should be more involved in determining them. As a group, scientists have a vast potential for influencing opinion, which has hardly been tapped.

President Nixon's 1972 message to the Congress on science and technology was excellent in content, though not in follow-through. Of particular relevance to science policy was his declaration that

we must appreciate that the progress we seek requires a new partnership in science and technology—one which brings together the Federal government, private enterprise, state and local governments, and our universities and research centers in a coordinated cooperative effort to serve the national interest.

Such a partnership is a laudable objective—one, indeed, toward which the U.S. Congress was reaching when, in 1968, it expanded the authority of the National Science Foundation to permit it to engage in applied research.

Many scientists were apprehensive that this expansion of authority might contaminate basic research. Yet, to be perfectly frank about the matter, most Congressmen are not overly concerned about the distinctions between basic science and technology. They tend to accept the convictions expressed by the leaders of the scientific community that the vigorous pursuit of basic research is a prerequisite to the proper application of science and technology; and for that reason, if for no other, they have supported basic research in the past. Congress also believes, however, that science and technology would be applied more rationally if the connections between them were examined more closely. It was to this end that the restructuring of the NSF in 1968 was addressed.

Glenn T. Seaborg put the question in perspective:

Scientists must shoulder another responsibility that will not be nearly so popular among them. As the complexity and depth of understanding of scientific information has increased, some scientists have been content to remove themselves from the frontiers by simply rationalizing that their research projects should be continued until all facets of a problem have been explored and all data refined to a greater and greater number of significant figures. We can think of the frontier of science as being a perimeter surrounding our amassed knowledge, and working on the boundary between the known and the unknown. This perimeter has grown enormously, but there are too many scientists still working at the boundary as it existed 20 years ago, even though there is great opportunity to carry out research at the new frontiers. In science as well as in other fields, there is a law of diminishing returns, a point at which the accumulation of additional information may not be worth the effort expended.

I am not advocating by this argument that some scientists be dragged grudgingly from their benches to the line, but rather that an enlightened science policy will offer to such scientists the opportunity to branch into more speculative directions with assurance that their past accomplishments are recognized. If the public and its elected representatives had the comfort of being assured that this nation would indeed structure the direction of science without interfering with intellectual freedom, I believe that a new era, the likes of which we have never seen, would dawn for science.[4]

This outline of the legislative events which culminated in the expansion of NSF authority in 1968 illustrates the difficulty surrounding any government reorganization. The authorizing legislation for the expansion, which grew out of the studies and hearings of my former Subcommittee on Science, Research and Development of the House Committee on Science and Astronautics, took some five years to complete, even though it dealt with a relatively small agency. The story also serves as a reminder that the U.S. Congress can initiate reorganizations without the impetus of a Presidential message. It further illustrates that that legislative process is a desirable means of involving the scientific and technical community in the formulation of science policy.

We should not neglect the legislative approach to science policy making, either now or in the future. Consider, for example, the recommendation of my former Subcommittee that a National Institute of Research and Advanced Studies be established.[5] In my view, a greater infusion of science advice into the legislative branch of the government can only be beneficial. Part of the genius of a legislature lies in its institutionalization of adversary procedures. As members of the Congress have become more sophisticated about science and technology, several have come to recognize that there can be honest differences of interpretation regarding science policy between two groups of equally competent scientists. This has been a difficult lesson to learn in a society which, in its dependence upon expert knowledge, tends to assume that expertise wipes out differences in opinion. Yet it is a point which must be grasped if efficient science policy making is to be established in an open society.

How a scientific adversary system can be institutionalized at the highest levels of the executive branch is not at all obvious. But it seems likely that if Congress were to establish the means for evaluating responsible scientific advice from many diverse quarters—not simply from the leaders of the scientific community who are too often regarded as its spokesmen on behalf of some imaginary unanimous constituency—then a good beginning would be made in at least one branch of the government toward institutionalizing scientific adversary procedures.

IV

An informed public which understood the full potential as well as the limitations of science and technology to affect social values and choices would be science's greatest ally. At the 1951 Arden House meeting, the American Association for the Advancement of Science issued a statement of purpose which said, in part,

It is clearly recognized that to define, for the general public, knowledge about science and its methods is a difficult, slow and never ending job. . . . In our modern society it is absolutely essential that science, the results of science, and the spirit of science can be better understood by government officials, by business men, and indeed by all the people.

When a second Arden House Conference was being suggested by the AAAS to look again at this question, the emphasis in the Conference plan was

. . . on the improvement of the effectiveness of science in the promotion of human welfare and improving public understanding of science, rather than on furthering the work of scientists and the fostering of cooperation among them.

I am convinced that this kind of commitment on the part of scientists can restore the faith and confidence of the public in science, whereas mere pronouncements, however true, on the virtues and intrinsic values of science will only evoke skepticism. Scientists have an obligation, during this period of reflection on the rise and fall of science policy in the executive branch, to speak out constructively on national issues and to help lay the groundwork for a new relationship of mutual respect and understanding that may someday be reestablished between science and the President.

REFERENCES

1. Hearings on Science Policy before the House Subcommittee on Science, Research and Development, July 7, 1970.

2. *Ibid.*, July 8, 1970.

3. "Public Papers of the President, Dwight D. Eisenhower," Document 230, pp. 789-790.

4. Glenn T. Seaborg, "Review of the National Science Foundation," Statement at the U.S. Atomic Energy Commission Hearings before the Subcommittee on Science, Research and Development of the House Committee on Science and Astronautics, 89th Congress, 1st Session, July 6, 1965.

5. Report of the Subcommittee on Science, Research and Development, April 15, 1970. That proposal is as valid today as it was in 1970 when we studied it.

DAVID J. ROSE

New Laboratories for Old

GOOD SCIENCE AND technology are arts of the rarest kind, though often given over to mediocre practitioners, and sometimes prostituted for daily pay. The art and the success come mainly in choosing which path to follow among many possible ones: far more technical and scientific directions always exist than can be followed. As in all things, initial choices must be made largely by intuition, before the facts or the consequences are fully known or understood; therein lies the art. To see the truth in this, one has only to look at bad science—journals full of dull elaboration on fashionable themes—or bad technology with its hackneyed or short-sighted responses to crises touching such matters as energy and the environment.

Until the late 1960's, there was cheap energy and very little concern for environmental protection. When concern for the environment became fashionable, it was often unaccompanied by any study of cost consequences. After 1970, when a growing awareness of impending energy problems led to new activities for the provision of energy, there was some recognition of how closely energy and environmental activities connect; there was, however, little concern for rational energy utilization or conservation until mid-1973. Even now, in 1974, proposed mixed menus of energy and environmental policies (more energy coupled with lower environmental standards, for example) remain unbalanced. Meanwhile, scientific and technological disciplines flourish, but with growing unease about their roles amid these larger issues.

Certainly something went wrong, or was never quite right in the first place. The tragedy is not one of having switched on lights of technology when we ought to have stayed in darkness—rather, with the chance to illuminate our ills and perhaps ameliorate some of them, we have made neon-lit strips and all-night drive-ins.

A significant fraction of these unfortunate results come from the narrowness of the vision in the institutions involved. Even now, not one of them is very good at working on anything either broadly or for the long term. Energy and the environment demand just those qualities of breadth and duration; hence, we are and continue to be in trouble, with no institutions designed or charged to develop broad alternative approaches, or sum up the costs and benefits over the many sectors of society. This paper sketches the features that more holistic working groups ought to possess; it analyzes the present institutional arrangements to find out why they do not now exist. Roots of both the energy crisis and the environmental crisis lie not in some stony ground of technological intractability, but in a morass of irresponsibility and dereliction.

New Environmental Laboratory Attempts

Between 1969 and 1971, the new awareness of environmental needs, combined with an uneasiness about the adequacy of existing institutions, led to several serious attempts to create new institutions. Most of the proposals advocated laboratory-like organizations, which were designed to have the following features:

1. They sought to incorporate a broad spectrum of abilities, not simply those of natural scientists and technologists.
2. They were intended to fill needs not met by private laboratories such as the Bell Telephone Laboratories, or consortium associations such as the Electric Power Research Institute.
3. They were large and extensive in outreach.
4. They were usually closely connected with the federal government.
5. They sought high visibility.
6. They had no executive power; they sought a license to explore, giving in return a promise not to usurp decision-making powers that properly belonged elsewhere.
7. They had substantial independence and internal disciplinary power.
8. They aimed to develop new technological and other options; to present credible policy alternatives, they took into account the often conflicting goals that exist in real life: conservation and exploitation, short and long time horizons, private and public sectors, and the like.

The tasks to be faced transcended regions and sectors, and could not be adequately contained—"internalized" in the present jargon—at less than the national level. The aim was to replace myriad sets of disconnected and unassessed challenges and responses by more logical processes, where the separate parts of problems were to be re-integrated; it was assumed that an illumination of the resulting options to decision-makers and the general public would lead to better decisions. In short, the advocates aimed at a sort of technology assessment on a grand scale, though few used that phrase at the time.

The literature of the early 1970's contains numerous suggestions for change.[1] Four principal proposals arose from groups that were prepared to proceed if and when adequate support materialized. The first was submitted to the Environmental Studies Board of the National Academies of Science and Engineering by a Committee under the co-chairmanship of Marvin L. Goldberger and Gordon MacDonald[2]; a second was prepared for the Ecological Society of America,[3] a third came from the Argonne Universities Association (AUA), the group formed to oversee the long-range activities of the Argonne National Laboratory;[4] the fourth came from the Oak Ridge National Laboratory (ORNL).[5] I mention these studies not so much to indicate their merits, as to examine the political and bureaucratic responses they evoked; they illustrate certain of the difficulties that will be described later more theoretically. The response by the federal government to the AUA and ORNL proposals makes for an interesting and instructive comparison.

A large semipublic debate on the AUA proposal took place in July, 1969; several high officials of the Atomic Energy Commission and members of the Congressional Joint Committee on Atomic Energy attended. Because the ideas were then only partly focused, and because some time would certainly elapse before any firm decisions were taken, comments could be made without much danger of any political commitment being made.

The then chairman of the JCAE, Representative Chet Holifield (R., California), enthused over these venturesome ideas:

In my opinion this awareness of the contamination of our environment is fully justified and long overdue. The rapid urbanization of our society and the burgeoning population require that we get control of the present problem and that we develop plans to cope with it in the decades to come. . . .

We [the Congress] are humble enough to know that we do not have within our own experience the specific knowledge necessary to provide the answers to the many and varied problems that cross our desks. . . .

It is in this field that the elected officials must have help from the specialists. Not only must we have their professional advice and services, but we must also have their willingness to do their part as informed citizens in their respective fields of expertise to bring to the public an awareness of a problem, such as pollution, which has devastating potential for harm.

He went on in a slightly cautionary vein:

Today I wish to sound a note of concern about a growing tendency on the part of specialists in a particular discipline of science and technology to assume omniscience in every field of knowledge. . . . As an elected representative in Congress, I must hear from experts who are specialists on both sides of a question. . . . It only confuses the issue for me to be faced, for example, with statements from a Nobel prize winner in biology on a subject in which he is not an expert.

His conclusion was, however, enthusiastic:

It is vitally important that we bring to bear the combined resources of our universities, our laboratories, our industries, and our federal, state and local governments in a coordinated effort to improve man's environment. We have seen the wondrous achievement of landing a man on the moon which was brought about by an integrated effort of our whole society at great expense. We must have similar success much closer to home.

Those words stand in stark contrast to what actually happened on February 6, 1970 when Senators Muskie (D., Maine), Baker (R., Tennessee) and some forty co-sponsors introduced Bill S.3140 into the U.S. Senate to establish National Environmental Laboratories, which was the result of the study prepared at the Oak Ridge National Laboratory on the request of the two Senators. The JCAE, still chaired by Rep. Holifield, instituted immediate inquiries to discover the perpetrators of this work. Feeling threatened by growing complaints about nuclear power policies and the general unresponsiveness of the AEC to the general public, the JCAE verbally ordered no public discussion of the idea by the AEC or its laboratories. Tentative plans for Senate hearings on the proposal were shelved. In April, without prior warning or discussion, the JCAE inserted several paragraphs into its authorization hearings, including the following:

The Joint Committee wishes to sound a note of caution in connection with the activities of AEC's national laboratories. These laboratories are major national assets. They were created, they exist and they are needed for AEC's nuclear missions. In recognition of the potentially valuable assistance that these laboratories could provide in environmental and other health and safety problem areas, regardless of any nuclear relevance, the Congress, on the initiative of this Committee, changed the law in 1967 . . . to permit AEC to assist others.

During the discussion in the House (about this amendment) it was made clear that the added authority was not to be used for "empire building." . . . Rather, the amendment clearly built in conditions. . . .

The Joint Committee sees signs that ambition to acquire new knowledge and expertise in fields outside the present competence and missions of an AEC national laboratory, in order to attain and provide wisdom which this country needs in connection with the totality and its non-nuclear environmental and ecological problems, is spurring at least one laboratory to solicit activities unrelated to its atomic energy programs and for which it does not now have special competence or talents. The Committee cautions the AEC and its national laboratories to stay within the bounds of this portion of [the section of law relating to work by the AEC's installations for non-AEC Agencies]. . . ."

The immediate target of those remarks was, of course, the Oak Ridge National Laboratory, whose imaginative Director, Alvin M. Weinberg, had already on occasion made himself unpopular at USAEC headquarters because of his unconventional but usually constructive views. The AEC amplified the JCAE hint, and again prohibited publication on the topic; in August 1970, it sent a guideline letter to all Area Managers and Laboratory Directors ". . . so that there may be close adherence."

Referring both to the spirit and the substance of the JCAE statement, it ruled that when a laboratory had no existing competence in connection with a proposal or request for work from other organizations, such work should not be undertaken. New National Laboratory roles, in other words, were to be restricted to those specifically initiated by the AEC. There were three additional specific "notes of caution." First, work for others ought to be within a laboratory's "existing competence." The letter specifically stated that the AEC laboratories traditionally lack significant competence in the social sciences; social science work, therefore, even in economics, should be clearly incidental to any new undertakings. Second, there ought to be no "brochuremanship" or promotional activities. Third, the letter indicated that the AEC intended that nothing could interfere with the paramount nuclear mission of the AEC laboratories.

[They] and other facilities, as far ahead as can be foreseen, are needed as atomic energy laboratories, and it is inappropriate to undertake empire building. . . . In addition, public pronouncements regarding major new non-nuclear missions and other activities should be strictly avoided whenever these pronouncements or activities tend to lead to misunderstanding of the primacy of the nuclear work and the importance of its continuation at AEC laboratories.

Finally, the letter stated that the AEC laboratories belonged to the AEC, and might not act independently; there was no right to submit a proposal to another Federal Agency, even by invitation.

When Senator Baker became a member of the JCAE in early 1971 this internal warfare subsided to a very considerable extent. A bill to create National Environmental Laboratories, modified to respond to certain AEC objections, actually passed the Senate, though its companion bill died in the House in 1972. Had Senator Muskie, one of the bill's sponsors, been the Democratic candidate for President, the House action might have been different. President Nixon, having threatened to veto the bill, gave the House Democratic majority an advantage that it might have used in support of Senator Muskie, had he been the presidential candidate. Meanwhile, in 1970, the National Science Foundation established a program at the Oak Ridge National Laboratory to try out, on a modest scale, certain of the approaches to environmental problems that had been suggested by the ORNL proposal. The program has been successful, despite various difficulties experienced by administrators at several levels in understanding either its goals or its role.

The Vacuum at the Top

In a way, one can understand the AEC, the Joint Committee, and many others who have been caught on these issues; the conventional wisdom they defend has many commendable features; it was once new and untried itself. "Look at all we have done for you," the AEC and the JCAE could have said, "Twenty-five years of service, and of very complicated decisions. It was never easy to convince Senators from coal-producing states to support a billion-dollar nuclear power program, nor was it easy to foresee some years ago even the approximate state of present affairs. The sudden appearance of ecofreaks, with unsettling and divisive challenges on environment and pollution—all unassailable by normal rhetoric—which it is impossible to be against, and which may be seized upon by the public and dangerously oversimplified, so that it is appropriated by every opportunist, creates the prospect of self-appointed messiahs arising, with no responsibility and infinite license. Many persons and groups have, in their novel words about institutional connectedness, failed to connect their tongues with their brains, as any detailed inspection of certain of their proposals will show." Then again, the AEC and JCAE may remark, each proposal coincidentally foresees need for a group of humble and revered oracles, who turn out to be none other than the proposers themselves.

It seems clear that the proposers of the Oak Ridge plan did not demonstrate to the satisfaction of the JCAE the existence of a real need for a new type of laboratory. Was there in fact a need? Or if there was, did present institutions satisfy it? An answer to these questions requires a brief foray into the R and D roles played by groups whose interests are considerably narrower than the interests of the proposed laboratory might have been.

At the narrowest level of institutional organization, individual companies work to develop specific items, from the sale or use of which they expect to benefit. An individual manufacturer of fossil fuel combustion equipment, for example, may develop a line of advanced burners, judging that appropriable benefits will accrue within an acceptable time horizon, say five to ten years. Under the proper external incentives, such a manufacturer may agree to engage in more socially oriented work; he may, for example, work on pollution control of effluents to meet imposed environmental restrictions.

At the next level of institutional organization are the industry-wide consortia and research groups. Their interests are broader than those of individual companies; their range of activities is greater; their time horizons longer. Developing hardware is not their principal task, in part because doing so could place them into competition with their members. Such "research institutes" are generally semipublic, though they need not be. The private Bell Telephone Laboratories, for example, long dominated the development of technological options for the American communications industry. The recently formed, semipublic Electric

Power Research Institute (EPRI), however, is a better example. Created to fill a void that existed in guiding and assessing the development of societally optimal electric power options for the future, the EPRI, if it carried out its mission, would encourage balanced research and development of advanced fossil fuel cycles. It would also consider both nuclear and fossil fuel options, and do economic analyses of various new possibilities. External incentives would help to shape programs at this level. EPRI, for example, might well be stimulated to study the long-term social consequences of various energy options, or of electrical energy conservation; if it failed to conduct such studies, Congress might be disposed to create a competing group that would do so.

At the third and highest level of institutional organization, one finds the principal difficulties that need to concern us. Before examining the national institutions that work in this domain, it is important to look at the properties of national problems here that are different from those encountered at lower levels.

(1)   They are broader and have much greater social content than those considered by groups at lower levels. For example, problems relating to cheap energy policy, rational land use, and environmental quality involve a larger number of organizational and social issues than those that are purely technological.

(2)   Many of the problems on this level interact and intersect. Thus, for example, cheap energy policy leads to the burning of high-sulfur coal that was strip-mined, without provision for land restoration; this raises questions of rational land use and of environmental quality.

(3)   Besides interacting with one another, the problems usually contain issues that are inherently in conflict. Recent environmental quality debates provide numerous examples of this. Should exploitation be stressed to the detriment of conservation? How can short-term and long-term advantages be balanced?

(4)   These problems require both long-range planning and massive coordinated efforts; they cannot be resolved simply by executive action or standard-setting decisions. Thus, for example, a national energy policy cannot be formulated without reference to detailed knowledge about synthetic fuels and nuclear reactors. But the development of such technologies is expensive, and so is the research that must be carried out to determine their impacts on health and environmental quality.

(5)   They usually have longer time horizons than the private sector is willing or able to accommodate. For example, air and water quality, or energy resources, are thirty year topics (at least); the private sector is limited, by economic considerations, to five or ten year time horizons.

(6)   Dealing with such problems often involves substantial present costs, but the benefits are usually delayed in time, and spread diffusely through society. For example, reducing urban pollutants by a factor of two will, according to Lester Lave and Eugene Seskin,[6] add three years to the life span of urban dwellers.

The constellation of these large issues—energy, environmental quality, transportation, public health, public education, and so forth—is closely related to our style of life; it is relatively indivisible, in the sense that changing one part affects most of the others. Thus the constellation starts to become more than just the sum of its parts, and the Europeans have even invented a name for it: the *problematique*.[7]

The question arises: Who prepares options at this level for the country's putative decision-makers? The answer: practically no one. As the Oak Ridge case history suggests, there may even be penalties exacted for trying.

One may expect that in an orderly national family the largest and strongest members will work on developing new options for the *problematique*. In the United States, for example, it offers a role for such bodies as the National Bureau of Standards, certain of the research laboratories of the Department of Health, Education and Welfare, and the National Laboratories of the Atomic Energy Commission. What have these laboratories done? What have they been permitted to do? The Oak Ridge and Argonne National Laboratories (of the USAEC) each have an annual budget of about $100,000,000. Surely these were appropriate places for the development of broad energy options. Until about 1973, however, the possibility of doing so hardly occurred to them. Apart from their activity in the basic sciences, which has little bearing on our discussion, they worked almost not at all on broad questions of energy technology and its consequences. They were not even concerned with nuclear energy as a whole; they concentrated on specific tasks related to specific nuclear reactor concepts.

This myopia had several causes and many consequences. First, the AEC and the Laboratories pretty much determined their own programs until the late 1960's; during that period of generally rising research and development funds, there was more than enough money to go around, and no one questioned either allocations or priorities. Second, the AEC and the Laboratories gravitated bit by bit into a symbiotic relationship; they concentrated on topics not likely to arouse public controversy that might threaten their continued existence; they worked on controlled fusion, various kinds of advanced reactor concepts, large high-energy physics projects, and the breeder reactor itself. Each of these studies was defensible *per se,* but together they did not constitute a program in the main stream of contemporary energy needs, let alone social needs. Interestingly, the oil companies had almost nothing to do with these matters. They had little stake in holding back the development of nuclear energy, which would be useful only for generating electric power, a task performed mainly by coal through the 1960's. Furthermore, by the time nuclear power came into its own, they foresaw that petroleum would be in tight supply.

Thus, by exercising comfortable stratagems and well-tuned arts of survivorship, the Agency-Laboratory combinations produced options based on unbalanced technological assessments; sometimes they produced nothing at all. Those working on the technology of breeder reactors (or on microturbulence, or on some other such matter) toiled on with the secure feeling of providing "contributions to the solution." The presence of so many individual contributors implies the presence somewhere of a contributee who reintegrates the material that has become so fragmented at various bureaucratic levels, hopefully converting it into sets of broad societal options understandable both to the public and to decision-making groups. For the most part, that role remains unfilled; the result is that reintegration languishes, and no balancing assessments are made between major technological options, in energy or elsewhere.

One might expect that universities and similar groups would assist in developing such broad-based choices; they do so, in fact, from time to time. But for many of them the role is too often unnatural. The large resources required to develop, for example, a new supertanker with engineered safeguards against oil spills is totally beyond what a university, given its present structure, is able to devote to the enterprise. Besides, the university, as a corporate educational enterprise, rarely, if ever, anticipates sufficient intellectual benefits to make the effort seem worthwhile. Indeed, with regard to the *problematique*, the universities have usually failed to distinguish themselves even in carrying out their traditional roles of generating new ideas and providing critical assessments.

This academic timidity is evident in the fields of both environment and energy. The federal government, university students, and conservation groups, assisted by a few far-sighted academics, were interested in the problems during the critical years, 1967-1970, but almost every main-line university group concerned with environmental questions was formed after the passage of the National Environmental Protection Act on January 1, 1970. To the best of my knowledge, no university in the United States established a center for the holistic study of energy problems before 1971, when many of the main issues were already well defined (to be sure, often by academics, but generally working counter to discipline-oriented departments).

What ought one to conclude? First, that a need exists for developing options with technological content at many levels, particularly with regard to national issues, and especially in cases where the benefits are not appropriable to any one part of the private sector. Second, that the task is not a trivial one which can be affected merely by promulgating regulations for the private sector to follow. Finally, that no one has been minding the shop, so to speak (at least in the energy and environmental areas, and probably nowhere else either), although a good number of workers have been busy in the store. It is as if the research department of a corporation were failing to carry out its prime purpose of providing new options for the corporation as a whole. The research department does not "decide" for the corporation; that, obviously, is quite another activity, involving all the departments. However, with the understanding that it has no executive role, the research department has—or ought to have—both the duty and the freedom to explore new options, even though some of them may presage trouble for certain operating sectors.

By analogy, it is almost self-evident that the nation and the federal government which leads it require new types of institutions that would collectively constitute a national research department. Such institutions need to be sensitive to broad private and public needs; all of these have both social and technical components. Lacking any sort of executive power, they ought to be free to explore ranges of technological and social options, even though some of these may be unsettling to certain parts of the private sector or to departments within the federal government.

As has already been suggested, some see substantial dangers in the establishment of national research institutions with broad mandates for exploration. A principal fear is that new elitists will be created, who will manipulate technology and its consequences, carefully constructing a climate where the only decisions that appear rational to decision-makers are those that are already tacitly decided on by the organization. This is a serious danger and raises the question put by Juvenal, "Quis custodiet ipsos custodes?" Indeed, who watches the watchers except society as a whole—"Nec quisquam unus, sed cunctus populus custodiet." These new institutions must obviously live in full public visibility; they must find their rewards in securing public participation and approval. Such public involvement will, on the one hand, insure that the work of the new institutions remains real and balanced; on the other, it will provide a check on both their actions and their motives.

Systematic Failures

Of the many difficulties involved in working successfully on the *problematique*, two stand out: natural tendencies toward excessive reductionism; incommensurate time horizons.

Reductionism versus Holism: The example of energy clearly shows the need for holistic approaches. The United States spends 125 billion dollars a year on energy, one-tenth of its GNP; in addition, energy affects many other sectors. Yet, until recently, it was treated fragmentarily. One cannot make a judgment between nuclear and fossil fuel generation of electric power without carefully determining the social costs and benefits of each, including those which involve the quality of the environment. Similarly, at a higher stage, one cannot judge the tradeoffs between various electric and nonelectric strategies without exploring each. At higher levels still, strategies for energy provision vie not only with each other, but also with those of energy conservation; sensible decisions at this level depend on the outcome of assessments at lower levels. At each stage, questions become less technical and more social.[8]

In this example, as in others, advances in understanding come at the synaptic points where various disciplinary specialties interact; optimizing along any single disciplinary line is no solution to the problem. Major problems cannot be properly defined in terms of single specific tasks; nor is it enough to think of the sums of such separate tasks. If one attempts to divide a problem into its separate parts—by whatever rules—some connective tissue will be left over, which cannot be ignored or separately assigned.

Environmental problems have suffered historically from sub-optimization. For example, environmental restoration is directly related to water quality; the Federal Water Quality Agency was created in the 1960's to have suzerainty over that particular aspect of the problem. Similar thinking led to the creation of a separate National Air Pollution Control Agency. Of course, the two are connected; the problem, to oversimplify, is whether we decide to burn our wastes or flush them down the river. Recognition of the inadequacy of these separate approaches, institutionalized in different federal departments, was part of the reason for the establishment of the Environmental Protection Agency in 1970.

The problems are not only connected; they are also nonlinear. A solution to any one part, or even a combination of solutions to separate parts, is not necessarily the optimum solution to the whole. For example, the coupled problems of providing more energy and maintaining environmental quality involved, at the outset, issues of fossil fuel provision and technologies of pollution control. Attempts at resolution now bring in issues and opportunities that were not present in the original problem statement: energy conservation, switch to nuclear power, limits to growth. Moreover, introducing such issues changed the character of the problem itself; it could not be properly defined *ab initio*, but only through the process of seeking solutions. Intermixing the problem and its solution leads to partly deferred problem definition, plus a natural and healthy uncertainty about what is to be done during the early work stages. Program administrators find their tasks more onerous, carried out for no obvious reward; they risk being accused of encouraging wooly headedness.

Despite the need for holism, one finds, in most cases, some form of reductionism. It is natural and comfortable to divide the task into bite-sized pieces; it is fatal to leave it that way. Respectable, clear-cut disciplines like nuclear engineering flourish at the expense of the large and difficult sociotechnological problem of how best to respond to the country's energy demands. The reasons for the mismatch are obvious: first, specific disciplines are certainly necessary, and it is possible to defend each one; second, specific disciplines are susceptible to neat organization; peer groups can judge performance, administer rewards, and provide prestige that is recognizable both inside and outside the discipline. Finally, disciplinary groups and their patrons form mutually reinforcing systems. Although they may be professionally demanding and difficult, they are less fractious and more comfortable than interdisciplinary groups. By a convention of mutual respect, participants avoid stepping on each other's toes; uncritically, they leave to others everything that lies outside their own narrow area.

Many people are still inclined to ask, "If we can go to the moon, why is it not possible to clean up our environment (or do something else)?" The answer, quite simply, is that getting to the moon was a well-defined disciplinary problem, with no surrogate goals permitted. Our task is more complex; we have for too long substituted work in comfortable disciplines for the real tasks that need to be attended to.

Only a short step beyond putting excessive trust in separated disciplines lies the error of mistaking the professional disciplines for the problems themselves. In this context, the disciplines are more nearly the tools and the scaffolding necessary for the construction of something else, which contains the real social worth. Universities take delight in embroidering disciplines on the pretext of excellence; they wish to be at the forefront of the field in question. What is it all for? That question is rarely asked, and when it is, the answer has to do with the need for securing support funds or jobs for graduates.

Incommensurate Time Horizons: The second major problem that leads to systemic failure in attacking the *problematique* is that of incommensurate time horizons. If each participating group stated at the outset what its time horizon was, subsequent discussion would be wonderfully facilitated. If our whole present concern is only for the next ten years—and next year for only nine, and so forth—there would be no need for environmental protection, energy systems analysis, resource conservation, or concern about the color of the dawn that bestrides the eastern hills. The discussion following the appearance of the book, *The Limits to Growth*, by Meadows and Associates,[9] offers a striking example. To certain critics, 2000 A.D. was the year after forever; to Meadows and company, the year 2020 A.D. was the day after tomorrow. As a result, each talked past the other.

As has already been suggested, natural time horizons in the private sector tend not to fit those of the *problematique;* the difficulty arises in part from how the private sector gets its money. At 14 percent per year rate of return on investment (before taxes), twenty-five cents invested today yields one dollar ten years from now, two dollars fifteen years hence. But conversely, one dollar of benefit ten years hence is worth only twenty-five cents now, and one dollar benefit fifteen years away is only worth thirteen cents now. Such considerations often lead to the desire for a short-term payoff, in five or ten years; environmental and other long-term problems tend quite naturally to be underemphasized.

The national program to develop clean fuels from coal or oil shale in environmentally and socially acceptable ways may be offered as an example. For simple economic reasons, the optimal strategy envisaged by the private sector lies in the direction of less environmental care than the public appears to want; we see also that the private sector is generally poorly motivated unless a payoff can be seen in something like ten years. Thus, the pattern arises of relatively little being done privately until it is too late for either the private or public sectors to undertake fifteen to twenty year programs of studying ecological efforts, of deciding how much land restoration is enough, and of developing new patterns of land use and new technologies for effluent control. The recent unforeseen increases in fuel prices exacerbate the kinds of difficulties described, since the private sector plans principally for even shorter time periods. Only the public sector, had it been operating properly some years ago, might have forestalled such difficulties.

Another time problem relates to the past versus the future. Solving present environmental problems and restoring the environment are essentially retrospective tasks related to past mistakes, of which the forward equivalent is the arranging of future technology in ways least likely to cause uninvited trouble. We have two outlooks—like the two faces of Janus—one showing foresight, the other retrospective correction. The first makes little sense without the second. Forward directions being unclear, option preparation for the future must be necessarily broad, and, inevitably, more expensive.

Preparing options for the far future often fails, for many reasons, to arouse great enthusiasm. As has already been suggested, the private sector has a limited time horizon. Past mistakes demand attention precisely because they are visible; possible future ones are not. Established groups often feel threatened by change or even the possibility of change. Because the future-oriented task is harder to see, it is much more difficult to do well. Finally, performance in activities defined by conventional wisdom determines rewards; often, the most able individuals (judged by their outstanding performances in scientific or humanistic disciplines) cannot be coaxed into untried ventures. Partly for these reasons, present federal government programs often lean toward policies designed to cure demonstrated ills. There is, however, some cause for optimism. Environmental Impact Statements, dealing with the future consequences of proposed actions, are increasingly being used for rational planning. For example, EIS's provide a more solid basis for public debate than have the previous documents; also, their adequacy may be challenged in the courts.

Even with these improvements, however, patronage of work on the *problematique* is both inadequate and inappropriate. Administrators at higher bureaucratic levels, themselves subject to excessive reductionism and wrong time horizons, tend to visit their faults on subsidiary organizations, sometimes by *force majeure*. This was all too evident in the Oak Ridge example. As seen by the working group, being too broad invited budgetary assassination for meddlesomeness.

Bureaucratic support must be not only passively permissive, but both understanding and positive, as the following illustration shows. A well-meaning group at a national laboratory might agree to develop the art and science of regional modeling, providing for growth, environmental impact, likely change in social structure, and so forth. It can proceed for a period measuring relatively timeless things: rainfall, slope of land, vegetation cover. But it requires more sensitive data, having to do with industrial activities and emissions for example. If it merely "accepts" data freely available and postpones analysis until the data are in some sense "complete," it realizes after a while that (a) many critical data are not forthcoming, often because local officials see adverse connotations; (b) even if reams of data are available, they apply only to the past. The group is obliged to think and analyze; then the trouble starts in earnest. The laboratory cannot do all its thinking in private; to succeed it must have detailed interaction with some region of the universe it seeks to understand. Inevitably, it is compelled to pass through some phase of external visibility, at a time when its sophistication and expertise are generally poorly formed. At that moment, during its days of heady anticipation, opponents will often attempt its assassination, complaining to the laboratory's overlords that it meddles where it has no competence. The charge of incompetence will be true and unanswerable, at least at that stage; the only defense could be on societal grounds, and not on professional or technical claims. No escape exists from this difficulty. To assume a low profile and say nothing controversial is to adopt disciplinary surrogates for motives that can only be described as self-preservation; the problem remains. Alternately, to labor privately toward perfection is almost certainly to labor in vain.

Conclusion

Dangers of unbalances appearing in the discussion of new institutions created to work on problems that are at once technological and social, such as the environment, are probably inevitable. Such discussions have been carried on principally by scientists and/or technologists, with social scientists being dragged in reluctantly, a circumstance both surprising and disturbing. The danger of qualitative misunderstanding is real; we propose solutions as if we understood how science, technology, society, and the environment all interact. In truth, no one understands at all well how the system works. By joining the social and the natural sciences, technology, and new ways of thinking about large problems, there is some hope for progress.

It should be possible to analyze the difficulties related to organizing effective work on the so-called *problematique*, which is the union of environmental, energy and other major national problems. It ought to be possible to indicate how newly structured institutions could work much more effectively than do present organizations in the several relevant areas; it may even be possible to point to certain modest experiments that confirm these opinions.

But the analyses, the roles and the new required institutions must be based on a very sophisticated outreaching; and, perhaps surprisingly, on a spirit that can only be defined as charitable.[10] The *problematique* requires that such attitudes be common. Unfortunately, the system that requires attention is too often dominated by more primitive attitudes, at least partly related to the reductionism, narrow constituencies, and preservation of institutional position, that are all too common. This is a pity, though by no means a surprise. The hope is that continued attempts to establish better institutions will themselves lead to a broadened understanding of the need for them, and that this will make for revised attitudes and the recognition of new opportunities.

REFERENCES

1. *Science* is a rich source of material, particularly volume 177 of 1972. The following articles are especially pertinent: Richard A. Cellarius and John Platt, "Councils of Urgent Studies" (25 August), pp. 670-676; the authors suggest holistic, problem-oriented institutions much like those I describe here, and include a shopping list of likely projects. J. E. Goldman, "Toward a National Technology Policy" (22 September), pp. 1078-1080; the concept here is too regimented to suit the real need, but helpful nevertheless. John Walsh, "Willow Run Laboratories: Separating from the University of Michigan" (18 May), pp. 595-596 and (25 August), 677-688.

2. M. L. Goldberger and G. MacDonald, "Institutions for Effective Management of the Environment" (Washington, D.C.: National Academies of Sciences and Engineering, January 1970).

3. Ecological Society of America and Peat, Marwick, Mitchell & Co., "National Institute of Ecology: An Inquiry," 1, supported by National Science Foundation Grant GB-6890-001, March 25, 1970. Available from Ecological Society of America.

4. "A Report on an AUA Conference, 27-29 July 1969," Conf. 690705 (Springfield, Va.: Clearinghouse for Federal Scientific and Technical Information, National Bureau of Standards, U.S. Dept. of Commerce).

5. "The Case for National Environmental Laboratories," Report ORNL-TM-2887 (Oak Ridge Tennessee: Oak Ridge National Laboratory, February 1970). An abridgment appears in *Technology Review* (April 1971).

6. Lester B. Lave and Eugene P. Seskin, "Air Pollution and Human Health," *Science*, 169 (August 1970), pp. 723-732.

7. Alexander King and Umberto Colombo, "Science and the Decision-Making Machinery of Society," Colloquium of the Committee on Science and Technology, September 13-14, 1973, Council of Europe, Strasbourg, France.

8. This point and others are excellently made in an article by Joseph Haberer, "Politicalization of Science," *Science*, 178 (November 1972), pp. 713-724.

9. Donella Meadows, Dennis L. Meadows, Jorgen Randers, and William W. Behrens III, *The Limits to Growth* (New York: Potomac Associates Universe Books, 1972).

10. In the sense of I Cor. 13 (Αγαπη).

RUSSELL MCCORMMACH

On Academic Scientists in Wilhelmian Germany

IN RECENT YEARS American scientists have been widely criticized by students, political activists, cultural and social analysts, and fellow scientists. The criticisms are well known. Scientists are accomplices in legal criminality, and uncritical of their patrons—big industry, the military, and government—powers that perpetuate Western imperialism, racism, and economic injustice, and despoil the planet. They abnegate their moral responsibility to make the world better, which it is in their power to do through political action and responsible research. Their rational, quantitative view of the world excludes as nonobjective, and thus nonexistent, all that is spiritual, magic, and poetic, distorting man's consciousness, alienating him from himself, society, and nature, and abetting the antihuman forces of the world. By their overspecialization, they bury us in facts, contributing nothing to our yearning to know the interconnectedness of things. Science should close shop; its vigor is spent, its problems exhausted. Science should declare a moratorium: it is too dangerous; it brings too rapid change for man and his institutions to cope with. Besides, the day of the great discoverer is past; science must turn practical to earn its keep. There is talk of new consciousness, of new culture. Whatever one may think of these judgments and predictions, they express attitudes that are important to some scientists, .as one can easily see by glancing at editorials and correspondence columns of *Science* and at certain presidential addresses of the American Association for the Advancement of Science from the mid-1960's.

America has been deeply divided by foreign war and internal inequalities, and American scientists and the American public have questioned the social and political responsiveness of science, as they have of all major institutions. Germany in the Wilhelmian years, from 1888 to 1918, was deeply divided by its powerful, quickening industrialization, and scientists and the public faced similar questions about the place of science in a society in change and conflict. Wilhelmian physical scientists found their work identified with a scientifically intensive technology that underlay much of Wilhelmian industrialization and that nourished a one-sidedly practical view of the world; they found it, too, identified with a one-sidedly rational view of the world that offended a segment of educated Germans who formed an avant-garde of an antimodernist culture imbued with artistic, mystical, and nature-worship values. By and large, Wilhelmian scientists steered a middle course. They upheld the values of nonutilitarian research and rational and empirical modes of inquiry; at the same time, they promoted industrial technology in limited contexts, as in the creation of applied scientific disciplines, and they recognized nonscientific components in the collective culture and in the worldview of the individual.

There are parallels between Germany at the turn of the century and America today. For one thing, important links between science, big industry, and government—links that in their modern form have elicited searching criticism in America in recent years—were emerging in Wilhelmian Germany. For another, Wilhelmian complaints about the specialization of science and its materialist and alienating tendencies resemble those that are still very much with us. Yet the historical situation of science is vastly different now from what it was then, and so are the practical questions surrounding it, both those of its practitioners and those of its administrators and public. We cannot look to historical parallels to give us specific answers to our contemporary questions about science. What we can learn from a study of Wilhelmian scientists is something about the functions of the scientific role in a time of rapid social change. Such a study illuminates an aspect of the situation of American scientists today. I return to this point at the end of the article, where I contrast popular conceptions of the scientific role in Wilhelmian and contemporary American societies.

II

In late-nineteenth-century Germany, one of the dominant justifications of the role of the academic researcher was that science was one of a number of distinct but interdependent cultural factors, and that its contribution was essential to a harmonious intellectual and material culture and to a harmonious worldview for the individual. The cultural justification rested on the self-appointment of Wilhelmian academic scientists to the class of culture-bearers. "Culture-bearer"[1] (*Kulturträger*) was a value-laden term denoting those who were considered well educated and qualified to judge matters affecting the quality of culture. The term was used mostly by humanists, but for the concerns I discuss in this article it is appropriate to extend it to scientists as well.[2] The culture-bearer—scientist and humanist alike—was vitally concerned with specialization, unity of knowledge, worldview, creativity, and the institutions that served culture, especially the universities that conferred on him his culture-bearer credentials and supported and transmitted his cultural work.

Scientists qua scientists saw as their responsibility the advancement and diffusion of a certain kind of knowledge that transcended national boundaries.[3] They regarded it as right and important to honor all major discoveries and their discoverers, whatever their national origin, and they expected to be so honored themselves. It was deeply satisfying to many of them to contemplate the universal objectivity of scientific knowledge. It was so to the physicist Max Planck, who spoke of the power of certain scientific results to command agreement from all experimenters, all nations, all cultures—even, he was sure, extraterrestrial cultures if there should be any. The price of universal truth was a de-anthropomorphic and abstract scientific world picture with its drab hues and alienating force; the attainment of universal truth was for one like Planck more than worth the price.[4]

As culture-bearers, Wilhelmian scientists saw their responsibility as honoring the whole of high culture and advancing its scientific parts. Although Wilhelmian scientists and scholars sometimes thought of themselves primarily as bearers of European, Western, or even world culture, in many contexts they conceived of culture with an explicit German reference. Wilhelmian culture-bearers had practical career interests that necessarily referred to the national context, but these did not lie at the emotional core of their self-image. That image derived from national cultural heroes, men who had realized in the highest degree the potentialities of the German mind with its penchant for unity, idealism, and abstraction.

Wilhelmian scientists had a dual commitment. As members of a scientific discipline, their intellectual being was wedded to the universality of scientific knowledge; the fruits of their work were gifts to all mankind. As culture-bearers, their intellectual motivation derived in the main from a national-cultural idealism; their work was in the first instance a gift to German culture, and only through German culture was it a gift to mankind. Their way of resolving the potential conflict in their academic-scientific role was through ideology. To give an illustration: culture-bearers believed that their work was the *raison d'être* of the nation and, since the nation was thought to derive prestige from their cultural attainments, they were, as scientists, able to pursue with easy conscience the pure scientific research that their scientific role encouraged.[5]

In this article I examine the ideological responses of Wilhelmian academic scientists—especially physical scientists—to industrialization and to its cultural repercussions. For this purpose, I draw heavily on invited addresses at ceremonial occasions and at scientific meetings. The speakers were eminent scientists who were accepted by the educated public as spokesmen for Wilhelmian science. Usually older than the average scientist, their formative years often fell in the early industrial period, during the time of heightened aspirations for political unification. Their ideologies reflected strong continuities with the cultural attitudes of pre-unification Germany, and they undoubtedly differed somewhat from those of scientists just beginning their careers in the period of intensive industrialization. By largely restricting my discussion to the views of scientific spokesmen, I deal with what I take to be dominant ideologies of Wilhelmian academic scientists. There is not space in this article to do justice to the intellectual ferment in Wilhelmian science: I do not take up the difficult problem of determining the attitudes of younger, lesser known scientists; nor do I discuss the League of German Monists, the Keplerbund, the Social Democratic Party, and other organizations with quasi-scientific ideologies that on certain points differed from the more dominant ideologies of academic scientists.

III

The condition of culture was an absorbing subject among Wilhelmian intellectuals, who studied it from every side, even organizing whole disciplines such as cultural history and cultural philosophy around it. For the purposes of this article, Wilhelm Lexis' comprehensive essay on culture is particularly fitting.[6] Although he was a humanist, Lexis had strong sympathies with the pure and applied sciences. His essay was the introductory statement in *Kultur der Gegenwart;* many of the twenty-two volumes that appeared—before the war put an end to the project—were devoted to the individual sciences and included contributions by the leading spokesmen for Wilhelmian science. For Lexis, "culture" meant the "rise of man above the natural state through the development and use of his intellectual and moral forces." He recognized that inherent in his understanding of culture was the threat that eventually a state of "overculture" might arise in which man would live in artificial conditions, wholly estranged from nature.

Lexis stressed that although culture was created by individuals, it could not arise outside society and the state. In addition to providing the conditions for cultural work the state had to take on cultural tasks, leaving, however, cultural decisions to the educated opinion of the culture-bearers. Unlike social progress, the aim of which was to level inequalities, cultural progress depended on a small elite, and only a minority could fully enjoy its highest achievements. Since culture sprang from the whole of man's complex intellectual nature, it had several interrelated aspects—practical, scientific, artistic, and ethical. Lexis believed that scientific culture evolved from a prior artistic culture, the earliest expression of the worldview of mankind's leading intellects. In recent times, scientific culture had fed into the practical—the economic and technical—culture, and was itself now dependent on its practical applications. However, in its essence, scientific culture was independent of the applications of science, for it was dedicated to a knowledge of nature, man, and humanity—a knowledge desired for its own sake—and to a worldview based on that knowledge. Like artists, scientists sought immaterial, ideal "cultural goods," deriving self-satisfaction from creative drives; by contrast, the driving force behind practical culture was human need, and its results were material "cultural products." Scientific culture shaped and ennobled the intellectual life of an educated people, for although few intellects were capable of advancing scientific knowledge, everyone was affected by it.

Lexis explained that since culture arose from different peoples and their different historical experiences, "cultural goods" bore specific national characteristics. In different cultural nations and at different times, different aspects of culture were emphasized, resulting in a fruitful international division of cultural labor. This specialization, however, entailed the danger that one aspect of culture might predominate and destroy the harmony of the whole. Lexis warned that measures had to be taken in time to prevent an imbalance of cultural factors from arising.

IV

German scientists had long placed their scientific ideology in the service of their greatest political cause, the unification of Germany. They regarded their largest, multidisciplinary national organization—the Society of German Natural Scientists and Physicians—as a model for Germany's future, and one of its sustaining purposes was that of furthering the unification of the German states. In 1871, at the meeting of the Society following unification, an official proudly recalled that formerly German "national unity" had existed only in its "scientific life," which had refused to allow the "beautiful myth" of unity to be forgotten. His boast that the "conquest of German arms is a conquest of the German mind"[7] expressed the culture-bearers' faith that Germany's scientific success contributed to her emergence as a great power. Into the 1880's, the Society might still be reassured that Germany's importance in the world was due primarily to her pure scientific achievements and only to a very minor extent to her material.[8] The vision of Germany as still predominantly a land of thinkers was the nostalgic vision of the traditional culture-bearers, and one that was to be increasingly challenged in the period following unification. The physical chemist Walther Nernst lectured the Berlin Academy in 1905 that whereas Germans used to look to "poetry" and "philosophy," they now looked to "grain fields" and "factories," and that if a scientific institution like the Academy did not adjust to that reality it would lose its influence in the nation.[9] The background of Nernst's observation was the recent, impressive scientific and technical development of German industry and agriculture.

German industrialism developed rapidly from the time of unification—increasingly so from about 1890—transforming Germany from an industrially backward power to Europe's leading industrial power. The scientific-technological drive of Wilhelmian Germany is illustrated by the electric industry, which was based almost wholly on recent developments in pure science. The grand commercial vision of Emil Rathenau, head of the giant combine Allgemeine Elektrizitätsgesellschaft, was to convert the entire German economy to electricity.[10] In the Wilhelmian era, whole German cities were electrified,[11] and Germany produced one-third of the world's electrical manufactures.[12] Indeed, the great prosperity of Germany from the end of the depression in 1895 to World War I owed much to its electrification.[13]

It was soon clear that Germany's industrial revolution promised to work at least as great a change in German life and consciousness as the political unification that had made much of it possible. In particular, Germany's industrialization challenged the educated Germans' comfortable assumption that the nation's values were identical with their learned and artistic values.[14] Many of them publicly deplored the noise, stench, and overpopulation accompanying industrial technology,[15] as well as the "nervous haste of our modern world"[16] and the whole complex of changes—urbanization, mechanization, and class conflict—engendered by the "soulless" machine age.[17]

For culture-bearers, however, much more than sensibility was at stake. Emerging with industrialism was a new cultural complex, controlled by the wealth of the propertied middle class, managed by a new professional class of engineers, and imbued with the politics of the working class. That the traditional culture-bearers felt a resulting sharp loss of authority in the nation is evident from the gloomy assessment of their situation by the psychologist Ernst Meumann.[18] Like many culture-bearers, he responded to the modern world with nostalgic yearning. He liked to believe that until the middle of the nineteenth century, the German princes and public took an avid interest in their great scientists and philosophers. By contrast, in his day the truly talented were obscured by a crowd of scientists with small talents, busy with small problems. The worldview of the public was shaped by scientific popularizers writing for the flood of half-scientific papers and reviews. In the rise of organized interests, scientists, scholars, and artists had been replaced by businessmen, press barons, and politicians, and Meumann concluded that the culture-bearers had all but lost their intellectual influence in the cultural life of the nation.

The alleged eclipse and degradation of the public conception of science in the emerging mass culture was not the only, or even the most severe, challenge to the Wilhelmian scientists' cultural place in the industrial age. Because of the widespread belief in the dependence of Germany's economic success on the scientific basis of industrial technology and because of the encouragement by industry and government of the expanding contacts of the physical sciences with technology, Wilhelmian scientists witnessed the increasing association of their disciplines with technical and industrial achievement. That association was a source of disturbing ambivalence for them, for they did not regard their national function as economic in the first instance, but intellectual, often priding themselves on their independence from productive forces. Their sharp distinction between science and technology, or more generally between theory and practice,[19] was rooted in part in their understanding of the condition of their intellectual freedom and creativity and, thus, of their cultural contribution to the nation. At the same time, however, they could welcome, along with the physicist Wilhelm Wien, the understanding that their work created the "firm foundations on which the pillars of our industry stand, which support a great part of our economy."[20] That understanding was one of the grounds for their unstinting support of Germany in her rise to great-power status, for they were convinced that an economically and militarily powerful Germany was a precondition of a science that reflected well both on themselves and on the nation.[21]

Wilhelmian scientists had no intention of denying that their work had valuable applications, even when they denied that they had direct responsibility for them. What troubled them was the possibility that people had come to see the applications and nothing else. They heard on every hand that "in recent times technology has given the natural sciences so many valuable stimuli that a certain displacement of the scientific center of gravity toward the technical side may hardly fail to be recognized";[22] that although scientists still sought basic natural truths, in the selection of their research problems they thought "technologically";[23] that although science brought technology into being, technology and not science determined the character of the present age;[24] even that technology now determined the science of the time rather than the reverse.[25] Within the Wilhelmian academic world a movement was launched to reorient academic science toward applications; its premise was that modern industry and the scientific technology that guided it were the dominant forces in contemporary culture.[26] That movement provoked the fear—fed by the frequency with which scientists saw their discoveries equated with telephones, electric lights, and the steam-powered conquest of space and time—that if technology were admitted into the universities the "trumpet sounds of technical success [would] drown out the soft music of natural laws."[27] Wilhelmian scientists were fond of arguing, from the principle of the division of labor, that they benefited technology most efficiently by ignoring it and leaving it to the practical specialists, resting their case on the classic scientific ideology: it was in the "proper advantage of state and society" to support a "free science, free from the proof of direct utility, free of course from signposts and warnings of forbidden paths, just as it originated in the time of past cultural conditions."[28]

V

A leading source of cultural discontent in Wilhelmian Germany was the specialization of modern research, a condition that was sometimes associated with the technological division of labor. Scientists shared in the discontent, feeling that excessive specialization produced something less than science; they spoke contemptuously of "mere facts" and "mere details." It was not that they denied the need for specialization; indeed, they understood that the strong development of specialized disciplines was at the root of German scientific hegemony and that the institutions that housed them were the model for much of the scientific world. Rather, they were concerned that the intellectual significance of science was jeopardized by research that was more compartmentalized than it need be. Wilhelmian scientists repeatedly reminded one another that they were more than specialists; they were committed to the unity of science and beyond that to the unity of intellectual culture. To themselves as educated men and to the culture that shaped them and whose guardian and developer they were, they felt a responsibility to oppose the unwanted consequences of specialization. In their general writings and addresses delivered to audiences inside and especially outside their own specialties, they often touched on the dangers of specialization.

Appropriate occasions for Wilhelmian specialists to address a wide audience were the annual meetings of the Society of German Natural Scientists and Physicians. The meetings evoked expressions of hope and despair from scientists; in principle they united members of all specialties in the entire German cultural spectrum like no other occasion, yet at the same time they exposed, in their specialist sectional organization, the growing fragmentation of modern research. The number of general sessions in the meetings decreased over the century; more important, the number of specialist sections increased, from three at the founding of the society in 1822, through seven in 1828, fifteen in 1871, to thirty-eight at the end of the century. In 1901, the Society reorganized to "counteract the too extensive splintering of the scientific interests of the meetings which had occurred over the years," reducing slightly, by combining neighboring disciplines, the number of specialist sections to twenty-seven. The President that year interpreted the Society's move as a promising sign that the peak of specialization had passed, that the different scientific specialties were interacting more and were approaching rather than receding from one another.[29]

The formal restructuring of the specialist sections of the meetings, however, was to provide little lasting comfort. Wilhelmian scientists continued to lament the uncontrolled growth of specialization. Associating narrow specialization with an unhealthy growth in what once was living, organically unified learning, they railed against the "measureless unhealthy splintering [of science that] carries within itself the seed of death," and the "excessive detailing [that] leads to the degradation of knowledge to the level of mere bread-and-butter study, to industrial training instead of true university education, to an abandonment by researchers of the ideal side of thought."[30] The President of the Society in 1912 put it succinctly: specialization was the "sickness of our time."[31]

Wilhelmian scientists repeatedly commended the Society for acting as a counterweight to the specialist scientific societies and congresses that constantly sprang up,[32] reaffirming its function of compensating for specialization. They commended it, too, for its original plan of bringing together in fruitful scientific relations natural scientists and physicians.[33] The unity for which German scientists praised their Society, however, was largely symbolic; coming into being only once a year, it could scarcely maintain "living contact between closely related neighboring specialties."[34] Having deplored the trend of specialization in their opening addresses at the meetings, scientists spent most of their remaining time in the sections of their specialties where the serious work was done. Yet the cultural values they expressed in ceremonious rhetoric were an essential element in their scientific commitment and elicited from them a powerful emotional response. The continuing existence of their national society reaffirmed the collective commitment of German scientists to the goal of the unity of knowledge. They spoke with pride of the accomplishments of their specialties, but as scientists imbued with holistic values they expressed recurring discontent with their regime of specialization.

VI

In reaction to the "social and cultural strains" arising from industrialization, some Wilhelmian intellectuals proclaimed a cultural crisis.[35] Political unification had not realized the hopes that had accompanied it; within industrial society, new internal divisions arose, leading some intellectuals to seek a new cultural unity based on antimodernist principles.[36] They opposed science and technology together with the whole new industrial "civilization," calling for a return to tradition, organic community, and Germanic "blood."[37]

Large numbers of Wilhelmian middle-class youth believed that the condition of overculture had arrived with the new industrial urban society. They rebelled culturally, not politically. The youth movement, founded in the 1890's, accented irrationality and instinct; its members embraced a romantic ideal of nature,[38] and opposed the "meshes of the mechanized modern life"[39] and, in general, the forces of modernity, including, often, science and technology.

The impatient, contemptuous attitude toward science of rebellious Wilhelmian youth was reflected in the play *Ithaca* by the young Expressionist Gottfried Benn.[40] The play is set in the laboratory of a professor of pathology, a monster of specialized pedantry. At the close of a course, the professor describes the precise tint of the stained cells from a rat's cerebrum. He knows that his students will understand the importance of this point, since a publication has recently appeared from a rival institute contesting his thorough researches on it. He boasts of the "enormous perspectives" he has opened up and which deserve a publication:

from this fine staining procedure, one could also differentiate rats with long black fur and dark eyes from those with short rough fur and light eyes, assuming that they are the same age, nourished with candy sugar, have played a half hour a day with a small puma and have all been allowed to evacuate their bowels twice in the evening at a temperature of 37.36°.

These perspectives bring man "one step nearer to the knowledge of the great connections that move the universe." The professor's students and young laboratory assistant are unmoved. For them, the universe that science depicts contains no unity, faith, feeling, or meaning, but only fragments, numbers, abstractions, words, and the brain; they demand instead myth, the transcendental, ecstasy, and dreams. They reject wholesale what they hold to be the scientific vision of the new man—"Homo faber" in place of the earlier "Homo sapiens"—that has resulted from the last hundred years of science and the technology that grew out of it. Outraged, they murder their professor. In the final speech, a student declares: "We are youth. Our blood cries out for heaven and earth and not for cells and worms." In rising "up in arms against the scientific century" and invoking in its place a "primitive way of life,"[41] Benn, through the students and laboratory assistant in his play, expressed the feelings of many Wilhelmian youth. The intellectuals, in turn, felt their criticisms, their rejection of specialized learning,[42] their worship of Nietzsche, a presumed enemy of science,[43] their longing for metaphysical world-views.[44]

Especially influential in the literature of cultural criticism was Julius Langbehn's *Rembrandt als Erzieher;* published in 1890, it went through thirty-nine editions within two years.[45] Detesting science and celebrating the mystic and irrational, Langbehn rejected the ideal of a balanced culture, proposing in its place a culture that was wholly artistic. He believed that the vocation of science was to "submerge itself in art,"[46] and he looked to an age of art to succeed that of science. Science, he said, had to become wholly subjective: false science was objective, dealing only in facts, whereas true science delivered value judgments.[47] For him, the epitome of the odious German science professor was the Berlin physiologist Emil Du Bois-Reymond, whose materialism was only to be expected from one who bore a French name.[48]

The physicist Paul Volkmann criticized *Rembrandt* for its "conscious hostility" to science, but soon concluded that the book was not a serious threat.[49] Hermann Diels, Secretary of the Berlin Academy, thought that the book may have contributed to attacks on the Academy, but that by 1902 it was "already forgotten."[50] It is no doubt true that *Rembrandt* was in itself unimportant for science, but its great popularity pointed to a cultural condition that was. Volkmann discounted the exclusively artistic education that Langbehn advocated on the grounds that since artistic talent was as rare as scientific such education would produce a crowd of flighty dabblers,[51] but the proposal had no real prospect anyway and only appealed as cultural fantasy. Langbehn played on the widespread anxiety that culture had become one-sidedly scientific and technical and that it had lost its idealistic and creative bearings as a result. *Rembrandt* was only one of many expressions of Wilhelmian longings for cultural rebirth that involved antimodernist attitudes; others included pan-Germanism, the cults of the peasant life and the Middle Ages, and the revival of Sturm und Drang in the arts.

VII

Although Wilhelmian scientists were familiar with attacks on what they regarded as their value-free, nonutilitarian research, even on their commitment to a rational, objective view of the world, they felt relatively secure in their scientific role. They tended to attribute the great success of science to the understanding that the first responsibility of the scientist was the unfettered search for natural knowledge. They believed that the exclusion of extra-scientific concerns had insured steady scientific progress through past times of social turmoil as it would through present ones.

American scientific spokesmen talk somewhat less than their Wilhelmian counterparts about unified worldviews and other lofty scientific and cultural ideals, but they too like to emphasize the cultural significance and motivation of their work. For many of them, science is a part of creative "culture,"[52] benefiting art and the humanities.[53] It belongs to the essential "intellectual infrastructure of society," like politics in Greece and religion in the Middle Ages.[54] It is a "way of looking at the world,"[55] capable of influencing the "concepts of man's place in the order of things."[56] American scientific spokesmen talk of the "curiosity" that has motivated the great discoverers,[57] of their "deep commitment to knowledge for its own sake,"[58] of their response to the "beauty and intellectual austerity" of scientific ideas.[59] They associate their motivation with the "traditional" meaning[60] and "highest allegiance"[61] of science—the search for truth.

American, like Wilhelmian, scientists attribute much of the discontent over science to broad cultural forces, only they tend to react more strongly. They see as a principal source of discontent the belief that science has failed as a cultural force for the good. Their spokesmen point to evidence of popular disillusion with scientific thought. That disillusion applies to the rationality of science,[62] to its ideal of quantitative, ethically neutral knowledge,[63] to the very possibility of objective knowledge,[64] and to the belief in the inexhaustibility of the intellectual challenge of science.[65] The critics of science charge that science has not satisfied man intellectually and emotionally. With its dangerous tendencies, which scientists concede, of "overspecialization"[66] and "Balkanization,"[67] science has failed to produce the unified picture of the world its critics ask of it. Scientists range in their responses to the cultural criticism from those who hold that science has much to learn from its philosophical critics[68] to those who hold the more traditional view that scientists should show the public the strongly "unitary nature of science,"[69] emphasizing the "connections between its different fields."[70]

In America today, as in Wilhelmian Germany, scientific spokesmen acknowledge that their work influences society not only through culture but also through technology,[71] and that attitudes toward technology, like those toward culture, are a principal source of the criticisms of science.[72] In language reminiscent of the Wilhelmian era, they may regret that today many Americans confuse the "high culture" of theoretical science with a technical "popular culture," a mix of applied science, technology, and materialism.[73] The contemporary criticism of the link between science and technology differs from Wilhelmian criticism in that it relates above all to the inhumane scientific technology of the war industries[74] and to the technological origins of the ecology crisis.[75] It relates, too, to certain applications of the new biology,[76] as well as to computers, automation, and other recent technological developments whose origins can be traced to advances in the physical sciences. American scientists, like Wilhelmian scientists, respond variously to criticisms of the technological implications of their work. Their responses include affirmations of the traditional view that scientists should be free to work on what they choose and not be compelled by funding policy to work on "relevant" problems.[77] They include, too, urgings that scientists redirect their interests to social problems[78] and mobilize to attack the critical problems for mankind's survival.[79]

Spokesmen for American science differ from Wilhelmian spokesmen most strikingly in their professions of social concern. They see the increasingly integral part that science plays in society and politics as a basic reason why criticisms of science are more severe today.[80] They talk of bringing science and politics closer together,[81] of developing a national science policy,[82] and of the role of science in helping to forge an ecologic, economic, and ethical world unity.[83] They look on the "asocial" ivory-tower scientist[84] as a phenomenon of the past; for having gained large social support, today's scientist has come to recognize the social responsibility that accompanies it.[85] Scientists point out that not only does their work have an influence on values,[86] but that science itself is not a "value-free activity,"[87] and that scientists themselves may create values in times of crisis.[88] Many accept that science, like other cultural activities, must expect criticism,[89] and that scientists themselves should exercise a critical function by pointing to the risks as well as the benefits of their activity.[90] They argue that even if scientists are not responsible for the applications of science, they are responsible for making known the implications of their work,[91] thereby serving as a "watchdog"[92] on applications and helping to avert the disasters that may result from modern technology.[93]

VIII

Although the unprecedented importance of science in World War I assured that science would be implicated in the moral outrage provoked by the war, just as it was in Walter Hasenclever's 1915 drama *Der Retter* likening modern war to the "child which atavism has begotten upon modern science,"[94] the most pressing moral perplexities of the scientific life were still largely in the future. Hiroshima, the Oppenheimer case, the space age, big science, the ecology crisis, and the Vietnam war have shaped our moral, social, and political concerns in ways that contrast with the predominantly cultural concerns of Wilhelmian academic scientists.

Today one sometimes hears science referred to as merely a problem-solving "game," a way of speaking that would have been unthinkable to most Wilhelmian scientists and which, I believe, points to a basic difference between then and now. It is not that Wilhelmian scientists did not know the pressures of our professional scientific life; they knew them well—the scramble for place and prestige, the constant, often harsh judgments, the overwork. Nor is it that they were less sophisticated about science then we; most of them had already given up the idea that science tells us about a real world beyond the senses. They did not conclude, for these or other reasons, that science had diminished intellectual or moral significance, and both their rhetoric and their actions were directed to substantiating their faith. The assurance that their work was widely prized as an essential cultural good provided a measure of insulation against hostility toward science, and militated against the scientific skepticism and indirection we sometimes see today. It is the absence, indeed the impossibility, of any comparable assurance today that makes science seem to some of its commentators and practitioners little more than a game, perhaps even a "dirty" game, interesting enough, and reasonably well paid, but without further significance. That is a reason why young scientists may feel anomie, like that of the industrial physicist in *The Principles of American Nuclear Chemistry: A Novel*, who, having spent his youth in Los Alamos where his father worked on the atomic bomb, quits his own defense job, wondering if "it was ever proper to feel love for science," certain only that he could never feel his father's robust commitment to the scientific life.[95]

We should not be surprised when we see that some scientists today respond to the moral issues that have attached to the scientific life since the Wilhelmian years by questioning the scientific role and, often, by combining their professional scientific work with political and social commitments. Such commitments may give meaning and direction to their work, serving in that sense much the same function that the commitment to a culture-bearer role within high culture did for Wilhelmian scientists. It would be valuable to study further the social and psychological reasons why some American scientists feel greater vulnerability than their Wilhelmian predecessors.[96] Here I will observe only that the scientific role continuously evolves, and that the recent great complexity of the involvement of science in national and international affairs has made a redefinition of the scientific role a matter of utmost urgency for some scientists.

REFERENCES

1. For an explanation of the term ànd a thorough study of a group to whom it applied, the German academic humanists and social scientists, see Fritz K. Ringer, *The Decline of the German Mandarins: The German Academic Community 1890-1933* (Cambridge, Mass.: Harvard University Press, 1969).
   For their helpful comments on and criticisms of this article, I wish to thank Joseph Caggiano, Paul Forman, Loren Graham, Christa Jungnickel, and Theodore Roszak, among others. The research was supported in part by a grant from the National Science Foundation.

2. My findings on Wilhelmian scientists lend support to Ringer's "impression that, in their attitudes toward cultural and political problems, many German scientists followed the leads of their humanist colleagues" (*ibid.*, p. 6). In his analysis of the role of the academic researcher in Germany, Joseph Ben-David, too, calls attention to certain shared values of scientists and humanists (*The Scientist's Role in Society* [Englewood Cliffs, N. J.: Prentice-Hall, 1971], pp. 108-138).
   In this article, where I write "scientists" I mean "natural scientists."

3. For a perceptive analysis of this and related aspects of German scientific ideology, see Paul Forman, "Scientific Internationalism and the Weimar Physicists: The Ideology and Its Manipulation in Germany after World War I," *Isis*, 64 (1973), pp. 151-180. My understanding of Wilhelmian science is heavily indebted to this and other writings by Forman.

4. Max Planck, "The Unity of the Physical Universe" (1908), in Planck, *A Survey of Physical Theory*, trans. R. Jones and D. H. Williams (New York: Dover Publications, 1960), pp. 18, 20; see also pp. 1-26.

5. *Ibid.*, p. 152.

6. Wilhelm Lexis, "Das Wesen der Kultur," *Die Kultur der Gegenwart*, 1. Teil, 1. Abt. (Berlin and Leipzig: B. G. Teubner, 1906), pp. 1-51.

7. Opening remarks at the first general session, *Verhandlungen der Gesellschaft deutscher Naturforscher und Ärzte*, 44 (1871), p. 17.

8. Opening remarks, *ibid.*, 56 (1883), p. 4.

9. Walther Nernst, "Antrittsreden," *Sitzungsberichte der Köinglichen Akademie der Wissenschaften zu Berlin* (1906), p. 551; see also pp. 549-552.

10. Gerhard Masur, *Imperial Berlin* (New York: Basic Books, 1970), p. 125.

11. Hajo Holborn, *A History of Modern Germany, 1840-1945* (New York: Alfred A. Knopf, 1969), p. 379.

12. Koppel S. Pinson, *Modern Germany* (New York: Macmillan, 1954), p. 227.

13. *Ibid.*, p. 220.

14. Ringer, *Decline of the German Mandarins*, pp. 1-13.

15. See, e.g., the report of an address by Werner Sombart, "Schattenseiten der Kultur," *Umschau*, 12 (1908), p. 317.

16. For a characteristic lament, see Walter König, "Goethes optische Studien," *Physikalische Zeitschrift*, 1 (1900), p. 454; see also pp. 454-463, 467-470.

17. Ringer, *Decline of the German Mandarins*, p. 3.

18. Ernst Meumann, "Wilhelm Wundt. Zu seinem achtzigsten Geburtstag," *Deutsche Rundschau*, 152 (1912), pp. 193-224.

19. Karl-Heinz Manegold, *Universität, Technische Hochschule und Industrie* (Berlin: Duncker & Humblot, 1970), p. 8, and *passim*.

20. Wilhelm Wien, *Die neuere Entwicklung unserer Univesitäten und ihre Stellung im deutschen Geistesleben* (Leipzig: J. A. Barth, 1915), p. 17.

21. Forman writes that Wilhelmian scientists believed in the "indispensability of Germany's political great-power status to her position as a scientific great power" ("Scientific Internationalism," p. 162).

22. "Vorbemerkung des Herausgebers," *Handbuch der Elektrotechnik*, 1, ed. C. Heinke (Leipzig: S. Hirzel, 1900).

23. Eberhard Zschimmer, "Naturwissenschaftliches und technisches Denken," *Naturwissenschaften*, 2 (1914), p. 44; see also pp. 412-414.

24. Hans Arnold, "Die Entwicklung unserer Naturerkenntnis," *ibid.*, 1 (1913), p. 865; see also pp. 835-839, 862-865.

25. Ulrich Wendt, *Die Technik als Kulturmacht in sozialer und geistiger Beziehung* (Berlin: G. Reimer, 1906).

26. Manegold, *Universität*, p. 105. The leader of the movement, the mathematician Felix Klein, upheld the "idealism of a pure scientific life" as a defense against the materialism that many university people thought inseparable from technology (p. 106).

27. Woldemar Voigt, "Ansprache Seiner Magnificenz des Herrn Prorectors, Geheimrat Professor Dr. Voigt, 1911/12" (Voigt Nachlass 5, Universitätsbibliothek Göttingen).

28. Otto Wiener, "Uber Farbenphotographie und verwandte naturwissenschaftliche Fragen," *Verhandlungen der Gesellschaft deutscher Naturforscher und Ärzte*, 1. Teil, 80 (1908), p. 137; see also pp. 112-137.

29. *Ibid.*, 1. Teil, 73 (1901), pp. 15-17.

30. *Ibid.*, 1. Teil, 81 (1909), p. 10.

31. *Ibid.*, 1. Teil, 84 (1912), p. 15.

32. *Ibid.*, 1. Teil, 83 (1911), p. 19.

33. Julius Wolff, "Ueber die Wechselbeziehungen zwischen der Form und der Function der einzelnen Gebilde des Organismus," *Ibid.*, 1. Teil, 72 (1900), p. 114; see also pp. 82-114.

34. *Ibid.*, 1. Teil, 81 (1909), p. 10.

35. Ringer, *Decline of the German Mandarins*, p. 3.

36. George L. Mosse, *The Crisis of German Ideology. Intellectual Origins of the Third Reich* (New York: Grosset & Dunlap, 1964), pp. 3-5.

37. Pinson, *Modern Germany*, pp. 271-272.

38. Walter Z. Laqueur, *Young Germany. A History of the German Youth Movement* (London: Routledge, & Kegan Paul, 1962), p. 6.

39. Pinson, *Modern Germany*, p. 272.

40. Gottfried Benn, "Ithaca," *Gesammelte Werke*, 2 (Wiesbaden: Limes Verlag, 1958), pp. 293-303.

41. Walter H. Sokel, *The Writer in Extremis: Expressionism in Twentieth-Century German Literature* (Stanford: Stanford University Press, 1959), pp. 91, 95.

42. Klaus Schwabe, *Wissenschaft und Kriegsmoral. Die deutschen Hochschullehrer und die politischen Grundfragen des Ersten Weltkriegs* (Göttingen, Zurich, and Frankfort: Muster Schmidt-Verlag, 1969), p. 207, note 224.

43. Paul Volkmann, *Erkenntnistheoretische Grundzüge der Naturwissenschaften und ihre Beziehungen zum Geistesleben der Gegenwart*, 2nd ed. (Leipzig and Berlin: B. G. Teubner, 1910), p. 293.

44. Hermann Diels, "Die Einheitsbestrebungen der Wissenschaft," *Internationale Wochenschrift für Wissenschaft, Kunst und Technik*, 1 (1907), col. 6; see also cols. 3-10.

45. Julius Langbehn, *Rembrandt als Erzieher*, 49th ed. (Leipzig: C. L. Hirschfeld, 1909), p. 103.

46. *Ibid.*, p. 164.

47. *Ibid.*

48. Fritz Stern, *The Politics of Cultural Despair: A Study of the Rise of the German Ideology* (Berkeley: University of California Press, 1961), p. 132.

49. Volkmann, *Erkenntnistheoretische Grundzüge*, p. 292.

50. Hermann Diels, "Festrede," *Sitzungsberichte der königlichen Akademie der Wissenschaften zu Berlin* (1902), p. 31; see also pp. 25-43.

51. Volkmann, *Erkenntnistheoretische Grundzüge*, p. 292.

52. Victor F. Weisskopf, "The Significance of Science," *Science*, 176 (1972), p. 143; see also pp. 138-146.

53. Homer E. Newell and Leonard Jaffe, "Impact of Space Research on Science and Technology," *Science*, 157 (1967), p. 39; see also pp. 29-39.

54. William Bevan, "The Welfare of Science in an Era of Change," *Science*, 176 (1972), p. 994; see also pp. 990-996.

55. *Ibid.*

56. René Dubos, "Science Critics," *Science*, 154 (1966), p. 595.

57. Alan T. Waterman, "The Changing Environment of Science," *Science*, 147 (1965), p. 15; see also pp. 13-18.

58. Herbert A. Simon, "Relevance—There and Here," *Science*, 181 (1973), p. 613.

59. Harvey Brooks, "Physics and Polity: Are Physics and Society on Divergent Courses?" *Science*, 160 (1968), p. 399; see also pp. 396-400.

60. Waterman, "Changing Environment of Science," pp. 14-15.

61. Harvey Brooks, "Can Science Survive in the Modern Age?" *Science*, 174 (1971), p. 29; see also pp. 21-29.

62. *Ibid.*, p. 24.

63. Don K. Price, "Purists and Politicians," *Science*, 163 (1969), p. 26; see also pp. 25-31.

64. Brooks, "Can Science Survive?" p. 23.

65. *Ibid.*, p. 25.

66. *Ibid.*

67. I. I. Rabi, *Science: The Center of Culture* (New York: World Publications, 1970), p. 92; quoted in Weisskopf, "The Significance of Science," p. 145.

68. Thomas R. Blackburn, "Sensuous-Intellectual Complementarity in Science," *Science*, 172 (1971), pp. 1003-1007.

69. William Bevan, "The General Scientific Association: A Bridge to Society at Large," *Science*, 172 (1971), p. 350; see also pp. 349-352.

70. Weisskopf, "The Significance of Science," p. 145.

71. Arthur Kornberg, "The Support of Science," *Science*, 180 (1973), p. 909.

72. Weisskopf, "The Significance of Science," p. 138.

73. Brooks, "Can Science Survive?" p. 23.

74. *Ibid.*

75. Robert S. Morison, "Science and Social Attitudes," *Science*, 165 (1969), p. 152; see also pp. 150-156.

76. *Ibid.*, p. 152.

77. Philip H. Abelson, "Science and Immediate Social Goals," *Science*, 169 (1970), p. 721.

78. René Dubos, "A Social Design for Science," *Science*, 166 (1969), p. 823.

79. John Platt, "What We Must Do," *Science*, 166 (1969), pp. 1115-1121.

80. Brooks, "Can Science Survive?" p. 29.

81. Price, "Purists and Politicians," p. 30.

82. Bevan, "The Welfare of Science," p. 995.

83. Glenn T. Seaborg, "Science, Technology, and Development," *Science*, 181 (1973), p. 13; see also pp. 13-19.

84. Bentley Glass, "The Ethical Basis of Science," *Science*, 150 (1965), p. 1261; see also pp. 1254-1261.

85. *Ibid.*

86. Price, "Purists and Politicians," p. 27.

87. Morison, "Science and Social Attitudes," p. 156.

88. Weisskopf, "The Significance of Science," p. 144.

89. Dubos, "Science Critics," p. 595.

90. Glass, "Ethical Basis for Science," p. 1260.

91. Brooks, "Can Science Survive?" p. 29.

92. Philip H. Abelson, "Social Responsibilities of Scientists," *Science*, 167 (1970), p. 241.

93. Brooks, "Can Science Survive?" p. 21

94. Sokel, *The Writer in Extremis*, p. 172.

95. Thomas McMahon, *Principles of American Nuclear Chemistry: A Novel* (Boston: Little, Brown, 1970); reissued as *A Random State* (London: Macmillan, 1970), p. 245

96. To understand the vulnerability of American scientists, which antedates World War II, one would need to study historically the place of scientists in American society.

SALLY GREGORY KOHLSTEDT

The Nineteenth-Century Amateur Tradition:
The Case of the Boston Society of Natural History

IT WAS NOT UNTIL the nineteenth century that the term amateur was used to
identify avocational scientific enthusiasts. Never fully defined by contemporaries and loosely grouped together, the amateurs held a vague position somewhere between the general public, which paid minimal attention to science,
and tne regularly practicing researchers. They differed from the general public
by their active interest in science as observers, collectors, and informed patrons
in the process of data gathering. As science became professional, such individuals
also found themselves estranged from full-time researchers and described negatively for a variety of faults—lacking specialized training or a collegiate degree
in science, having no contracts (or inappropriate ones) in the professional world,
or producing too little in the way of publications in recognized journals. In the
interests of establishing standards and status, the professionals emphasized the
gulf or separation between full-time and avocational science studies implying
higher and lower orders. Meanwhile, as the public became aware of major
figures in science, it became less impressed by local "experts." This shift left
the amateur uncomfortably in the middle, not fully understood by either end
of the continuum. The purpose of this paper is to consider how the amateurs
have been viewed and how they organized themselves in the nineteenth century,
and to conclude with some observations on the functions which they filled.
Much of my discussion will relate to the Boston Society of Natural History
[BSNH] between 1830 to 1880 as a case in point.

As a field, the history of science has become increasingly interested in
questions often noted as "external"—the study of science in relationship to a
particular social setting. This expression is meant to underline the contrast with
internal investigations that trace developments in a specific discipline. Exter-
nalists investigate the nature of the community of scientists, the impetus for
pursuing particular lines of research, and the importance of public interest and
support, both moral and financial. By implication they argue that the environment of science affects its effort and product. Together with sociologists of
science, historians of science now frequently consider the nature of scientific
communities, comparing them over time and across national boundaries. Two
recent symposia analyzed learned societies and institutions as one way of assessing how political, social, and economic considerations influenced research and
individual activities.[1] Nearly every presentation emphasized how the particular
agency studied advanced (or failed to advance) modern, professional science.
But, despite interest in the context of science, almost no attention has been

G. Holton and W. A. Blanpied (eds.), Science and its Public, 173–190. All Rights Reserved.
This article Copyright © 1976 by D. Reidel Publishing Company, Dordrecht-Holland.

paid to the individuals designated as amateurs. When mentioned by historians they are noted in passing as important supporters of particular scientific activity, as beginning professionals in a temporary situation or as time-consuming nuisances to the researchers. But almost nothing has been done to investigate the nature of the amateur tradition. Because little historical attention has been paid to amateurs, there are only vague answers to fundamental questions: Who were they? What were their motivations? What functions did they fulfill for science and the public? Attention to the amateur tradition to date has been primarily intended to determine when and how a change occurred that established modern, professional standards.[2] Often historians designate seventeenth and eighteenth-century researchers as "amateur" in terms of their training and expertise.[3] By the nineteenth century, given options for training and practice, the term amateur designates a more discrete class of individuals.

Several descriptive models of the amateur have been preferred. The image of the New World's "gentlement farmers" and physicians examining local flora and fauna in hopes of an invitation to membership in the Royal Society is frequently cited.[4] More recently Nathan Reingold has argued that many "cultivators" (a noun he selects because it is currently less pejorative than amateur) used science as a hobby, an extracurricular activity that had status and might hold applications useful for other careers. [5] Another historian studying natural history in Cincinnati argues that his group of interested but avocational collectors were disinterested scholars studying "science for science's sake"[6]—which would seem to distinguish them from their professional colleagues only in the time and energy they were able to devote to the enterprise. A fourth possibility suggested in the personal history of several researchers is that the amateur activity provided a gradual training in situations where no alternative education was available. While each of these descriptions is useful, there is still no explanation demonstrating why the activity of the amateur came to be so negatively regarded in the United States nor what role the amateur played prior to elimination from the scientific community.

The dismissal of the amateur by the historian of nineteenth-century American science is not surprising given the sources used—the manuscripts and publications of the most aspiring professional scientists. Caught up in cultural nationalism, a major goal of the scientists in America was to convince European friends that they were doing valid research and to insist that "pure science" needed steady support and should not be called upon for specific applications. As a result, a thesis arguing that Americans were indifferent to basic research was perpetuated by excellent scholars who accepted at face value the breast-beating and frustrations of the scientists who not infrequently used polemic about the low status of basic research in order to gain more support for science. Thanks to the work of Nathan Reingold and others, scholars now recognize that whether the measure is great scientists, the nature of the work performed, or European comparisons, American science by the middle of the nineteenth century was respectable and important.[7] Unfortunately the self-image of these aspirng professionals continues to influence historical assessment of the period. Because leading contemporaries dismissed the part-time practitioners as peripheral, no one has investigated the role played by the amateurs who were, in fact, an integral part of the process of normal science.[8] Their existence cannot be denied. Although their activity might be minimally acknowledged, both the public and knowledgable colleagues were aware of their interest and effort. The amateurs should be of interest to the social historians as well as the more traditional historians of science because as a subset they mirrored attitudes in science while being in close touch with contemporary conceptions of science. Their perspective on science is unique: they were more consistently interested in scientific investigation than the general public but had a less narrow vision of new research than did many researchers. The intermediate position which they occupied between the public and the scientific community was occasionally uncomfortable. The scientific establishment found them to be an important liaison, capitalizing on their interest and, in turn, driving support for the scientific enterprise.

The descriptive term "amateur" belongs to the nineteenth century. Entomologically linked to the French *amateur*, which is in turn from the Latin to love, by 1803 the expression was in frequent use in England. Originally it denoted a person who pursued any interest as a past-time, distinguishable from those who pursued an area for monetary or academic reasons. Its reference was to the arts and literature as well as to science. As the century progressed, the term was used as an adjective and increasingly connoted a dabbler or superficial student.[9] The shift in implication was gradual; explaining the change will be far more difficult than describing it. Two parallel developments were significant—the adherence to a "scientific way of knowing" that was associated with a specially trained group, and the popular enthusiasm for science that developed originally in Parisian salons and English sitting rooms but penetrated to the popular imagination.[10] From the mid-eighteenth century, public lectures on science indicated that science had an audience, although content ranged from the display of phenomenon like the two-headed calf to electrical demonstrations.[11] The eclecticism of public curiosity reinforced scientists' desire to distinguish themselves from the lecture-circuit promoter. The amateur lost visibility and purpose to the professional on one side and the popularizer on the other. By the end of the nineteenth century American science had patrons but very few real amateurs.

When and why did the image of the amateur shift? At the beginning of the century the educated savant was respected and the local collector sought after by interested zoologists and botanists.[12] The shift was not inevitable. Russians also developed scientific institutions and worked to establish a cultural and scientific identity but, according to Vuchinich, the amateurs managed to maintain an important role in Russia.[13] By the 1840s the Russian amateurs were more than a chorus of enthusiasts, they produced work of high quality, and were credited by full-time colleagues with linking professional and public response. As late as the 1860s they provided the forum and vehicle for the scientific community to state publicly their collective views on the social roles of science, educational policies, government research, and professional problems.[13] The amateur was not the counterpoint to the scientist but rather a valued link. In the United States this pattern was early in evidence but failed to persist. By the late 1830s, change was evident. Some of the outstanding younger scientists felt their efforts were hampered by curious but intermittently interested and poorly trained colleagues in local societies and either withdrew from or tried to control local societies. Concerned about the image of American science in Europe, the ambitious young men wanted to monitor all presentations and publications that represented American science. Joseph Henry, physicist and later head of the Smithsonian Institution, wrote to his friend Alexander Dallas Bache while traveling in Europe in 1838, "There is a great prejudice and perhaps in some respects a just one ... against Americans ... [that arises from aristocratical and institutions abroad and results in] the general low opinion which is maintained relative to American science and literature."[14] International credibility forced them to seek unambiguous authority. They were torn, however, in their efforts to distinguish themselves. Many of the amateurs were personal friends, former classmates, or even prominent political leaders. Such ties could not be severed with impunity.

Dependence on an interested public was a position bitterly resented by the emerging professionals. The scientists frequently bemoaned the problem of finding support for research in a democratic society, arguing that European patronage by royalty and the upper classes offered greater consistency and a higher level of understanding than did the less literate and less sympathetic mass public in the United States. Despite their complaints American scientists knocked on many doors—from those of private patrons through the arched Gothic portals of local, state and federal government offices, to the lecture halls of the general public—with remarkable success.[15] Assisting them in their claims and unifying their efforts were philosophical and scientific societies.

It was, ironically, the same societies which had first encouraged voluntary groups to discuss science casually and pioneered in establishing liaisons between the scientist and the community which became agents in formulating the distinction between scientist and amateur. While local societies by the middle of the century emphasized their professional role, all of them continued to hold meetings open to the public. Who should lecture to the public? Increasingly the prominent scientists resisted the call to publicize science, and into the gap proceeded the popularizer—reasonably knowledgable, interested, and with a sharper awareness of what the public wanted. The amateur member became at best the pre-professional. A frequent analogy held that just as science had come through an amateur stage, so individuals, too, had a naive period of enthusiasm before beginning serious work. The professional went on to further become identifiable and certified by college degree, by occupation, by publications, and by affiliations. Professional schools were serious about distinguishing their progeny. This was evident first in medicine and engineering, the former seeking licensing by the 1820s. Scientists by mid-century established collegiate programs in specific fields and used certificates or degrees as credentials.[16] They began to insist that they, too, were professionals and should be given good positions with opportunities for research. Their efforts should be published or presented to colleagues who could discuss results in detail. Their task, as they saw it, was to "advance" science, not promote it.

Amateurs in urban societies were similar to the pre-Civil War scientific community; like the professional scientists, most seem to come from middle or upper-class homes and had college educations. But they had not, for the most part, dedicated themselves to scientific research. Other amateurs (rural or of a lower class) remained in that status because they lacked the education, the leisure, or the contacts which could have given them the opportunity to become professional. Thus, the amateurs were either voluntarily or involuntarily part of the mainstream rather than the vanguard of science.

The amateurs were neither totally dismissed nor denigrated by the 1840s. But the credit given them was somewhat grudging and patronizing; thus, Alexander Dallas Buche, great-grandson of Benjamin Franklin and head of the U.S. Coast Survey, observed in his presidential address before the American Association for the Advancement of Science in 1851:

The world is made up of ordinary men, and it is part of common sense not to despise their doings. The specimens collected or the observations made by the humblest geologist who ever wielded a hammer, or the meekest astronomer who ever noted a transit, serve as part of the foundation of the superb structure raised by Von Bach, or by Leverrier. If the zeal of *second-rate* men is warmed into activity and directed in the development by such influence, the general level of science is raised by slow deposits, which may on occasion make mountains by upheaval.[17]

Bache's geological analogy could scarcely have been lost on the amateurs who generally attended AAAS meetings of the 1850s in substantial numbers: yes, they had a contribution, provided they remembered their usual place was on the bottom! Other leading scientists acknowledged that the amateurs assisted them specifically in gathering the raw data necessary for the "real work" of science. The unaffiliated and uninitiated amateur was sometimes handled intemperately, crudely put down before an audience.[18] Private correspondence expressed disdain for the naive comments and questions occasionally addressed to professionals—yet in a congenial local club the same query after dinner might be answered with enthusiasm.[19]

Public and peer approval as well as personal aspirations certainly contributed to the creation of boundaries and hierarchies. In addition, there was a concern that under the pressure of industrialization and materialism, the intellectual pursuit of basic science might be lost. The concerns of such major scientists as Joseph Henry and Alexander Dallas Bache when they contemplated the future of science mirrored the fears of such Europeans as Charles Babbage, whose "Reflections on the Decline of Science..." in England reverberated across the Atlantic.[20] This generalized anxiety that major breakthroughs were not forthcoming and that minor problem-solving distracted top thinkers, affected amateurs negatively. Amateurs seemed to require time and attention but to provide few concrete results. Moreover, as governments offered more support for science, and the colleges assumed more responsibility for educating and assisting scientists, the amateur groups no longer seemed so critical nor the amateur an essential liaison between scientists and the public. The professional could more readily dismiss the "mere amateur" as expendable when new support systems emerged and expertise was distinguishable. The loss of the amateur tradition was, however, more subtle in its effects.

The foregoing discussion, admittedly somewhat abstract, suggests that the negative attitude toward amateurs originated with nineteenth-century scientists and persists in current historical literature. A case study of the Boston Society of Natural History will reveal how the amateurs saw themselves and pursued their own goals in relationship to science. Founded in 1830 and celebrating its fiftieth anniversary in 1880, the BSNH spans an era of critical change in the membership composition, the activities, and the purposes of amateur societies. A longitudinal analysis of the Society also demonstrates that the amateur tradition persisted even as its functions changed. The Society, which had begun as a voluntary association of active amateurs became, during the course of its first hundred years, a public Museum and lecture hall located on one of Boston's most fashionable squares, and today remains one of that city's most important public educational institutions. A search through the rich, largely unresearched manuscripts of the now-deceased Society, housed at the Boston Museum of Science, reveals how the group viewed itself, and what functions it consciously or unconsciously fulfilled for science during its active years in the nineteenth century.

Boston may have provided a special climate, distinct from other major centers of scientific activity. Although Philadelphia challenged its hegemony and the nation's scientific practitioners were widely scattered, the Bostonians had shown a steady interest in science from an informal club of Cotton Mather through the foundation of the American Academy of Arts and Sciences (1780). Activities and interests varied over time, favoring either the natural or the physical sciences. At the beginning of the nineteenth century certain Bostonians, most of them with Harvard degrees, established a short-lived club, the Society for the Study of Natural Philosophy. The young, wealthy Federalists tried to study science by duplicating classic experiments and forming collections of artifacts. Their efforts were more serious than the philosophical amusements of the eighteenth century salon but with similar results—none. Demand of their individual professions caused the members to drop out. But a recent analysis of the group concludes that in the process of their study they gained a healthy respect for science and that "years after the SSNP had disappeared, its former members exercised substantial influence in favor of institutional support."[21] The Linnaean Society (1815-1823), founded shortly thereafter, was composed primarily of medical practitioners, who took field trips together in order to gather natural history and mineralogical specimens. During its few active years the Society concentrated on the collection and classification of regional specimens. Both efforts suggest that the Boston-Cambridge area was supportive of science. Local scientific practitioners usually became members of the American Academy of Arts and Sciences and brought science to other informal discussion clubs as well. Laymen and amateurs seemed committed to the proposition that science was essential to American progress: names like Appleton, Lowell, and Lawrence frequently appear on subscription lists for new monographs. Economic prosperity reinforced local belief that New England did, in fact, lead the nation.

The BSNH was, like its predecessors, founded to bring self-edification to its members but also to give focus to individual activities.[22] The stated goal was personal "improvement, and ... the cultivation of a taste for natural history in our community."[23] The group was avowedly avocational in its pursuit of knowledge, curious about natural science and assuming that its activity would advance its development. In an address in 1833, the Society's president noted,

It is very true that most of us are so connected, in our several professions, with those to whom our first and chief attention is due, that we cannot lawfully be absorbed in pursuits which are extraneous to our immediate obligations; ... but we can take a little from our leisure, and a little from our indulgences, and a little from our rest, and make our very amusement and healthful recreation contribute to the welfare and growth of this Society.[24]

Under the state-granted charter, members selected colleagues socially and educationally compatible; the Boston Athenaeum served as the Society's first meeting place. The members, as several examples indicate, were in harmony with the general cultural assumptions of the nation and felt that there were important possibilities in the amateur life. Founding members Walter Channing and George B. Emerson are typical. Dr Channing (1786-1876) had also been a founding member of the earlier Linnaean Society and held the offices of curator and vice president in the BSNH. His older brother was the eminent theologian Dr William Ellery Channing. Genial and witty, Walter Channing was once asked if he was the famous Dr Channing who preached, and he responded, "No, it's my brother who preaches, I practice."[25] What he practiced was medicine, education, reform, and religion. Channing entered Harvard in 1804 but left after participation in a student rebellion. He then studied medicine in Boston (and also travelled to visit medical schools in Philadelphia, Edinburgh, and London). In 1812 he established a local practice and became Lecturer in Obstetrics at the Harvard Medical College. Lively, socially conscious and intellectually curious, he was a typically active member in the early years of the BSNH.

Another founding member, George B. Emerson (1797-1880) was born into a medical family in Maine. Apparently interested in botany even as a child, he took courses with William Peck while at Harvard. When he graduated in 1819 he turned to education, becoming the first principal of the Boston English Classical High School and then opening his own school for young women. An educational reformer, he wrote the memorial to the Massachusetts legislature that resulted in Horace Mann's appointment as Secretary of the new Board of Education.[26] Emerson retained his interest in natural history, however, and was president of the BSNH from 1837 to 1843; during his presidency he persuaded the legislature to sponsor a scientific survey of the state. Not only did Governor Edward Everett concur with the legislation, but he accepted recommended appointments, including that of Emerson himself who, after nine summers of field work and research, reported on the trees and shrubs of the state. Emerson remained an active member of the BSNH through the 1870s.

The Society also could and did assist purposeful young amateurs interested in science. Augustus A. Gould (1805-1866) was born in New Hampshire (curiously, all the members mentioned thus far were born outside Massachusetts); he earned his own way through Harvard College and then studied medicine with Walter Channing and others to earn an M.D. in 1830. Although dependent on an income from practice, he found the leisure to become a collector of Cicindelae and published an account of them in 1834, after presenting preliminary results to the BSNH. In 1841 Gould published his results from the state survey; his analysis of the geographical distribution of shells systematically demonstrated for the first time the barrier that Cape Cod forms for some species. By the 1840s he was established as an authority on shells, working on the results of the Wilkes Exploring Expedition, and associated with Louis Agassiz in producing *Principles of Zoology*. Like Emerson, he was a religious person, active in his own profession (as the president of the Massachusetts Medical Society and as collector of the State's vital statistics), and almost continuously an officer in the BSNH.

Gould is on the line between amateur and professional; never devoting all his time to zoology, he managed to become a professional by his expertise on mollusca.[27] Undoubtedly he provided an inspiration to younger members who had few financial resources and no initial contacts in science. At the other end of the amateur spectrum was Henry David Thoreau. Elected a corresponding member of the Society in 1850, Thoreau rarely if ever attended meetings but he made frequent use of the library. Although Thoreau was unwilling to engage in technical investigation and eschewed "fact grubbers," his fellow Transcendentalist Ralph Waldo Emerson remembered that Thoreau was never "idle or self-indulgent."[28] Thoreau's journal documents his careful, even systematic interest in botany and ornithology.[29] The unconventional Thoreau was indifferent to rules and hours and at least once came through a window of the Society's residence in order to borrow books before the librarian had arrived. Other members seemed to find him interesting and readily allowed him use of the Society's resources. Basically membership was open and eclectic. The boundary on membership seemed to be sex and, less explicitly, class. Leadership rotated within a small circle of the membership.

The botanical organization had characteristics that were also found in similar societies located in Philadelphia, Albany, and New York. The BSNH's yearly activities are not pertinent here. But, in summary it should be noted that they collected a substantial reference library enriched by periodicals that were obtained through an international system of exchange. Equally important for study was an extensive collection of botanical, geological, and zoological specimens that related particularly to the New England region. Quite early the membership had established a geographical boundary to their interests which enabled them to concentrate their skills and to complete a systematic collection. A regional emphasis meant exhibits, lectures, and tours could complement the local research interests of members. The BSNH was a cooperative enterprise, with the active members serving as volunteer curators, reporting on new books, and presenting recent research. All of this activity, together with the more publicly visible effects of the State's natural history survey reinforced the members' assumption that amateurs had a contribution to make to science.

From the outset, however, they felt community responsibility. Buoyed by the contemporary movement toward self culture, enthusiastic educational reformers like Emerson and Channing turned their attention to promoting interest in science. Certainly the committees were aware that:

A large collection has the effect of attracting great attention, and the wondering thousands who are drawn by its exhibition to visit it daily or weekly, enjoy an innocent pleasure that is well worth providing for in all large communities, especially as the influence may often go far beyond gratifying curiosity.[30]

Clearly the Society hoped that visitors would provide additional income, but the implied financial considerations were tangential. The Society's officers were genuinely interested in arousing the curiosity of at least a sector of the public and the collections were open for viewing without charge one or two days a week. Yet for some of the invited lectures by prominent scientists a charge of three dollars per person was made—evidently a select audience was anticipated. An effort was made to present the best and most up-to-date interpretations in science. Lyceum-level lectures or public spectacles had no place.[31] In fact, the Society was delighted when member Jeffries Wyman presented a paper discussing a purported skeleton of a Sea Serpent. A New York promoter had exhibited a "fossil" with head, teeth, ribs, paddles and vertebrae neatly cemented together which he called *Hydrarchus Sillimani*. The specimen measured about one hundred and fourteen feet and attracted a large crowd of visitors. Without much difficulty Wyman found that the vertebrae were not from one individual creature but belonged to many different ages and did not present any of the characteristics of an ophidian reptile. Its teeth were probably those of a warm-blooded, mammiferous animal. [32] Revelation of the hoax delighted members who pointed out that they were a responsible group and would introduce the public to scientific truths, not spectacles and speculations.

Clarifying science for the public became an increasingly important function. By the 1840s the research efforts were being complemented and even superseded by research and informal study groups at Harvard and by the drive toward professionalization. Members, however, did not initially fear specialization; in 1841 one member told his fellows:

The process is clearly this: the great accumulation of facts in any sciences causes an absolute necessity for arrangement into divisions and subdivisions; the more extensive the more clearly defined and simple are these divisions, and each becomes the object of a separate study; hence, the subject is more easily mastered, more easily grasped by the mind, while the man of comparatively little leisure can undertake a single division and not only keep pace with discovery, but soon add something to what is already known.[33]

Quite clearly and directly he outlined the aspirations of most amateurs. Naive? Perhaps. In the 1840s, however, the BSNH had a specimen cellection among the best in the nation, and all local, prominent natural scientists, including the internationally known Louis Agassiz and William Barton Rogers, were members. When Lewis R. Gibbes came north from South Carolina, he studied the Crustaceans at the BSNH and made a full catalogue of the collection.[34]

The Society's daily activity, however, was conducted by the amateur members; professionals had other priorities. In an effort to document research done primarily by members or on the Society's collection, a *Journal* was established in 1830. Gradually it specialized in botany and zoology, parallel to the Philadelphia Academy's efforts in geology. After 1840 the Boston society also published *Proceedings*, hoping that a summary of their activity would be read by the general community. Members believed,

as the subjects which present themselves for the discussion of the Society embrace the whole region of Nature, there is no want of a variety nor of interest; those who have attended them [the meetings] never seemed to separate but with reluctance, and a desire to continue their investigations...regular publication of the discussions, and notice of the objects presented, is absolutely necessary to let the world know that the society is in existence:...they bring before the public, in a form which is convenient and agreeable, facts which they would not become acquainted with from scientific papers in the journal.[35]

Despite efforts to coordinate and unify the Society's own activity, however, the mid-century decades were difficult ones. Attendance dropped and finances were strained. The BSNH felt the strain of professionalization which turned attention from the local organization toward national societies like the American Association for the Advancement of Science (1848) and the National Academy of Sciences, which was founded in 1863 and had a carefully selected membership.[36] In addition, emphasis on systematic training of specialists in the colleges and increasing reliance on graduate training aboard minimized the possibilities of the casual, mutual-help training of older members. Nonetheless the BSNH proved resilient and gradually shifted its goals. Unable to compete with emerging national societies with professional goals, the BSNH sought alternative ways to assist science.

Designing the Society's program so as to attract the public became an important priority—but with the end still the representation to the public of the best of scientific expertise. Initially the Society hoped to maintain a place for active practitioners and established special study sections in entomology and microscopy.[37] These proved short-lived. Practically as well as philosophically the Society shifted in terms of its perceived base of support, its educational goals, and the self-image of its members. The old assumption had been that the public would seek out the society; but increasingly the BSNH actively promoted itself. By the 1860s it reported regularly to local newspapers about its activities.[38] In 1857 the members petitioned for land in the fill area of the Back Bay with the object of erecting a museum.[39] Simultaneously, they sought prominent scientists to give addresses open to the public. The intention was, as a committee informed William B. Rogers,

in accordance with the very general sentiment that it is especially desirable at this time to attempt to give a more prominent place in the public estimation to the Boston Society of Natural History by calling attention to its character and purposes...[40]

The purpose was avowedly public in contrast to the small, mutual-help Society of the 1830s.

In 1867 the new Museum of Science opened on Berkeley Street, housed in an impressive brick and stone structure just off Copley Square. The Museum like the Public Library, suggested a kind of social reorganization in Boston. A proud secretary reported that in the first year 3600 persons had come to see the exhibits. The appeal was now to the general public—they were invited to visit, not to become members—but making this appeal was not without difficulties. A disgruntled curator noted the high cost (in both money and energy) of being the keepers of "scientific Truths instead of the rubbish which is so often forced upon the public under the name of natural history".[41] His frustations were specific:

The dust brought in and kept in motion by hundreds of feet, penetrating the cases and injuring the specimens; the defacement of the walls by the ignorant and foolish, the damage done to furniture and the glass of the cabinets by careless or malicious persons...while the hinderance to the skilled scientific labor continuously in progress here, by the crowd and noise, cannot be measured by dollars and cents.[42]

His conclusion was that this commitment was nevertheless important and that the conditions had to be endured.

The old enthusiasm for the sciences themselves seemed curiously displaced by a sense of obligation to educate in the sciences. But as a result of the more extensive activity or perhaps the new style of activity, attendance at meetings increased from an early average of less than twenty to thirty or forty persons at the biweekly meetings. The old cooperative efforts were taken over by a paid curator with a bachelors degree in science from Harvard, Alpheus Hyatt. An enterprising manager, Hyatt helped initiate a new lecture series in 1870, suggesting the Museum should be an "instrument of popular culture."[43] The Lowell Institute trustees were persuaded to provide some support for a series of lectures expressly directed at local elementary teachers.[44] The public was obviously interested and more than seven hundred teachers responded.[45] Similarly a general lecture series—as contrasted to earlier individual topical lectures—emphasized scientists who were local experts: the speakers included Frederick W. Putnam on archeology, Nathan Shaler on geology, explorer John Wesley Powell on enthnography, and Samuel Scudder on butterflies. In 1880 Hyatt reported on the results of a decade. To him the Museum embodied the results of researchers and the message to the public had been

that no love of the merely curious, no love of facts for facts sake alone, no love of nature, even solely for nature's sake...induced the busy men who founded this institution, or the naturalists who helped them, to give up their hard earned hours of rest to its service. Our museum must make the public appreciate that these men believed in self-culture.[46]

The message was historically accurate, but no longer applicable. The amateurs in the Society had begun to create a new hierarchy (a corps of elite *within* the amateur tradition). Only after tense meetings did they accept the new implications of amateur status and expand their vision of appropriate membership for the BSNH.

Professionalization in science depended on higher education and informal certification by peers; barriers to either or both of these routes necessarily resulted in exclusion. As late as the third quarter of the nineteenth century the Boston Society of Natural History provided access to research facilities and to the publications that in turn permitted certification. Everywhere in America women had found themselves blocked along both avenues, but by the 1870s they were finding educational opportunities (through undergraduate science curricula) in the newly established women's colleges and in certain state land-grant colleges. The trend propelled the BSNH to consider a situation it had successfully ignored for years—its policy of the exclusion of women. Women had been allowed membership in the American Association for the Advancement of Science prior to the Civil War and the American Academy of Arts and Sciences had initiated Maria Mitchell in 1848. The use of natural history texts in young women's preparatory schools had acquainted women with the sciences, and by the late 1860s the BSNH itself had hired women as assistants in the library and used two women to assist the curator. Still it resisted efforts to make them members.[47] Women were welcome to visit the museum, as were children, during public visiting hours. In 1870 a card listing regular meeting times for the year indicated that for all sessions, except those at which nomination for new members occurred, "members might invite attendance of ladies."[48] But the invitation-only policy seemed increasingly inappropriate.

Perhaps only by chance, the Centennial of 1876 brought a revolution to the Society over the question of whether women should be admitted into regular membership. The issue was among the most hotly debated in the entire set of handwritten minute books. Taken by surprise at a motion to admit women, the members in attendance concurred but cautiously voted to postpone action until a special committee could report on the "expediency of the proposed measures."[49] At the following meeting the assigned committee recommended the "unconditional admission of women to membership." Although the Society had widely expanded its membership in other ways, Samuel H. Scudder suggested that the subject of women needed further consideration and recommended "the formation of grades of members."[50]

Scudder, son of a wealthy merchant and a graduate of the Lawrence Scientific School (1862), made his reputation as an entomologist and later became a paleontologist with the United States Geological Survey. Reasons for his opposition to women were not clear but in an effort to limit the move he contacted associates by letter, asking their opinion about whether women should be admitted. Several responses reinforced his position and articulated the private fears of male members and contemporary social concerns. Alexander Agassiz replied, "I have no special reasons against it, except the extremely rare chance of offending their ears by some embryological fact..."[51] Another correspondent also worried about chaste ears but suggested that the difficulty could be handled by sending out cards with the title of the talk. He recommended that if there was some possibility of an embarrassing discussion "some special mark, as a skull and crossbones, or other object indicating 'danger', might be put after the subject."[52] Member Charles Sprague, however, articulated two hesitations that reflected on his belief that the Society still had standards to maintain which might be undermined by the admission of women:

There have been some very learned women. But the great majority, even of the highest of female minds, have not that peculiar logical exactness which is characteristic of the scientific minds of men. They have not that unimpassioned, pertinacious, comprehensive and self-denying impulse which leads some men to devote themselves to a dry detail of study in a field totally uninteresting, except to the votaries of each specialty.[53]

Moreover, they would undermine the seriousness of the society by tempting the men away from serious work:

The presence of *woman* is inevitably a hindrance to a calm, passionless, self-forgetting study. Some motives are so constituted as to withstand this influence; but I question whether science would not suffer in a mixed assemblage of young men and young women brought together not by the affinities of social life and station, but in the promiscuous intercourse of a society of comparative strangers... I refer in all this to *real science*, not to that smattering of knowledge, beyond which you and your coadjutors have long since passed.[54]

Sprague feared the human nature of both men and women, apparently, but placed the burden of exclusion on women. Still another correspondent suggested that it was fine to admit some women, but the Society should be cautious not to include "a heap of ranting spiritualists, suffragists, dress reformists and worse."[55]

Others, however, approved and felt equal recognition was important. One respondent argued that they should be regular members and not pay "full price for a half a loaf."[56] A letter from the Academy of Natural Sciences in Philadelphia expressed surprise at the question, pointing out that women were admitted as full members of their Academy, and that they had been allowed to attend all meetings without problem. Scudder's specific questions elicited an additional comment:

I should consider that ladies attending scientific meetings ought to leave at home all false modesty and enter into discussions intellectually and without regard to the apparent delicacy of their situation. Ladies have male physicians and attend medical lectures upon delicate subjects; therefore why should they not run their chances of being shocked at the society meetings?[57]

J. Ellen Cabot of Brookline was perhaps the sole woman consulted; she replied:

My feeling is in favour of the admission, or rather, against exclusion; which is always bad unless necessary, and should always place the burden of proof on the excluders.

The objection I suppose is that it will tend to restrict the freedom of discussion, by limiting the topics and giving a bias to the treatment, towards what may be considered as suitable or as attractive, to women, —I suppose there may be some danger of this sort, but I do not believe it will be any serious obstacle to anybody who has anything to say which has seeming importance.[58]

She concluded by noting that probably there would not be many women interested, so that the Society would not be overrun. She added that if, in fact, including women as members proved a "dangerous policy", the Society could easily stop selecting women.

Apparently persuaded that the inclusion of women was inevitable, Scudder took the recommendation of several correspondents that two levels of membership be created.[59] While the two levels, Associate and Intermediate, were not specifically sex stereotyped, the implication was clear. Debate became so heated that in the midst of an argument advocating full membership for women, the session was adjourned.[60] The two categories were established in a newly written constitution, along with yet another resolution to insert the word "male" before "Associate members."[61] But this motion was defeated two months later.[62] Significantly, the constitution also changed the wording of the requirement for membership from "contribution to science" to "aid to science."[63]

Fears of domination or negative influence proved to be unfounded. Fifteen women joined almost immediately and most had Back Bay, Beacon Hill, and Jamaica Plain addresses. They attended meetings but were rarely mentioned as discussants; none gave papers, or were elected to office, or were promoted to associate membership.[64] They were more active in the section meetings, where, perhaps, they felt less hostility and more interest in and reinforcement for their work.

Considering the barriers erected against women was not a digression but clearly suggests something about the nature of the later amateur tradition.

Aware that their own contribution was uncertain and frequently denigrated, the members of the BSNH worried about maintaining credibility with the professionals. The arguments against women were basically defensive, and the result was the creation of levels of membership which implied greater status (and perhaps closer identification with professionalism) to certain members. The price was another kind of exclusion with the new subset identified as more "amateur-like." The amateurs, by implication, admitted that a hierarchy was appropriate and that culture bearers were important but of lesser status.

Throughout the period, the Society maintained an "aristocracy of spirit." They formulated their goals in the rarified atmosphere of Beacon Hill and could not follow the logical conclusion of their program—that everyone should not only have access to scientific knowledge but should be able to advance to the limits of personal interest and talent. Increasingly didactic, even patronizing, the Society became less effective as a coherent society of amateurs. Promotion of science became a central function and was carried out primarily by a hired staff. The active amateur participation in science was almost entirely eliminated and the professionals took little interest in the educational and museum activities of the local society.

This development should not, however, discourage scholars from investigating further the nineteenth-century amateur. The Boston Society of Natural History presents only one aspect of the amateur tradition, and a single case study is certainly not sufficient to make a good model. The amateurs within the BSNH showed remarkable ability to study independently, to define their own goals, and to shift both goals and functions. If the group of amateurs added relatively little to pure scholarship or even to the training of scientists, it did establish a research collection and an excellent library that served the scientists admirably. In addition, it provided an intellectual climate —demonstrated in the Gray-Agassiz debate over Darwin—as well as an opportunity to present papers and to have them published. Moreover, by forging links with a particular elite in Boston, the Society encouraged local financial support for science as well. To the greater Boston community it offered the best of science available in the mid-nineteenth century, the adventure of new discovery through recent research. If there was ascertainable "public conception of science," the BSNH was probably critical in its formation.

A common accusation or assumption was that the amateur was responsible for the tendency to seek the "usefulness" or applicability of science or to indulge in philosophical, sentimental interpretations. However, within the BSNH there was almost no discussion of applied science nor the religiosity so feared by the emerging scientific professionals. Rather the BSNH as a Society assumed the value of science implicitly and promoted the study and analysis of basic natural science.

Investigation of the amateur tradition is only beginning and much more data are required. The BSNH indicates that as a cohesive group in urban society, its member amateurs functioned rather well. But information is also necessary on those persistent amateurs who existed outside the comfortable sphere of like-minded friends—those in rural areas or those who were de facto barred from affiliation by sex or class during the active years of the amateur.

The evidence of these individuals' activity is found in the textbooks and field guides produced by women and other local experts and in the correspondence of members of the Boston Society with like-minded friends throughout the expanding United States. Who were these other amateurs? What were their goals and their function? How did they view themselves? Questions remain. One general conclusion, however, is likely to emerge from further study: the efforts and concerns on behalf of science were perhaps more diffused throughout society in the nineteenth century than in any other and the amateur tradition provides an important clue as to how that was possible.

REFERENCES

1. The two symposia were sponsored by the American Academy of Arts and Sciences in June, 1973 and June, 1975. The proceedings of the former are to be published by the Johns Hopkins University Press.

2. George H. Daniels, "The Process of Professionalization in American Science: The Emergent Period, 1820-1860," *Isis*, 58, No. 2 (1967), pp. 151-166.

3. Although Dorothy Stimson's *Scientists and Amateurs: A History of the Royal Society* (New York: H. Schuman, 1948) makes use of the term throughout her volume, she never defines it; nor does she demonstrate that it was in contemporary use during the Royal Society's early years.

4. Raymond P. Stearns, *Science in the British Colonies of North America* (Urbana: University of Illinois Press, 1971).

5. Nathan Reingold, "Definitions and Speculations: The Professionalization of Science in America in the Nineteenth Century," unpublished manuscript for the American Academy of Arts and Sciences' Conference on the History of Learned Societies in America, June, 1973.

6. Henry Shapiro, "The Western Academy of Natural Sciences of Cincinnati and the Structure of Science in the Ohio Valley, 1810-1850," unpublished manuscript for the American Academy of Arts and Sciences' Conference on the History of Learned Societies in America, June, 1973.

7. This is especially clear in an article by Richard H. Shryock, "American Indifference to Basic Science in the Nineteenth Century," *Arch. Int. d'Hist. des Sci.*, 2, No. 5 (1948), pp. 50-65.For a fascinating unraveling of the myth see Nathan Reingold, "American Indifference to Basic Research: A Reappraisal," *Nineteenth-Century American Science*, ed. George Daniels (Evanston: Northwestern University Press, 1972), pp. 38-62.

8. There has been important and well-researched analysis of the pre-Civil War scientific community but almost all scholars looked exclusively to the establishment of science as a profession.

9. *Oxford English Dictionary* (Oxford: Oxford University Press, 1961), 1, p. 265. Professional is given as an antonym.

10. Roger Hahn, *The Anatomy of a Scientific Institution: The Paris Academy of Science, 1666-1803* (Berkeley: University of California Press, 1971), pp. 86-89.

11. The range of public curiosity is demonstrated in two excellent studies of popular culture during this period; see Carl Bode, *The American Lyceum: Town Meeting of the Mind* (New York: Oxford University Press, 1956) and Neil Harris, *Humbug: The Art of P. T. Barnum* (Boston: Little, Brown, 1973).

12. John C. Greene, "Science and the Public in the Age of Jefferson," *Isis*, 49 (1958), pp. 13-25, and "American Science Comes of Age," *Journal of American History*, 55 (1968), pp. 22-41.

13. Alexander Vuchinich, *Science in Russian Culture* (Stanford: Stanford University Press, 1963 and 1970), 1, p. 380 and 2, p. 80.

14. Quoted in *Science in Nineteenth-Century America: A Documentary History*, ed. Nathan Reingold (New York: Hill and Wang, 1964).

15. A. Hunter Dupree, *Science and the Federal Government* (Cambridge, Mass.: Harvard University Press, 1957) and Howard Miller, *Dollars for Research: Science and Its Patrons in Nineteenth-Century America* (Seattle: University of Washington Press, 1970) demonstrate that the societies were heavily dependent on private support.

16. Daniel H Calhoun, *The American Civil Engineer: Origins and Conflict* (Cambridge, Mass : Harvard University Press, 1960).

17. Alexander Dallas Bache, "Address," American Association for the Advancement of Science, *Proceedings* (1851), p. xlii.

18. Sally Gregory Kohlstedt, "The Formation of a National Scientific Community: The American Association for the Advancement of Science, 1848-1860," unpublished dissertation (University of Illinois, 1972), Chapter VI; Daniels, "Process of Professionalization," p. 164, see reference 2.

19. Evidence of attitudes toward the non-professionals and the double standard are evident in the manuscripts reprinted by Nathan Reingold, ed., *Science in Nineteenth-Century America*, see reference 14.

20. For a useful discussion of the concern see *The Papers of Joseph Henry*, 1, ed. Nathan Reingold (Washington, D.C.: the Smithsonian Institution Press, 1973), pp. 342-43.

21. Linda K. Kerber, "Science in the Early Republic: The Society for the Study of Natural Philosophy," *William and Mary Quarterly*, 29 (1972), pp. 279-80.

22. The only histories of the Society were completed by members on anniversary occasions. The best is Thomas T. Bouvé', *Historical Sketch of the Boston Society of Natural History* (Boston, 1880), which simply summarizes year-by-year activities and offers sketches of leading members. See also *The Boston Society of Natural History, 1830-1930* (Boston, 1930). The Boston Museum of Science (BMS) has an extensive collection of the printed and manuscript records of the Boston Society of Natural History (BSNH) including the correspondence files from 1837.

23. BSNH, *Journal*, 1 (1831), p. 11.

24. *Ibid.*, pp. 12-13.

25. Bouvé, *Historical Sketch*, p. 84.

26. *Dictionary of American Biography*, 6, pp. 127-128. The Emerson Manuscripts at the Massachusetts Historical Society are almost entirely personal family correspondence and indicate that Emerson nearly went into medicine because he was so attracted to science. Letter to his father, Harvard, December 12, 1819.

27. *Dictionary of Scientific Biography*, 5 pp. 477-479, The extensive Gould manuscripts at the Houghton Library, Harvard, indicate the ambitious Gould was not always similarly supportive of his rural correspondants, as shown in debates with James G. Anthony of Cincinnati and Charles Baker Adams at Middlebury Colleges; also see a letter from Augustus Mitchell to Gould, Portland, June 25, 1855, BSNH, BMS.

28. Ralph Waldo Emerson, "Thoreau," *Atlantic Monthly*, 10, No. 58 (1862), p. 239.

29. Kenneth W. Cameron, "Emerson, Thoreau and the Society of Natural History," *American Literature*, 24, No. 1 (1952), pp. 21-30. The library journal recording book withdrawals shows that Thoreau was interested not only in zoology and geology but also in the anthropological research of Henry Schoolcraft and Samuel G. Morton.

30. Bouvé, *Historical Sketch*, p. 13.

31. This continuous fear was expressed abroad as well as in the United States. See Hahn, *Anatomy*, p. 235.

32. BSNH, *Proceedings*, 2 (1845), p. 65.

33. J. E. Teschmacher, *Address delivered at the Annual Meeting of the Boston Society of Natural History, Wednesday, May 5, 1841* (Boston, 1841), p. 5.

34. BSNH, *Proceedings*, 2 (1945), pp. 68-70. The Harvard faculty also formed an informal scientific club that was deliberately designed to include those with amateur interests. See A. Hunter Dupree, *Asa Gray, 1810-1880* (Cambridge, Mass.: Harvard University Press, 1859), pp. 121-22.

35. Quoted in John C. Warren, *Address of the Boston Society of Natural History* (Boston, 1853).

36. Henry Shapiro (see Ref. 5 above) argued in his paper on Cincinnati that the establishment of a more professional national group, far from broadening local enthusiasm for science, seemed to undermine the energy of local amateurs by the discouraging comparisons of effort and expertise evident in the new organization.

37. Lawrence Badash, "The Completeness of Nineteenth Century Science," *Isis*, 13, No. 216 (1972), pp. 48-58. Badash presents (based on work with physical scientists) another explanation for the shift of interest away from the theoretical by noting that many scientists believed that the fundamental research was already completed and specialists were now taking problems to the "next decimal place."

38. "Scrapbook, 1866-1871," BSNH MSS, Boston Museum of Science Library [hereafter BMS]. The Scrapbook contains newspaper clippings, principally from *The Christian Register* but occasionally from the *Advertiser*, the *Journal*, and the *Post*.

39. "Council Minutes, 1840-1861," December 7, 1857, BSNH MSS, BMS.

40. Theodore Lyman, James E. White, and Amos Binney to Rogers, n.d., n.p., Rogers MSS, Massachusetts Institute of Technology Archives. A letter declining an invitation is dated April 8, 1860.

41. *Boston Transcript*, March 1, 1872, from "Scrapbook, 1866-1871," BSNH MSS, BMS.

42. *Ibid.*

43. *Boston Society*, p. 46.

44. The series was initiated with thirty-three lectures on mineralogy in the winter of 1871-1872. After 1874 lecture series on botany and zoology as well as physics, chemistry and physical geography were regularly sponsored by the BSNH but financially supported by local citizens. In May of 1876 Professor Shaler reported that the practical teaching was "giving to the public opportunities which are probably unequalled in any other city and its effect, on the advancement of science cannot fail to be felt." BSNH MSS, BMS.

45. BSNH, *Proceedings*, 16 (1871-1872), p. 26. "Qualified scientific workers and lecturers are too few, and too much absorbed by strictly professional duties to act of themselves and directly upon the scholars; they must depend upon the teachers."

46. *Ibid.*, 21 (1880-1882), p. 2.

47. Women had given donations through the years to the BSNH, including Sarah Pratt's bequest of 4000 conchology specimens and $10,000. The minutes record that a woman was

hired in the library in 1864 and that by 1873 Miss Carter and Miss Washburn were working with the Museum curators, identifying specimens and organizing exhibits.

48. "Scrapbook, 1866-1871," BSNH MSS, BMS.

49. March 1, 1876, "Record Book for 1875-1883," BSNH MSS, BMS.

50. *Ibid.*, March 15, 1876.

51. To S. H. Scudder, Cambridge, March 3, 1876, BSNH MSS, BMS.

52. Samuel Kneeland to S. H. Scudder, Institute of Technology, March 21, 1876, BSNH MSS, BMS.

53. Charles Sprague to S. H. Scudder, Boston, March 20, 1876, BSNH MSS, BMS.

54. *Ibid.*

55. Edward Turse to S. H. Scudder, Salem, March 21, 1876, BSNH MSS, BMS.

56. Kneeland to S. H. Scudder, Cambridge, March 3, 1876, BSNH, BMS.

57. Clarence Burrage to S. H. Scudder, Brookline, March 23, 1876, BSNH MSS, BMS.

58. J. J. Ellen Cabot to S. H. Scudder, Brookline, March 23, 1876, BSNH MSS, BMS.

59. J. A. Allen, Cambridge, March 23, 1876 and Charles Sprague, Boston, March 16, 1879; also see March 15, and April 5, 1876, "Manuscript Minutes, 1875-1883," BSNH MSS, BMS.

60. *Ibid.*, April 5, 1876.

61. *Ibid.*, April 19, 1876.

62. *Ibid.*, May 3, 1876

63. *Ibid.*, April 19, 1876. A circular for that meeting noted specifically: "Judging from the limited number of contributions to science hitherto made by women, it is probable that very few women will attain immediate membership, at least for many years to come; and therefore your Committee believes that the sentiments of those who have opposed the unrestricted admission of women to the Society's meetings, will be practically regarded, while at the same time, as is desired by others, women are actually placed upon the same footing as men. At present, one half of the general meetings are open, by invitation, to women; under the proposed plan, probably nearly all would be open, by right, to women who had obtained Associate Membership."

64. "Constitution and By-laws" with signatures from the period between 1864 and 1884. At least forty-two women joined by 1884; some other signatories give only initials and might also be women.

65. A version of this paper was first presented at the Conference on *The Public and the Professional: Critics and Partisans of Science and Medicine in Modern History*, Radcliffe Institute, April 26, 1974. My thanks to Margaret Rossiter, Gerald Holton, and Barbara Rosenkrantz for their comments on an earlier draft. The library staff of the Boston Museum of Science, especially Barbara Wiseman and Edward D. Pierce, were exceptionally cooperative with me in efforts to find the various manuscript materials currently under their care. A grant from the Simmons College Fund for Research has assisted me in gathering materials for a larger study of the amateur tradition.

JOHN A. MOORE

Creationism in California

THERE IS AN ELEMENT of pathos in Santayana's proverb to the effect that those who cannot remember the past are condemned to repeat it. The implication is that knowledge of the past will prevent error in the present; unfortunately, there is very little evidence to support such a hopeful philosophy. The generals and kings whose actions lead to disastrous ends are not unfamiliar with history. Their behavior depends not so much on ignorance of history as on the hope that history is not going to happen to them.

Those evolutionary biologists who engage in cyclical debates with Biblical creationists on the origin and evolution of life are usually aware that these debates have occurred before. For more than a century, there has been a chronic level of activity that, at intervals of roughly a generation, becomes acute or even fulminating. We are in such a period of increased activity today; biologists and creationists are saying essentially what was being said in the mid-nineteenth century by Huxley and Wilberforce, or by Gray and Agassiz.[1] Neither group is convincing the other; indeed, an accord is impossible. The strict creationist rejects the statements and procedures of the scientist because he feels they do not lead to God. The evolutionist rejects the statements and procedures of the creationist because they do not lead to a scientifically acceptable understanding of nature.

Creationists search nature for evidence for conclusions they have already accepted; evolutionists, on the other hand, use observations and experiments on natural phenomena to help them reach their conclusions. Thus the two groups are using wholly divergent and incompatible systems of thought. Yet the lines between the two camps were not always so sharply drawn. The eighteenth and early nineteenth centuries saw little conflict between Biblical statements and the slowly emerging sciences of biology and geology. Indeed, the beautifully adaptive structures and processes of animals and plants were viewed as evidence of the Creator's mode of action, and of His loving concern for the living world. In 1802, William Paley, Archdeacon of Carlisle, codified this point of view in his *Natural Theology; or Evidences of the Existence and Attributes of the Deity, Collected from the Appearances of Nature*.[2] During the 1830's, the eight volumes of the Bridgewater Treatises expanded this theme.[3]

Within decades of the publication in 1859 of *On the Origins of Species*, the terms "evolutionist" and "creationist" began to be taken as synonymous with "scientist" and "fundamentalist," so that by the end of the century most of those in the former camp would have accepted Thomas Huxley's statement that

The only rational course for those who had no other object than the attainment of truth, was to accept "Darwinism" as a working hypothesis, and to see what could be made of it. Either

it would prove its capacity to elucidate the facts of organic life, or it would break down under the strain.[4]

Yet, as Jacques Monod has aptly noted, acceptance of the proposition that "objective knowledge is the *only* authentic source of truth" is of necessity a matter of
ethical choice.[5]

Fundamentalist theologians, possibly less sure of their faith than the more
enlightened nineteenth-century natural theologians, have continued to insist that
Biblical statements about nature really mean what they seem to be saying: for example, that the earth is of recent origin and the organisms of today are essentially
the same as those formed on the fifth and sixth day of creation. It is this fundamentalist strain in religion, so strong among certain Protestant sects, that has led to recent vitriolic conflicts with science. The Darwinian hypothesis, with its strong implications that *Homo sapiens* and all other creatures are the products of
evolutionary change and that the origin of life could be studied as a scientific
problem and need not be accepted as a divine *fait accompli*, was and is anathema
to the fundamentalists.

Although the creationists' antagonism toward organic evolution has not abated,
their strategies for undermining the influence of evolutionary biology in the public
schools have changed. Whereas the main thrust of the creationist movement of the
1920's, which culminated in the Scopes Trial, was to get evolutionary theory out of
the schools, they are now demanding that their own theory be given equal time.

A generation ago (and, to a considerable extent, today), local and state
pressures were sufficient to cause publishers to omit from textbooks topics deemed
objectionable to lay groups. As far as biology books were concerned, human
reproduction and evolution were the only topics likely to offend the sensibilities of
anyone. Thus, biology was usually limited to the study of species like frogs whose
sex life and evolution were hardly likely to be threatening to anyone. Even in cases
where a large fraction of the citizenry preferred to have their children taught these
subjects, objections from a few individuals in a few localities were often sufficient
to cause the books for the entire nation to be censored. The alarm of publishers and
authors at the prospects of losing profit and causing trouble kept the central theory
of biological science out of the biology books. For their part, many teachers then
and now have had to deal with social situations where to espouse evolution is tanta-
mont to professing a disbelief in God—a position unlikely to enhance the prospects
for renewal of a teaching contract.

The book situation began to change in 1960, when a group of scientists from
across the nation, working on the Biological Sciences Curriculum Study, published
three versions of a high school Biology textbook.[6] The BSCS contract with the
publishers contained a critical clause which gave it complete control over the
content of the books. A national organization, unlike an individual author, carried
sufficient clout to insist that evolution be included in its text even though sales
would suffer. So finally not only evolution, but also reproduction, was fully discussed in high school books—although a salesman for one of the three versions of
the text allegedly used as a selling point the fact that *his* version had many fewer
listings in the index under "evolution" than either of the other two. Some school
boards (in the State of Texas, for example) reacted in the time-honored manner and
demanded that the books be altered. But the BSCS maintained the position that

the only basis for change would be scientific error or inefficient pedagogy and refused to alter the books. Subsequent events showed that nearly every objecting school board ended up adopting the books—evolution, sex, and all.

Thus, in the mid-1960's the censoring power of the school boards proved to be less strong than had previously been assumed. Word was spreading that BSCS biology was the "new thing," and there were community pressures on school boards to be up to date, even if a little wicked, rather than behind the times and fully virtuous. Once this situation was understood, nearly every newly published biology book included an explicit discussion of evolution.

Despite the fact that many teachers continued to ignore what was clearly a contentious subject in their communities, many professional biologists interpreted these brave doings as "victory." In 1959, the well-known geneticist H. J. Muller had written a forceful article, "A Hundred Years without Darwinism Are Enough."[7] But now, apparently, biology teaching was ready to emerge from its century in darkness.

But what was perceived as victory has, as we shall see, turned out to be just another movement of a pendulum that oscillates between times of greater and lesser toleration for scientific and rational thought. Conceivably there will always be some who are content with natural explanations of natural phenomena and others who will accept only supernatural explanations, at least for some natural phenomena. In the first century of the Darwinian Age, power has alternated between the two groups. As a scientist, I should like to believe that there has been steady progress toward an acceptance of naturalistic explanations for the phenomena of nature, but I have no hard evidence to support this hope. On the contrary, as a result of my own work on the BSCS books, I have received many letters, even petitions signed by whole classes, asking me to mend my ways. Indeed, there seems to be a marked swing away from public willingness to accept naturalistic explanations. One obvious case in point is the vigor of creationist campaigns to gain "equal time and emphasis" for their point of view in the nation's schools.

The Contemporary Creationist Position

California has been a center of activity for the current creationist movement. The state is the home of some of the most vigorous individuals and organizations promoting creationism, and it is here that these groups have perfected their newer strategies. Since the state's public school textbook sales account for about 10 percent of the total market in the United States, it is obvious that these strategies, if successful, would have a decided effect on science books for the entire nation. California is, in short, a bellweather state. Certainly any success the creationist movement might have there would give courage to the creationist crusaders elsewhere.

The intellectual leadership of the creationist movement in America is provided by two closely related organizations, which have several leaders in common—the Creation Research Society, and the Institute for Creation Research, a division of the Christian Heritage College of San Diego.[8] Since all full or voting members of the Creation Research Society must have earned an M.S. or Ph.D. degree in science, or some equivalent, one can assume that they are all professional people.

Both full and associate members subscribe to the following statement of belief:

1. The Bible is the written Word of God, and because it is inspired throughout, all its assertions are historically and scientifically true in all the original autographs. To the student of nature this means that the account of origins in Genesis is a factual presentation of simple historical truths.

2. All basic types of living things, including man, were made by direct creative acts of God during the Creation Week described in Genesis. Whatever biological changes have occurred since Creation Week have accomplished only changes within the original created kinds.

3. The great Flood described in Genesis, commonly referred to as the Noachian Flood, was a historic event worldwide in its extent and effect.

4. We are an organization of Christian men of science. . . .[9]

Elsewhere, these creationists have stated that the notion of a young earth is part of their creed: "Most creationists believe that the age of the earth can be measured in thousands rather than millions or billions of years."[10] Not surprisingly, the M.S.'s and Ph.D.'s who are able to ascribe to these beliefs are seldom biologists and geologists whose studies relate to the origins and diversity of living creatures. I am not aware of any member of the Creation Research Society who is doing research that any biologist or geologist who is not also a member would regard as pertinent to either evolution or creation. Members are more likely to be professionals in the fields of space science, engineering, physics, mathematics, chemistry, and science education.

There are also a number of individuals active in the creationist movement who are not affiliated with either of these groups. One of the most prominent, politically, is Vernon L. Grose, an engineer with an aerospace background who is Vice President of the Tustin Institute of Technology, a Santa Barbara firm that conducts management seminars. According to Nicholas Wade, Grose feels that the Institute of Creation Research staff are "not fully scientific in their selection of the evidence."[11] Grose is concerned that "schoolchildren, brought up to believe there is a God, are now told in the name of science that God has conclusively been shown to be out of the Picture. I want that to be withdrawn and a neutral or *pro-theistic account to be given.*"[12]

The resurgence of creationism in California has been encouraged by the conservative outlook of the State Board of Education (in 1963 it ruled that evolution was to be taught as theory rather than as scientific fact), which was strongly influenced by Max Rafferty, who was Superintendent of Public Instruction until 1970. Rafferty's outlook is set forth in a booklet called *Guidelines for Moral Instruction in California Schools,* prepared under his direction and issued by the State Department of Education in 1969.[13] The flavor of this document can be gathered from these excerpts:

I always think that America was built upon the Bible and we have as a result the highest civilization the world has known [page 7].

The crisis of our time is that these people [Margaret Mead and others] have not bothered to examine the guides which history and experience offer to us. Their rejection of our traditions begs the questions: Can a child in a school system dedicated by law to the affirmation of a religious and moral heritage be taught to question the substance of that heritage? Can children be taught to judge "right or wrong" as the unsteady product of their individual consciences? Is this not in violation of Education Code Section 13556.5 [formerly Section

7851]? Is it not also in violation of more recent legislation to protect the child's (and parent's) morality from attack by secular Humanists [pages 49-50]?

Recent revelations about the successes of "sensitivity training" in the colleges, and now in the high schools, suggest that those dedicated to this goal, however wellmeaning they may be, are in fact aligned with revolutionary groups acting contrary to public policy; that is, they intend to use the schools to destroy American culture and traditions [page 51].

The teaching of evolution as a part of the religion of Humanism, therefore, is yet another area of concern to parents and teachers alike who wish to abide by the mandates of the laws and of the State Board Resolution that "Christian parents . . . are protected by law against any attempt to destroy or weaken their children's faith in their particular church." In this instance, as with other areas of controversial instruction, it is how the subject is treated by the teachers, what materials the teacher uses that matters. If the origins of man were taught from the point of view of *both* evolutionists and creationists, the purpose of education would be satisfied. By concentrating on only *one* theory and ignoring others, it is tantamount to indoctrination in one special religious viewpoint [page 64].

Thus Rafferty and the authors of *Guidelines* exhort us to return to the perceived morality of the Founding Fathers, a morality they feel is being undermined by progressive education, sex education, behaviorists, antitraditional social scientists, Marxists, evolutionists, and Margaret Mead. As Paul Kurtz, editor of *The Humanist*, has remarked:

The *Guidelines* seem to want to turn the public schools into an extension of the conservative wing of the Republican party at prayer and to break down the constitutional principle of separation of church and state.[14]

The Battle of the Books

From 1965 to 1969, even while Rafferty's *Guidelines* were being prepared, the California State Advisory Committee on Science Education was at work on a very different type of document, eventually published as *Science Framework for California Public Schools Kindergarten—Grades One Through Twelve*.[15] Among the distinguished scientists, educators, and teachers who worked on the book were Wallace R. Brode (American Chemical Society), Jacob Bronowski (Salk Institute), Robert M. Gagné (University of California, Berkeley), Ralph W. Gerard (University of California, Irvine), Paul DeHart Hurd (Stanford), Ray D. Owen (California Institute of Technology), and Henry Rapoport (University of California, Berkeley).

When it was presented to the State Board of Education in the fall of 1969, the Advisory Committee's draft of *Science Framework* provided the creationists with an opportunity for a frontal attack. As its title implied, the draft was intended primarily as a framework rather than as a compendium of detailed suggestions. In Appendix A, the Committee listed seven major conceptual systems, with subdivisions, that "provide a foundation for the understanding of how certain facts are related" and "offer a perspective by means of which future discoveries may be correlated and understood." Conceptual system G was "Units of matter interact," and the subdivision G-2 was "Interdependence and interaction with the environment are universal relationships." Examples of interdependence and interaction were given from the fields of astronomy, physics, chemistry, and biology. The biological examples included mention of the integration and interaction that occur at the levels of the components of cells, entire cells, organisms, populations, communities, and ecosystems. It was noted that some of these interactions are cyclic—such as those involving food, water, and the stages of the individual's life. The final example of interactions was that of evolution, involving both living and nonliving phenomena. The two paragraphs of the original draft that provided focus for the ensuing controversy read as follows:

> Perhaps the best known evolutionary product of interactions are changes which occur in living organisms over long periods of time. From the origin of the first living particle, the evolution of living organisms was probably directed by environmental conditions and the changes occurring in them. A soup of amino acid-like molecules, formed in pools some 3 billion years ago, interacted with oxygen and other elemental constituents of the earth, probably giving rise to the first organization of matter which possesses the properties of life.
>
> Evidence indicates that nearly 2 million living species, and millions of extinct species, are descendants of this early form of life. This diversity among living organisms is the result of natural selection, preserving characteristics which have allowed adaptation to the many kinds of environments on this planet. Long-term adaptation is evolution. Evolution results from mutations and genetic recombinations in the organism which, through natural selection, have produced a more efficient relation with the changing environment than less successful ancestors.[16]

Although much responsibility for education in California is delegated to local school districts, the State Board of Education, whose members are appointed by the Governor, has the responsibility of establishing overall educational policy and of approving textbooks used in the elementary and junior high schools. The State Curriculum Commission, whose members it selects, assists in examining textbooks. The State Superintendent of Public Instruction, an elected official, is Executive Officer of the State Board of Education and was the Chairman of the Curriculum Commission.

When the draft of *Science Framework* was presented to the State Board of Education, several of its members urged that it not be accepted unless "Creation Theory" was given a place in it. At a meeting on November 13, 1969, Vernon Grose, acting in the capacity of a concerned citizen, submitted his objections to the draft. The Board was most sympathetic to Grose's concern and voted unanimously to throw out the two paragraphs on evolution prepared by the Advisory Committee on Science Education and to substitute Grose's ideas about creation and evolution. They were summarized in three remarkable paragraphs.[17]

> All scientific evidence to date concerning the origin of life implies at least a dualism or the necessity to use several theories to fully explain relationships between established data points. This dualism is not unique to this study but is also appropriate in other scientific disciplines, such as the physics of light.
>
> While the Bible and other philosophic treatises also mention creation, science has independently postulated the various theories of creation. Therefore, creation in scientific terms is not a religious or philosophic belief. Also note that creation and evolutionary theories are not necessarily mutual exclusives. Some of the scientific data (e.g., the regular absence of transitional forms) may be best explained by a creation theory, while other data (e.g., transmutation of species) substantiate a process of evolution.
>
> Aristotle proposed a theory of spontaneous generation. In the nineteenth century a concept of natural selection was proposed. This theory rests upon the idea of diversity among living organisms and the influence of the natural environment upon their survival. Fossil records indicate that hundreds of thousands of species of plants and animals have not been able to survive the conditions of a changing environment. More recently, efforts have been made to explain the origin of life in biochemical terms.

The State Advisory Committee on Science Education vigorously opposed this change in its original draft. It finally concluded that it could live with the third paragraph, but regarded the first two as wholly unacceptable. The State Board of Education, however, overrode the scientists and science educators on its own Advisory Committee and on this important issue accepted the opinion of Grose, an individual who could claim no professional expertise either as a scientist or as a science educator.

In rebuttal, on December 4, 1969 the State Advisory Committee on Science Education issued a statement on the yet unpublished *Science Framework* which included the following:

The State Advisory Committee on Science Education takes strong exception to the alteration of content made by the State Board of Education on page 125 of the Committee report, Science Framework for California Public Schools. It also takes exception to the manner in which these changes were introduced.

The Committee, composed of over a dozen competent educators and scientists appointed more than four years ago by the Board, has worked faithfully, diligently and carefully to forge a guide for the science education of the children of this state. Its only recompense would be in contributing to a public service. It called on additional experts for advice, utilized the efforts of national science and education groups that had earlier devoted much time and money to the problem of science education, circulated preliminary drafts of its Framework to teachers and education administration in this state, and held meetings around the state with these groups to solicit criticisms and suggestions. All these contributed to the version officially submitted to the Curriculum Commission (which strongly endorsed it) and to the Board at its September meeting. The Board asked for certain changes, essentially for injecting some extraneous religious or metaphysical views into the document, and the Committee moved as far in the requested direction as its integrity would permit. (Some members felt it had even then gone too far.) Our revised version was considered by the Board on November 13, 1969 and the crucial changes were then excerpted from a document offered by a non-board member, with minimal discussion and without giving Committee members present any opportunity to comment.

The changes, though small in extent, have the effect of entirely undercutting the thrust of the 205 page document. Our concern has been overwhelmingly with the nature of scientific evidence and conclusions and the great care needed to avoid error and dogmatism. The topic of evolution entered almost incidentally; in fact, the content of science in general is underplayed. The Board, by pitting a scientific fact (and theory as to its mechanism) against a particular religious belief as if they are commensurate, has thus offended the very essence of science, if not also that of religion.

The Committee meeting at the San Francisco airport area today (December 4), therefore repudiates the changes introduced by the Board. Of the fifteen members of the Committee, fourteen are present at this meeting and unanimously voted the following resolution. Whatever the response of the Board, we request an opportunity to present our views before it at an early meeting. It is also our intent to make a prompt public release of this full statement.

Resolved: The State Advisory Committee on Science Education requests the California State Board of Education in the following order of preference, to:

(1a) Restore the Framework to the form submitted for its November 13 meeting.

(1b) Restore the Framework as in (1a) but add the statement on creation as a clearly separate item over the name of the Board.

(1c) Incorporate into the Framework as now altered a statement prepared by the Committee, explicitly repudiating the objectionable passage, with reasons. Further, the

Committee states its intent to recommend against teaching materials in science that follow the disputed guide line.

(2)   If the Board refuses to take any of the actions indicated above, the Committee members decline, individually and collectively, to have their names associated with the document in any way. We also suggest that others whose names are mentioned as contributors be offered the opportunity to withdraw their names. . . .[18]

The State Board of Education selected alternative 1c, the one that stopped just short of resulting in the full repudiation of *Science Framework* by its authors. Thus, in its final printed form, issued in 1970, *Science Framework* carried the following disclaimer of the first two paragraphs of the Grose statement:

This statement was prepared by the State Board of Education as an explanation of its position and inserted in this copy in lieu of two sentences which were deleted. This statement does not meet with the approval of the State Advisory Committee on Science Education, nor does its inclusion in this manuscript have the approval of the Committee.[19]

Clearly, the creationists had won the first round.

The initial reaction of the intellectual community to this creationist victory can hardly be described as vigorous. Practitioners of the social sciences and the humanities apparently concluded that the affair had nothing to do with them, a position which most have maintained to this day. The vast majority of natural scientists seemed equally unconcerned; many seemed to regard the controversy as just another example of the lunatic fringe of American life making a public display of its appalling ignorance. But a few, among them the distinguished geneticist-evolutionist G. Ledyard Stebbins of the University of California (Davis), and William V. Mayer, the Director of the Biological Sciences Curriculum Committee, thought otherwise. They and Jerry P. Lightner, Executive Director of The National Association of Biology Teachers, saw in the creationist challenge serious threats to science and education and sought to arouse the scientific and educational communities.

And they were serious. Had *Science Framework* been just another governmental report to be filed and forgotten, it might have been sufficient to give the creationists' insertion a condescending smile and forget about it. But the document represented official state policy on science education, and provided the basis for screening all science textbooks to be purchased by the state for use in its elementary and junior high schools (senior high schools select their own books). Therefore, the next critical stage of the battle centered on the selection of books. The procedure is complex. The State Board of Education invites publishers to submit books for consideration. These books are then referred to the State Curriculum Commission for evaluation, partly on the basis of whether they comply with the criteria set forth in the framework for the discipline. Thus, in the present case, if the books discussed biological evolution, they would also have to discuss creationism.

The State Curriculum Commission submits its list of acceptable textbooks to the State Board of Education. The Board considers the recommendations and tentatively adopts certain books, which are then made available to local school districts and interested citizens for further review. After this, the Board makes final decisions, and the local school districts can make their selection from among the approved books.

Books were screened approximately every five years; 1972 was the year when elementary science books were to be considered. Several of the biology books submitted for consideration were of the creationist variety. Some of these, apparently including those most favored by the creationists on the State Board of Education, were not selected by the Curriculum Commission. At its meeting on May 11-12 1972, however, the Board restored these rejected books to the list to be sent to the local school districts. Shortly thereafter the State Curriculum Commission was dissolved by the Board and replaced by a new Curriculum Development and Supplemental Materials Commission (CDSMC). The creationists, including Vernon Grose, were well represented on the new commission. In contrast with the original commission, it included but one lone professional scientist, Junji Kumamoto, a chemist at the University of California (Riverside). Kumamoto became the Chairperson of the Science Subcommittee, which, like its parent committee, included a strong contingent of creationists. The subcommittee was charged with preparing a list of acceptable elementary science books, which would be presented first to the CDSMC as a whole, and then to the State Board of Education.

It now appears that Kumamoto, more than anyone else, was responsible for keeping creationist theory out of the science books used in the California schools. He was forceful and effective in arguing that science books should restrict themselves to dealing with phenomena that can be studied by the usual procedures of science and that, by no stretch of the imagination, could the account of creation given in Genesis conform to that stipulation.

The Scientists Wake Up

By the summer of 1972, the scientific community was finally becoming aroused. The goings-on in California had been given wide publicity in the mass media and it was clear that the creationists were not simply going to fade away. In the fall of that year, the State Board of Education was deluged with resolutions requesting that it not mandate the teaching of creation as a scientific theory. The first of these resolutions was passed by the Commission on Science Education of the American Association for the Advancement of Science (AAAS). Shortly thereafter, statements were made by the National Academy of Sciences, the Board of Directors and the Council of the AAAS, the Board of Directors of the American Chemical Society, the American Association of Physics Teachers, the Council of the Academic Senate of the University of California, the nineteen Nobel Laureates living in California, the American Institute of Biological Sciences, and the Society of Vertebrate Paleontology, among others. In addition, many individuals wrote directly to the State Board of Education and to the Chairperson of the CDSMC. The National Association of Biology Teachers and its vigorous director Jerry P. Lightner prepared to challenge the Board in the courts.

The bulk of the resolutions and letters reached the State Board during the autumn months when it was making its final decisions about the science books. It is difficult to evaluate their effect on the actions of the Board. Some think that they were most useful; others that they may merely have antagonized the Board members. In any event, the Board could no longer continue to believe, as apparently some of its members had, that professional scientists who had studied the origins and diversity of life were seriously divided on the question of whether the biological theory of evolution or the Biblical account of creation was more valid.

In the maneuvering that led to the final decision, the Board was divided. In fact the forces were so equally divided that neither side could prevail; it could merely prevent the other side from prevailing. Normally the Board has ten members, but during these deliberations there was one vacancy. According to the rules that govern the Board's decisions, six votes are required to carry a motion. On most of the votes, the score was five for the creationist point of view and four against. Thus, a creationist motion to demand that the books discuss both evolution and creation would fail for lack of a sixth vote. Similarly, a motion to prohibit the teaching of creation as a scientific theory would also fail—by two votes.

During the summer and fall of 1972, there had been considerable public interest in the controversy and the press had carried many articles, editorials, and letters to the editor. With the approach of the public hearings by the State Board of Education, scheduled for November, the two sides marshalled their forces for the fray. It was rumored that no less a personage than Wernher von Braun was to appear for the creationists. He didn't, but he presented his views to the Board in a letter to Grose; his position was basically the old argument that, since there is design in the universe, there must have been a Designer. The scientists, of course, also had their say, but probably those most effective in urging the Board not to mandate the teaching of creationism as science were important representatives of Catholic, Jewish, Protestant, and Buddhist religions. These individuals made it abundantly clear that they did not regard the controversy as one between religion and science but as one between fundamentalism and science.

For example, the Very Rev. C. Julian Bartlett, Dean of Grace Cathedral in San Francisco stated:

You are fully aware that the creation myth-story set forth in the Book of Genesis was for many centuries considered by Christians and Jews alike as the reliable account, quite literally, of the origin of our physical environment and of the various forms of life, of whatever nature. That Biblical myth-story was but one of many such which were developed by primitive religions. Over 100 years ago modern science began to dismantle the superstructure of religious myth-stories of origins, and of the Genesis story in particular, by means of scientific investigation. In so doing, science rendered Biblical religion an inestimable service in that religion was thereby enabled to recover a simple truth about the Book of Genesis: i.e., that it is a religious and therefore theological document and not a scientific treatise.[20]

Furthermore, the Dean stated, "If at any time, *any* theological doctrine should be proven incorrect under the impact of scientific *knowledge,* I shall discard that theological doctrine." He requested that the Board delete the controversial Grose paragraphs in the *Science Framework,* paragraphs which, if allowed to stand, would be "incredible, appalling and preposterous."

The Rev. Hogen Fujimoto, speaking for the Buddhist Churches of America, stated, "It is my firm conviction that the school is not the proper place to teach Divine Creation. It belongs in the church or in the family."[21]

Rabbi Amiel Wohl, of Congregation B'nai Israel in Sacramento, pointed out that in Judaism,

we would never purport to place the Creation Epic as a scientific theory of creation. We understand it as a theological statement. . . . It would be confusing to call science, religion, or religion, science, or to confuse the two in the study of pure inquiry. . . . From the earliest period foreward, our Jewish faith has never been weakened or threatened by the new knowledge. . . . In the public domain and public schools, we have to be very careful to avoid any particular group sectarian ideas.[22]

Rita R. Semel, speaking for the San Francisco Conference on Religion, Race and Social Concerns, felt that the proposal to teach creationism was "based on a profound misunderstanding of the respective roles of science and religion," and that the Board should "choose science textbooks which deal with science and which do not venture into the fields of theology and religion."[23]

Dr. Conrad Bonifazi, Professor of Philosophy and Religion at the Pacific School of Religion in Berkeley, summed up the situation in a letter to Kumamoto.

Broadly speaking, then, the situation is thus: an extremely conservative wing of Christian sectarianism, which has little or no repute in the world of theological scholarship, adheres to a literal interpretation of the Bible, and is therefore committed to saying that evolution contradicts the biblical account of creation. Its belief in the "infallibility" of the Bible does not even permit it to recognize that in Genesis itself there are two accounts of creation, each differing from the other in background and in content. It is also true that the major *denominations* of Protestantism and the Roman Catholic Church in the United States recognize and condone the teaching of evolution in the disciplines of natural science. These denominations represent a large majority of Christians in this country who raise no objection to the teaching of evolution in schools.[24]

The effect the public hearings had on the Board is uncertain. The basic lineup seemed to remain much the same, but possibly the pro-creationist group became less sure of complete success, for during December and January a compromise of sorts was reached. The immediate issue before the Board was whether or not to accept the list of books that had been screened by the Curriculum Development and Supplemental Materials Commission. There was not one overtly creationist text on the list. (The publishers, to their credit, had so far largely ignored the creationist stipulations of *Science Framework*.) The Board had been assured by its chief counsel that it was legally bound only to submit books to its Curriculum Commission for evaluation, and not to follow the commission's recommendations (though other staff lawyers disagreed). The Board might also have demanded that the publishers of the listed books insert material on creation to conform with the statement in *Science Framework* specifying that science books must deal with both evolution and special creation. But again the creationists could not quite muster the votes to enable them to follow either of these courses of action.

The Board finally approved the list, but it accompanied its approval with several decisions which indicate that the affair is far from closed. Darwinian evolution was to be identified in the textbooks as "theory" to indicate that it had a speculative nature. In discussions of the origins of life it was required that "dogmatism be changed to conditional statements where speculation is offered as explanation for origins." Furthermore, the books were to emphasize that science may deal with the "how" of origins of life but it has nothing to say about the "ultimate causes." The Board then appointed a committee, heavily weighted with creationists, to see that authors and publishers made the desired changes.

In addition, the Board decided that, although science books will not be required to treat creation as a science, social studies textbooks will have to state that the biological theory of evolution is not the only way that people have sought to explain the origin and diversity of life—that there are other ways, including the accounts of creation given in Genesis. Thus the hot potato has passed to our colleagues in the social sciences, who may now view the controversy with more concern than they have exhibited heretofore. All that is certain at the moment is that there will be a fight.

Meanwhile, another group has entered the fray. Shortly after the open hearings of the State Board of Education in November, 1972, an organization styling itself "Creation Evolution Equality" was formed to pressure the California State Legislature into allotting equal time in the schools for creation and Darwinian evolution. Floyd L. Wakefield of South Gate, one of the assemblymen who will sponsor a bill to achieve this goal, stated, "If the legislature is fair, it will put creation and evolution on equal footing."[25] If his efforts fail in the legislature, he may sponsor a state-wide initiative to allow the voters to decide the question.

The question of whether or not creationism must be taught in science classes in California may now be answered—at least for a time. Kumamoto's Science Committee of the Curriculum Developmental and Supplemental Materials Commission revised the controversial paragraphs of the *Science Framework* and presented them to the State Board of Education. The Board, at its meeting on March 14, 1974, approved the following wording:

Interactions between organisms and their environment produce changes in both. Changes in the environment are readily demonstrable on a short-term basis (i.e., over the period of recorded history circa 5,000 years) and have been inferred from geologic evidence over a greatly extended period of time (billions of years), although the further back we go, the less certain we can be. Prehistoric processes were not observed, and replication is difficult. During the past century and a half the earth's crust and the fossils preserved in it have been studied intensively by scientists. Fossil evidence shows that organisms populating the earth have not always been structurally the same. The differences are consistent with the theory that anatomical changes have taken place through time.

This process of change through time is termed evolution. The Darwinian theory of organic evolution postulates a genetic basis for the biological development of complex forms of life in the past and present, and the changes noted through time.

The concepts which are the basic foundation for this theory are: (1) that inheritable variations exist among members of a population of like organisms; and (2) that differential successful reproduction (i.e., survival) is occasioned by the composite of environmental factors impinging generation after generation upon the population. The theory is used to explain the many similarities and differences which exist between diverse kinds of organisms.

The theory of organic evolution, its limitations notwithstanding, provides a structural framework upon which many seemingly unrelated observations can be brought into more meaningful relationships. Biologists also have developed hypotheses concerning the origination of life from nonliving matter (e.g., the heterotroph hypothesis) derived from experiments and observations.

Philosophic and religious considerations pertaining to the origin, meaning, and values of life are not within the realm of science, because they cannot be analyzed or measured by the present methods of science.

The Board is still considering whether or not it will require social studies classes to include creationism and evolution as parallel ways of viewing the natural world.

Creationism and the Public Understanding of Science

Surely one may ask, "Is there no place in the curriculum for a consideration of the diverse explanations that various cultures have given for their origins?" After all, these myths are an important part of the cultural history of all peoples. Indeed, they have been a part of our human heritage for millennia; Darwinism for little more than a century. Why shouldn't classroom students be given a review of the data which led to the rejection of special creation as a useful hypothesis for explaining the origin and diversity of living creatures? By the same token, might it not be useful for a young student to understand why most of us no longer believe that the earth is flat or that the heavenly bodies encircle the earth in awesome Ptolemaic procession?

These are complex questions involving the emotionally charged relations between science and religion which have persisted at least since the time of Bruno and Galileo. But I, for one, have yet to hear of a scientist who has campaigned to have Genesis thrown out of the churches, or who has demanded that preachers give equal time to Darwin and Moses. The reluctance of most scientists to allow biology books to discuss creation at all, even to show why it must be regarded as a useless hypothesis from a scientific point of view, is based on a reluctance to give some teachers an excuse to teach religious doctrine as a part of scientific material. This is not an unfounded fear, the Constitution of the United States notwithstanding, as former Superintendent Rafferty's *Guidelines* suggest. Scientists who have dealt with fundamentalists simply do not trust them; they rather imagine that, if the fundamentalists had the power, they would happily reinstitute an inquisition.

Personally, I find it very difficult to understand why creationists insist on having biology teachers discuss creation as a *scientific* theory. Possibly they regard this as the only possible action available to them. The Supreme Court in the case of *Epperson vs. State of Arkansas* (1968) made it clearly unconstitutional[26] to throw Darwinism out of the classroom. At least until the advent in 1960 of the Biological Sciences Curriculum Study, the strategy of bringing local pressures on teachers, school boards, publishers, and textbook committees often effectively kept evolution out of the classroom. Thus, it might have made some sense, as far as the creationists were concerned, to intensify these time-honored methods. But when they demand that creation, as described in Genesis and elsewhere, be discussed as a scientific theory, one can only conclude that they have not thought out the consequences of their request.

Let us suppose that there were an entirely neutral teacher who, to the fullest possible extent, gave equal time and emphasis to biological evolution and to special creation as competing ways to understand the origins and variations of living organisms. Let us assume also that this teacher dealt with the material in a purely scholarly and scientific manner. The scenario might well develop as follows.

Clearly it would be helpful to the student to understand, at the offset, the different sources of the data. Thus, the teacher would explain that the data in support of evolution derives from the procedures of geologists, anthropologists, biologists, and chemists and, to a lesser extent from those of astronomers, physicists, and mathematicians, and would show how all these types of data support each other. The teacher would then point out that the basic data for the Judaeo-Christian creation theory can be found in Genesis, and that many peoples have similar oral and written traditions. Some of them teach that mankind was formed out of clay in a world already inhabited by animals and plants. Some teach that he arose from the blood of a hero or god. Others make no suggestions about creation, and no predictions of an end. And still others teach that there are cycles of creation and annihilation. Furthermore, in the Old Testament, there are two conflicting accounts of creation known to biblical scholars as the J (for Yahweh) and the P (for Priestly) traditions—the former much older than the latter.

At some stage, as he listens to this evidence, a student is bound to ask, "Which of these contradictory accounts of creation is true?" The teacher will then have to explain that belief in any creation myth is based on faith, not on scientific evidence. Furthermore, in all honesty, the students would have to be told that essentially all scientists who have studied these matters have rejected all creation myths as useless in obtaining a verifiable understanding of the natural world. The teacher would also have to note that even most Biblical scholars long ago stopped looking upon Genesis as a source for scientific knowledge. Most would agree with Theodor Gaster, who, when discussing the cosmogony of Genesis, writes:

Hebrew thinking on the subject was mythopoetic rather than intellectual, issuing more out of the imaginative fancy than out of logical inference or disciplined inquiry. Accordingly all efforts to reconcile biblical cosmogony and modern science rest, in the last analysis, on a fundamental misunderstanding of its purport and intent and on a naive confusion between two distinct forms of mental activity.[27]

If the class devotes equal time and emphasis to the two points of view, they will have ample opportunity to investigate the claims of some creationists that there is scientific evidence for their theory. Thus, for example, they would consider the evidence for the creationist contention that, barring sampling errors, the fossils of all geological ages are essentially the same, as well as for their contention that the earth is very young. Inevitably, their study would show that overwhelming evidence from a variety of scientific disciplines indicates that neither of these contentions is even remotely plausible.

But surely the most telling argument of all has nothing to do with details about the age of the earth or the differences in its inhabitants of bygone eras. If a statement is to be treated as a scientific theory, one must follow scientific procedures to reach it. Basic to the whole scientific enterprise is a refusal to use supernatural forces to explain natural phenomena. Thus, the fact that creationists base their notions on the supernatural acts described in Genesis or elsewhere, destroys any pretext that creationism is scientific. By definition, no theory based on supernatural phenomena can be a part of modern science. The creationists are not dealing with science if they accept an "ultimate cause" fully able to suspend the laws of science.

In fact, this point might seem so elementary as to make it difficult to understand how the creationists themselves are able to believe that their theory could be treated as scientific, any more than an anger-of-the-gods theory of disease could be put in the same class of statements as theories of disease based on biological phenomena. Yet, as the data presented by Etzioni and Nunn elsewhere in this volume clearly indicate, an individual's toleration for the procedures and conclusions of modern science is a highly complex function of many psychological, sociological, and economic factors. Indeed, their data suggest that at different times the same individual manifests various degrees of toleration for rational explanation.

In terms of the public understanding of science, our study of creationism in California supports the view of many thoughtful educators that, over a century after Darwin published his theory, the triumph of evolutionary biology is still more apparent than real. Few people (including our colleagues in the social sciences and humanities) seem to understand clearly what is at stake. The sad fact is that most individuals who "believe" in evolution do so because they believe that scientists who have studied the question objectively believe in evolution. And, as the events in California suggest, it is quite possible to persuade such individuals that creationism and the evolution theory are equally useful ways of describing nature. Those swayed by such an argument demonstrate either a fundamental ignorance of the nature of science or a distaste for what it stands for.

By the same token, many people who are hostile to science for one reason or another might well find the arguments of the creationists appealing—if only because the creationists are challenging the monopoly on truth which such people allege that some mythical scientific "establishment" claims. Max Rafferty's *Guidelines* are a case in point. Duane Gish of the Institute for Creation Research also made this point succinctly when he stated that "The authoritarianism of the medieval church has been replaced by the authoritarianism of rationalistic materialism."[28]

As a biologist and educator, I would like to believe that the evidence in favor of evolution is so persuasive that any rational person must agree with Huxley that the acceptance of "Darwinism" as a working hypothesis is "the only rational course for those who have no other object than the attainment of truth."[29] Yet, if educated people—including high-school biology teachers—are prepared to accept or reject evolution according to the degree of their faith in scientists, then clearly we have not been performing our tasks as well as we might have. It may be true that the creationists could not carry the day if they were restricted to the procedures of science, but this supposition should not give us too much comfort. Few human actions are based entirely on scientific procedures. Neither can we hope to neutralize the creationists by legal means alone. If, as Jacques Monod suggests, we believe that "objective knowledge" is "the *only* authentic source of truth," then, we must demonstrate to the public the reasons why we consider such knowledge superior to any other.

REFERENCES

1. For a review of the early creationist controversy see C. C. Gillispie, *Genesis and Geology* (Cambridge, Mass.: Harvard University Press, 1951; rev. ed., New York: Harper and Row, 1959). The earliest American debates on evolution were held between Louis Agassiz and Asa Gray at the American Academy of Arts and Sciences in 1859, and have been reviewed by A. H. Dupree, "The First Darwinian Debate in America: Gray Versus Agassiz," *Daedalus*, 88 (Summer 1959) pp. 560-569.

2. William Paley, *Natural Theology; or Evidences of the Existence and Attributes of the Deity, Collected From the Appearances of Nature* (London, 1802).

3. The Bridgewater Treatises were published in London during the 1830's. The authors and abbreviated titles are:
Thomas Chalmers, *The Adaptation of External Nature to the Moral and Intellectual Condition of Man.*

John Kidd, *The Adaptation of External Nature to the Physical Condition of Man.*
William Whewell, *Astronomy and General Physics Considered with Reference to Natural Theology.*
Charles Bell, *The Hand, Its Mechanisms and Vital Endowments as Evincing Design.*
Peter Mark Roget, *Animal and Vegetable Physiology Considered with Reference to Natural Theology.*   ⚹
William Buckland, *Geology and Mineralogy Considered with Reference to Natural Philosophy.*
William Kirby, *The Habits and Instincts of Animals with Reference to Natural Theology.*
William Prout, *Chemistry, Meterology, and the Function of Digestion, Considered with Reference to Natural Theology.*

4. T. H. Huxley, "On the Reception of the Origin of the Species," *The Life and Letters of Charles Darwin including Autobiographical Chapter,* ed. Francis Darwin (London: John Murray, 1888), Vol. 2, p. 198.

5. Jacques Monod, *Chance and Necessity: An Essay on the Natural Philosophy of Modern Biology* (New York: Alfred Knopf, 1971), p. 169.

6. *Biological Science: An Inquiry into Life* (New York: Harcourt, Brace & World, 1963, 1968, 1973); *Biological Science: Molecules to Man* (Boston: Houghton Mifflin, 1963, 1968, 1973); *High School Biology* (Chicago: Rand McNally, 1963, 1968, 1973).

7. H. J. Muller, "One Hundred Years Without Darwinism Are Enough," *School Sciences and Mathematics,* 59 (March 1959), p. 304.

8. The Creation Research Society was organized in 1963. Its first president was the horticulturist Dr. Walter E. Lammerts. Dr. Henry Morris, a hydraulics engineer, was president for six years. It has approximately 400 voting members and 1200 associate members. It publishes a quarterly and a high school biology textbook, **Biology: A Search for Order in Complexity, ed. John N. Moore and Harold S. Slusher (Grand Rapids, Mich.: Zondervan Publishing Co., 1970).**
    The Institute for Creation Research is part of Christian Heritage College, near San Diego, California. Dr. Morris played a key role in founding the Institute and the College in 1970. Dr. Morris is the Director of the Institute and Dr. Duane T. Gish, who has a Ph.D. in biochemistry from Berkeley, is the Associate Director. Gish and Morris devote a very large amount of their time lecturing on creation and against evolution to church and college groups in the United States and throughout the world. These two, plus Dr. John N. Moore, a philosopher and professor of Natural Science at Michigan State University, are the most visible and vigorous creationists in the land.

9. Excerpted from a pamphlet entitled *Creation Research Society* that gives the history, activities, officers, publications, and requirements for membership.

10. The quote is from the creationist textbook, *Biology: A Search for Order in Complexity,* p. 416. In addition, Duane T. Gish has described the "creation model" in an article published in the *American Biology Teacher,* 35 (March 1973), pp. 132-140.

11. Nicholas Wade, "Creationists and Evolutionists: Confrontation in California," *Science,* 178 (1972), p. 727.

12. *Ibid.*

13. Max Rafferty, Superintendent of Public Instruction, *Guidelines for Moral Instruction in California. A Report Accepted by the State Board of Education, May 9, 1969* (Sacramento: California State Department of Education, 1969).

14. Paul Kurtz, *The Humanist,* 29 (September-October 1969), p. 1.

15. *Science Framework for California Public Schools. Kindergarten—Grades One through Twelve,* prepared by the California State Advisory Committee on Science Education and adopted by the California State Board of Education (Sacramento: California State Department of Education, 1970).

16. This material is quoted from the mimeographed draft of *Science Framework* that the State Advisory Committee presented to the Board.

17. *Science Framework*, page 106.

18. Mimeographed statement.

19. *Science Framework*, page 106.

20-23. These excerpts are taken from the written statements that the individuals presented to the State Board of Education. Each person who desired to speak at the public hearings was required to submit a prepared statement.

24. Letter dated November 2, 1972.

25. UPI report, *Press* (Riverside, Ca.), March 30, 1973.

26. *New York Times*, November 13, 1968.

27. *The Interpreter's Dictionary of the Bible*, ed. George Arthur Buttrick *et al.*, Vol. 1 (New York: Abingdon Press, 1962), pp. 702-709.

28. Wade, "Creationists and Evolutionists," p. 728.

29. Huxley, "On the Reception of the Origin of the Species."

DOROTHY NELKIN

Science or Scripture: The Politics of "Equal Time"*

THE PRESSURES of Biblical Creationism on the teaching of science have exercised remarkable influence on state legislatures and school boards throughout the country in the last few years. In a dozen states, "creationists" have been demanding that public schools balance the teaching of evolution with alternate explanations of man's origins.[1] They base their demands for "equal time" on the belief that *Genesis* presents a viable alternate theory of origins capable of evaluation by scientific procedures. The activities of these creationists must be understood in the context of more widely-based concerns with the teaching of science in public schools, reflected, for example, in the textbook controversy in Kanawha County, West Virginia, and in the protest against *Man: a Course of Study*, a federally-funded social science course for junior high schools.

While religion clearly sustains many of the groups who continue to oppose the teaching of particular scientific theories, the issue extends well beyond the old conflict between science and religion. One must ask why in this science-dominated age the old resistance to evolution theory has gathered new momentum. How have small groups of "believers" been able to intrude their ideologies into educational establishments? What are the issues that have forced public recognition of concerns that have long been ignored as merely the rumblings of marginal groups of religious fundamentalists?

Opposition to the teaching of evolution is taking place at several levels. The core activists call themselves "scientific creationists"; and are often trained as engineers or physical scientists, working primarily out of "research centers" and Bible colleges. They present themselves as scientists claiming to be engaged, not in a controversy between religion and science, but in a debate about the methodological validity of two scientific theories. Their organizations and activities reflect the image of science: titles and credentials are "proof" of legitimacy; research projects are designed to examine their hypotheses.

Another level of activity is represented by an organization of religious scientists, the American Scientific Affiliation, and by many individuals who share creationist beliefs but remain ambivalent about including these beliefs in biology courses. They argue that biology teaching is "dogmatic" and "exclusive," and they seek to effect a "more just and balanced approach." Finally, there is an anonymous but large group of people who are the financial, political and social base for the more active creationists. These are mainly followers of tradi-

* Editor's Note: Dorothy Nelkin's essay treats the "Creationism" controversy from a different perspective from that of the previous essay.

G. Holton and W. A. Blanpied (eds.), *Science and its Public*, 209–227. All Rights Reserved.
This article Copyright © 1976 by D. Reidel Publishing Company, Dordrecht-Holland.

tional fundamentalist ministries, but they also include people who perceive science as the source of a damaging materialist culture and the cause of a decline in public morality.

Creationists try to influence educational policy by working through local school boards, state curriculum committees and state legislatures. They have met with some success. In California the State Board of Education's *Science Framework*, developed in 1969 to guide the selection of biology textbooks for public schools, recommended teaching both evolution and creation as alternative viable scientific explanations for the origin of life.[2] Textbooks used in most parts of Texas make no mention of evolution, and in Tennessee legislation requiring "equal time" for creation theory in biology courses was actually passed in 1973.

California creationists eventually failed to implement the teaching of creation in public schools,[3] and even in Tennessee, the law was struck down as unconstitutional in 1975.[4] There remains, however, a solid group of fundamentalists who continue to oppose what they feel is a biased scientific explanation of the origin of life, and one that violates their religious beliefs and their First Amendment rights. Opposition to the teaching of evolution, for example, is the major theme in the six and a half million copies of *Awake* and the *Watchtower* distributed weekly by Jehovah's Witnesses, and in numerous tracts published by other missionizing sects.[5] The teaching of evolution was one of the concerns in Kanawha County, the seat of West Virginia's state government, where for five long and violent months in 1974 citizens protested against the selection of textbooks by professional educators in the public schools. There the reaction against "godless" teaching in the schools involved pickets and strikes, the closing of mines employing 5,000 workers and the fire bombing of school buildings.[6] Reflecting local commitments, the Board of Education in Kanawha County approved a district-wide adoption of creationist textbooks.

In 1972, William Willoughby, the religious editor of the *Washington Evening Star*, sued H. Guyford Stever, director of the National Science Foundation (NSF) "in the interest of 40 million evangelistic Christians in the United States" for using tax money to support education that violated religious beliefs.[7] The NSF had helped to finance the development of the Biological Sciences Curriculum Study, a series of textbooks that modernized the public school teaching of biology and included extensive discussion of evolution. While this case was dismissed by the courts, the issue of NSF support of science courses emerged once more in 1975 with the reaction against *Man: a Course of Study* (MACOS), a social science course used in about 1700 schools in 47 states. NSF had provided financial support for the development, dissemination, and implementation of the MACOS curriculum by the Educational Development Center of Cambridge, Mass. MACOS is explicitly based on assumptions about the relationship of man to animals. These assumptions and their evolutionary implications so provoked opposition that in 1975 Congressman J. Conlan (Arizona) introduced an amendment to the NSF appropriations bill that would prevent the NSF from funding secondary school courses altogether. The bill only narrowly failed to pass (by a vote of 215 to 196).[8]

To most biologists it seems preposterous that anti-evolutionists could still persist in their opposition to one of the most firmly established generalizations of modern science. They have labelled those who questioned the validity or limits of modern science as "crackpot", "ignorant", "irrational". But these explanations hardly shed light on the social and political roots that sustain the objections to science teaching—objections that seem to grow with the increased popularization and improved dissemination of scientific information. Objections are especially vehement as scientific research itself focuses more directly on human behavior, and they can be found among diverse groups ranging from rural folk in the mining regions of Tennessee to the middle-class, educated creationists of southern California, where membership in the Creation Research Society requires an advanced degree in natural sciences.[9]

Analyzing the various groups who are questioning the teaching of evolution in public school, one is struck by three themes that pervade their discussions and demands: 1) a decline in technological optimism; 2) a challenge to the authority represented by scientific "dogmatism" and the increasing professional determination of school curriculum; and 3) a defense of pluralist and egalitarian values that appear threatened by science, especially as it touches on human behavior. Following discussion of these themes, we will inquire into the problems of communicating science that are suggested by the persistent conflict over the teaching of evolution. This paper will not deal with the creationists' professional competence, nor with the quality of their "science." For the central issue is not the validity of ideas that clearly originate in religious fundamentalism, but the social basis of their influence in a "scientific" age. This analysis will, therefore, examine the political influence of creationism as reflecting the convergence of widely appreciated social concerns with misconceptions about science, and suggest how these are perpetuated by basic problems of communication.

The Social Concerns of Creationists

The Decline of Technological Optimism: The hostility to scientific rationality evident in the demands of creationists reflect considerable skepticism about the social effects of technology, and discomfort with the uncertainties and disruptions that are increasingly evident in a technological society.[10] The creationists have been most active in California, a state which experienced extraordinary economic fluctuations associated largely with science-based industry. Indeed, many creationists are engineers once employed in the aerospace industry.

Passing from the extremes of technological optimism, California came to witness the fruits of disillusion. Creationists associate hippies, campus revolts, drugs and environmental problems with the liberalism of a technological age. They talk about a decline in moral and religious values caused by the dominance of scientific perspectives. Vernon Grose, author of the section in the Board of Education's *Science Framework* that suggested the inclusion of creation theory in biology textbooks, associates evolution theory with "a campaign of secularization in a scientific, materialist society—a campaign to totally neutralize religious convictions, to destroy any concept of absolute moral values, to deny any racial differences, to mix all ethnic groups in cookbook proportions and finally the latest—the destruction of the distinction between male and female." All this, he claims, is nonsense. Differences were created by design; the source of our problems is the secularization of religious belief.[11]. Creationists also dwell on the misuse of science:

It seems to me that man is not getting better, but is developing more diabolical ways of hurting his fellow man. Moreover, because of his ineptness and his inability to see all of the implications of his actions, he misuses many of his scientific accomplishments ... he is hardly in a position to guide evolution.[12]

Two aspects of evolution theory are especially disturbing to creationists; the relationship between animals and man, and the implication of relativism. They argue that emphasis on the genetic similarity between man and animals is a socially dangerous concept that may encourage animal-like behavior:

If man is an evolved animal, then the morals of the barnyard or jungle are more natural ... than the artificially imposed restrictions of premarital chastity and marital fidelity. Instead of monogamy, why not promiscuity and polygamy? ... Self preservation is the first law of nature; only the fittest will survive. Be the cock-of-the-walk and the king-of-the-mountain. Eat, drink and be merry, for life is short and that's the end. So says evolution.[13]

One woman blamed the "streaking" craze on the theory of evolution. "If young people are taught they are animals long enough, they'll soon begin to act like them."[14]

Similar concerns underlie the objections to the MACOS program which uses examples of animal behavior to develop concepts about the nature of man. MACOS traces the life cycle of the salmon to suggest the role of parental protection in human life; the adaptive behavior patterns of gulls to suggest the relation of adaptation to survival; and baboon behavior to illustrate facets of interpersonal relationships. The premises underlying such examples created such antagonism that the NSF felt forced to place a moratorium on further dissemination of MACOS materials, and undertook a major review of its pre-college science curriculum activities.[15]

Creationists further object to the theory of evolution as "a religion of relativism" that "denies absolute standards." One is told that "it erodes everything we hold dear in this country." "Belief in evolution is morally destructive and harmful to family life." "Already we condone fetucide and abortion and devalue human life, all in the name of biological principles."[16] According to creationists, the concepts of relativism implicit in evolution theory also serve as a license to exploit nature, fostering the self-centered behavior that is responsible for environmental problems. Recognizing man as "God's steward" would help him to understand and live in harmony with nature—to preserve given forms.[17] The creationists' arguments have thus touched on some widely-held concerns raised about the social implications of modern science and technology.

Challenges to Authority: Another major theme in the creationist literature is resentment of professional authority. This is expressed in attacks on scientific dogmatism and in efforts to strengthen local control over the educational establishment. Creationists talk of "arrogance" and "absence of humility" among scientists.[18] A sympathetic journalist writes of his great joy "to see science humbled," to see a break in "the monopoly of truth."[19] Creationist Vernon

Grose observes in a speech to the California Board of Education that "after all, scientists put on their trousers one leg at a time every morning, just like the rest of the world." [20] And a Jehovah's Witness writes about the "arrogant authoritarianism" required by evolutionists to "sustain what they cannot prove" [21]

The California Board of Education voted against including creationist beliefs in biology textbooks, but they did appoint a committee to remove dogmatic statements from textbooks and to clearly indicate the conditional nature of the theory of evolution. This committee of four included two members of the American Scientific Affiliation and two school board members who had accepted the revised *Science Framework*. The committee chairman announced his intention to be as much on guard against "the religion of science" as he would be against other religious incursions, and he formulated the problem as "neither creation *vs.* evolution, nor design *vs.* chance, but the evidence of a Supreme Being, an issue that lies beyond the limits of science." [22] The examples in Table 1 suggest the nature of the subsequent textbook revisions approved by the Board of Education.

The creationists' concern with authority is also expressed in their demands for local control and their resentment of professionals who determine the content of school texts. The creationist movement in California was started in 1963 by two mothers from Orange County who were concerned that parents were losing control of the values taught to their children. A Texas organization has called for public participation in the selection of textbook material. "Unless local people take an active voice in assisting the authorized units of government in the program of selecting textbooks, the selection will continue to deteriorate." [22] Particularly irritating to such groups is the role of the NSF in supporting the development of textbooks. It was this "federal take-over of education" that provoked the 1972 lawsuit against NSF for its support of the BSCS materials and that later aroused special concern in the MACOS dispute. Congressman J. Conlan from Arizona, for example, reported the indignation of his constituents about

the insidious attempt to impose particular school courses...on local school districts, using the power and financial resources of the federal government to set up a network of educator lobbyists to control education throughout America...We Americans place a high value on local autonomy. Local school boards, reflecting the prevailing social norms of the community, should be the final arbiter of curriculum development. [23]

Similarly, Kanawha people wanted "to get the government down to where they'll listen to us little old hillbillies." [24] Education was a community issue, and the powerful professionals were not to be trusted. [25]

The system of public education has been one of the last grass roots institutions in America. School systems have traditionally been decentralized, run by local school boards composed of elected citizens (i.e., non-professionals). There has, however, been a gradual erosion of local control through court decisions, reliance on non-local funds, mergers of school districts, and the general trend towards professionalism in education. Those who retain the expectation of local control over educational policy are especially threatened by the growing power of state-wide curriculum committees and departments of education. Some claim that the professionalization and standardization of education favors only

Table 1

Changes in Biology Text Books Recommended by the California Board of Education

Changes in Definitions

Original Version	Changed Version
Science is the total knowledge of facts and principles that govern our lives, the world, and everything in it, and the universe of which the world is just a part	Science is one way of discovering and interpreting the facts and principles that govern our lives, the world and everything in it, and the universe of which the world is just a part. The scientific way limits itself to natural causes and to descriptions that can be contradicted, at least in principle, by experimental investigation.
Evolution is a central explanatory hypothesis in the biological sciences. Students who have taken a biology course without learning about evolution probably have not been adequately or honestly educated	Evolution is a central explanatory hypothesis in the biological sciences. Therefore, students need some knowledge of its assumptions and basic concepts.

Qualifications to "Reduce Dogmatism"

Original Version	Changed Version
Scientists can reconstruct the (prehistoric) animal	Scientists do their best to reconstruct the (prehistoric) animal
Scientists believe that these species were ancestors	According to the evolutionary view, these species were ancestors
A short description...	A short approximate description...
Modern animals that are descendents...	Modern animals that seem to be direct descendents...
Some fish began to change	Some fish began to change although we don't know why
The earth had spore bearing plants long before the first seed plants appeared	There is evidence that the earth had spore bearing green plants long before the first seed plants appeared
Paleontologists...have reconstructed past history...	Paleontologists...have done their best to reconstruct the past history...
How do we know...	On what basis has it been concluded that...

Original Version	Changed Version
The evidence that shows how...	Evidence that is often interpreted to mean...
They would have to change	They would have to be different
Fishes adapted...	Fishes were adapted...
Paleontologists have been able to date the geological history of North America	Paleontologists have assembled a tentative outline of the geological history of North America

Changes to Avoid Evolutionary Assumptions

Original Version	Changed Version
Slowly, over millions of years, the dinosaurs died out	Slowly, the dinosaurs died out
As reptiles evolved from fishlike ancestors, they developed a thicker scaly surface	If reptiles evolved from fishlike ancestors, as proposed in the theory of evolution, they must have developed a thick scaly surface
Many scientists believe that the universe had a beginning similar to that of a snow fort. They believe that the stars and the galaxies of the entire universe in the beginning were in the form of very small scattered particles	Science, by definition, cannot say anything about where the first matter and energy of which our universe is made came from. That is because there cannot be any science without matter and energy to deal with. When scientists speak of the beginning of the universe, therefore, they mean the first interactions of matter and energy. Many scientists believe that these first interactions were like those in making a snow fort....
Shortly after the flying reptiles took to the air, the early birds developed	Birds appear in the fossil record shortly after flying reptiles
The constant rate at which radioactive elements give off particles enables scientists to determine how long it will take for one ball of any sample of a radioactive element to form another element	Scientists know that radioactive elements give off particles today at a constant rate. If they assume that this rate has remained constant back in time to the date of interest, they can determine how long a ...
Scientists believe life may have begun from amino acids or viruses, neither of which is usually considered living. Scientists believe life may have been transported from another planet	Scientists do not know how life began on earth. Some suggest that life began from non-living material. Others suggest that life may have been transported...
Plants took to the land and conquered it	Plants appeared on the land

the middle class, others are concerned with increasing taxes, and still others oppose the loss of community control for ideological reasons. The decline in local control has, of course, been a major source of tension with respect to ethnic balance and busing as well as in matters of curriculum and textbooks. Thus any effort to oppose government directives can count on a wide base of support.[26]

Centralization of curriculum decisions does require some assumptions concerning public choice; for public school policy, as a collective decision, is binding on the individual. Similar in this way to fluoridation decisions, once a curriculum is established, every student is involved. State-wide policies are developed either on the assumption that the knowledge in question is objective and indisputable (e.g., mathematics), or that a given policy is of broad social benefit, reflecting dominant values and shared objectives as to the purpose of education (e.g., American history). The creationists' objection to evolution as "dogma" suggests that even with respect to science, unresolvable value differences preclude such consensus. For they argue that the teaching of evolution contributes to the difficulty of maintaining the religious commitment and value orientation of a minority group in a scientific society.[27]

Their demands raise difficult questions: Does the State have a responsibility to require that certain textbooks be used in all classes? What kind of constraints and standards can be imposed to balance academic interests with political and religious pressures? How can one determine if textbook material is anti-religious, or in some way biased? And, more generally, how can a policy-making commission respect diverse individual beliefs while making collective decisions?

The Ideology of Equal Time

Let us present as many theories as possible and give the child the right to choose the one that seems most logical to him. We are working to have students receive a fair shake.[28]

In January 1973 Henry Morris, the president of the Institute for Creation Research, wrote to the director of the BSCS and challenged him to a public debate on the topic: "Resolved, that the Special Creation Model of the History of the Earth and its Inhabitants is More Effective in the Correlation and Prediction of Scientific Data than is the Evolution Model."[29] Morris proposed that the winner of the debate would be determined by audience applause. The issue, claimed the creationist, is one of free public choice, of equality, of fairness. Creationists argue that since the Biblical theory of origins is scientifically valid, it deserves "equal time." When there are two "equally valid hypotheses" it is only "fair" that students be exposed to both theories and be allowed to choose for themselves. Creationists accuse those who deny the right to equal time of being closed-minded ("those narrow men, like people standing in the sunlight who argue there is no evidence for the existence of the sun"[30]) or exclusive ("using their authoritarian position to protect their own domain"[31]).

The concept of equal time originated with the FCC Fairness Doctrine, confirming the responsibility of broadcasters to afford "reasonable opportunity for the presentation of contrasting viewpoints on controversial issues of public importance."[32] The concept has been extraordinarily difficult to define and to implement, but, equated with fairness and justice and rooted in the democratic impulse, it has enormous appeal in American society. The concept has had an especially important influence on American education which has been described as an almost "evangelically egalitarian system." [33]

The wide appeal of "equal time" has helped to strengthen creationists' arguments, buttressing the claimed anti-authority appeal of the movement. Creationists take great pride in the fact that their constituents are not limited to fundamentalists. When William Willoughby sued the NSF for its support of textbooks teaching evolution, he reported that he received letters from allegedly non-religious people who sympathized with demands for equality from minority groups. It is simply not "fair," claimed his correspondents, that some groups are disadvantaged in this way. Interviews with teachers and school administrators also suggested widespread sympathy with the argument that it is only fair to include creationists views as an alternate explanation of man's origins and many of those interviewed said they saw no harm in presenting these views in science classrooms.

The creationists' demands have brought into the political arena questions that are normally resolved by professional consensus. The scientific merits of the theory of evolution do not remove it from political turmoil. On the contrary, the persistence of conflict reflects a tension between assumptions about uniformity in the teaching of a specific subject and the expectations of pluralism in an educational system that is directed far beyond simply providing factual knowledge to students. The intention of education includes "life adjustment" and preparation for a role as "moral citizen in a democratic society"—broadly-conceived functions that open public schools to considerable pressure. In the last decade, we have seen textbook changes instigated by several minority groups and by women who see the presentation of history as biased and who seek to have their interests more fairly represented. Their demands have formed the model for creationists. The concept of pluralism allows that minority groups have a right to maintain cultural and religious traditions in the face of pressures for conformity. The creationists have extended this concept to the teaching of science. Indeed, the struggle against cultural hegemony is paralleled by the struggle against scientific conformity, as creationists insist that public school teaching must respect beliefs that are outside the dominant culture.[34] And as in other controversies, the active participation of an articulate if extremist minority in a local issue can be highly effective[35]

The Communication of Science

Evolutionists were incredulous that creationists could effectively call into question the expertise that had developed California's science curriculum.

Consider the incredibility of the fact that the California Board incorporated a statement of opinion from a member of the audience into a document that had been laboriously prepared over a period of years by a group of experienced and knowledgeable members of the State Advisory Committee on Science Education.[36]

Incredulity was sometimes expressed in terms of amused disdain. It was

proposed that Bible publishers insert a paragraph in *Genesis* to indicate that
scientific method rejects supernatural approaches in explaining the universe.
Amusement, however, soon gave way to defense that appeared at times to
approach the dogmatic character of the creationists' position. As creationists
brought their social and political concerns to the discussion of the science curriculum, biologists found themselves extending their arguments well beyond
the limits of science. Each group claimed the other based its beliefs on faith,
and each claimed dispassionate objectivity while defending its position with
passionate commitment.

The Problem of Dogmatism: Scientists' Response to Challenge: The noted physicist Victor Weisskopf once remarked, "If I wanted to be nasty toward the
evolutionists I would say they are surer of themselves than the nuclear physicists—and that is quite a lot."[37] Indeed, biologists present the theory of evolution as demonstrated truth. T. Dobzhansky, for example, asserts that: "No informed person entertains any doubt of the validity of the evolution theory in
the sense that evolution has occurred."[38] Certainty is often stated in terms
designed to provoke a defensive response. Thus, George Gaylord Simpson
claims, it is obvious "to any informed, open-minded, and rational person that
the facts agree only with the hypothesis of evolution."[39]

Historically and methodologically, much of science developed in opposition
to the dogmatism of religion, and scientists themselves understand their work
as approximate, conditional and open to critical scrutiny. Yet this is in striking
contrast to frequent examples of the public representation of science as authoritative, exact, and definitive. The organized skepticism towards scientific findings that is tacitly understood by those who practice science contrasts sharply
with its external image. Perhaps the most difficult concept to convey to those
who are not scientists is the delicate balance between certainty and doubt that
is so essential to the scientific spirit. Textbooks in particular tend to convey a
message of certainty to the non-specialist, for in the process of simplification,
findings may become explanations, explanations may become axioms, and
tentative judgments may become definitive conclusions.[40] Few textbooks are
careful to emphasize the distinction between fact and interpretation, or to
suggest the intuition and speculation that actually guide the development of
scientific theory. The media communicate only the sensational aspects of scientific work—the great bio-medical accomplishments and the most dramatic
discoveries. Little is ever said about the scientific enterprise itself. The relationship of science to cultural and social attitudes, to the social organization of
scientific research, or to the personal characteristics of scientists themselves is
given little attention. One historian of science has suggested only partly tongue-
in-cheek that "the history of science be rated X," for a proper study of the
historical development of scientific concepts would often do violence to the
professional ideal and public image of science.[41]

The authoritarian public representations of science are reinforced by a
gospel of consensus that is shared by many scientists who deeply want to avoid
challenge and criticism from those outside the profession. Evolutionists are
particularly defensive, perhaps because of the long history of criticism in a

subject that is rich in philosophical and theological images. Unable to ignore persistent challenges to an authority which they believe derives from the weight of scientific evidence, many evolutionists have responded with a kind of scientific fundamentalism, emphasizing the neutrality and apolitical character of their work.[42]

Scientists, as most other professionals, are greatly concerned with maintaining autonomous control over their work, and control necessarily includes determining the dissemination of science; that is, the content of science education. It was the external threat to professional control that evolutionists found most irritating. "The State Board's repudiation of its own committee in favor of a lay opinion from the audience should ultimately become a classic example in textbooks on school administration of how *not* to proceed with the development of standards," claimed an evolutionist.[43] "Why are comments related to science made by high-priced technicians such as medical doctors and by persons in related fields of technology more readily acceptable as statements of science than those made by scientists themselves?" complains another.[44]

The creationists' public claims of scientific verity were especially embarrassing Scientists maintain internal control of work in their disciplines through informal communications, and through a peer review system that determines research funding and the acceptance of papers by journals. Careful of their external image and eager to avoid political interference, scientists usually try to avoid public exposés of arguments among themselves.[45] Creationists, however, claim also to be scientists and they adopt the language and forms of science. Yet they reject the constraints of scientific norms by seeking external political judgments of the validity and justice of their arguments.

In responding to the creationists, scientists too found themselves arguing, though uncomfortably, in political terms. Indeed, placing their arguments side by side with those of their critics, we find striking similarities (Table 2). The debate has aspects of a battle between two dogmatic groups. The anti-dogmatic norms internal to science faded as evolutionary scientists sought to convince non-scientists of the validity of their work.

Science and Personal Beliefs: The concept of equal time—the suggestion that questions of fairness should determine scientific merit and the substance of scientific education—left scientists utterly amazed: Can quacks be entitled to equal time? Should Christian Science be in health books, the stork theory in books on reproduction, and astrological lore in expositions of astronomy?

Concepts of pluralism, of equity, of "fairness," of wide-open participatory democracy as defined in a political context, are incongruous in science. The process of consensus-seeking in science is based on the existence of an organized body of knowledge, and on an intricate network of procedures that are accepted by a community of scientists who share norms concerning appropriate behavior and standards of acceptable truth.

To be a scientist is not only to do what scientists do, but to accept what other scientists accept... a professional scientist depends on this universe of discourse for an important part of his or her identity.[46]

Scientific theories are taught because they are accepted by this scientific com-

Table 2

Contrasting Arguments of Creationists and Evolutionists *

Creationist Argument	Evolutionist Argument

ON SCIENTIFIC METHODOLOGY

Creation theory is as likely a scientific hypothesis as evolution. Neither theory can be supported by observable events, neither can be tested scientifically to predict the outcome of future phenomena, nor are they capable of falsification. Evolutionists, while claiming to be scientific, confuse theory and fact. And it is unscientific to present evolution as a self-evident truth when it is based on unproven *a priori* faith in a chain of natural causes. Based on circumstantial evidence, evolution theory is not useful as a basis for prediction. It is rather, "a hallowed religious dogma that must be defended by censorship of contrary arguments". The situation is a trial of Galileo in reverse.[1]

Creationism is a "gross perversion of scientific theory." Scientific theory is derived from a vast mass of data and hypotheses, consistently analyzed; creation theory is "Godgiven and unquestioned," based on an *a priori* commitment to a six-day creation. Creationists ignore the interplay between fact and theory, eagerly searching for facts to buttress their beliefs. Creationism cannot be submitted to independent testing and has no predictive value, for it is a belief system that must be accepted on faith.

ON MORAL IMPLICATIONS

Man is a higher form of life made in the image of God. To emphasize the genetic similarity between animals and man is socially dangerous, encouraging animal-like behavior. As a "religion of relativism", evolution theory denies that there are absolute standards of justice and truth and this has disastrous moral implications.

"Tampering with science education by insisting on the priority of feeling over reason, of spontaneity over discipline, of irrationality over objectivity, the honorable man wrecks his own ideals. By attempting to redefine science for his own purposes, the honorable man finds himself in the company of a young hippie radical representing the counter-culture, who indiscriminately is throwing out a life of reason based on objectivity and thus gives himself license to live carelessly and dogmatically."[2]

ON POLITICAL IMPLICATIONS

Evolution is a scientific justification for "harmful" political changes. The evolutionary philosophy which substitutes concepts of progress for the "dignity of man" has been responsible for "some of the crudest class, race and nationalistic myths of all times: the Nazi notion of master race; the Marxist hatred for the bourgeoisie; and the tyrannical subordination of the worth of the individual to the state."[3]

Creationism is a form of right-wing conservatism, as evident in the role of Reagan appointees in the California Board of Education. "Attempts to legislate belief systems through controlling printed materials in the public schools have frequently been a part of fascism."[4]

ON LEGAL IMPLICATIONS[5]

Public schools cannot legally deal with questions of origin that are the domain of religion. They infringe on constitutional rights as guaranteed under the "establishment of religion" and "free exercise" clauses of the first amendment, the "equal protection" and due process clauses of the fourteenth, and constitutional guarantees of freedom of speech. Teaching evolution amounts to the establishment of "secular religion", interfering with free exercise of fundamentalists' truths and violating parental rights. Moreover, restricting the teaching of alternate theories violates the free speech right of teachers.

Exclusion of creation theory in science classes is justified by the first and fourteenth amendments. It is unconstitutional to teach children in a way that would blur the distinction between church and state. Creationism is non-scientific and religious, and therefore to include it would amount to "the establishment of religion." Imposing non-scientific demands would restrict the freedom of teachers to teach and students to learn. To require equal time for doctrines that have no relation to the discipline of biology would impose unconstitutional constraints on the teacher's freedom of speech.

ON RELIGIOUS EQUALITY

To select one set of beliefs over another is to suggest that one group of people is superior to another. Creationists are a persecuted minority. In view of the wide range of beliefs among Americans, teaching evolution is divisive and inequitable and reflects the dogmatism of an established group. "Science has been oversold in Western culture as the sole repository of objective truth... the authoritarianism of the medieval church has been replaced by the authoritarianism of rationalistic materialism."[6]

Creationist demands that their beliefs be taught in public schools represent the tyranny of a minority; a few people are using democratic protections to subvert majority interests. To teach creation theory would violate the beliefs of other religious groups. Justice, in this case, would also require teaching hundreds of other mythologies reflecting the belief of the American Indians, Hindus, Buddhists, Moslems, and so on. Religions can co-exist with science because they operate at a different level of reality.

ON SOUND EDUCATIONAL PRACTICE

Education in biology is "indoctrination in a religion of secular humanism." It is a breach of academic freedom to prevent the teaching of arguments that have withstood challenges for 6,000 years. "Science demands that our children be taught an unproven, undocumentable theory. There is neither a scientific nor moral base upon which to refuse our school children access to another, much documented theory—the theory of Genesis creation." Sound educational practice requires teaching creation as an alternate theory so that students can decide what to believe for themselves.

To include creation theory in scientific classes would be poor pedagogy leading to ridicule and rejection of both science and religion. If creation were presented as an alternate hypothesis, even less would be taught about science than is taught today. Furthermore, it would be a breach of academic freedom to require to teach what is essentially a belief system. In-depth studies of the relationship between science and religion are too sophisticated for public schools. It is sound educational practice to focus on an accurate presentation of scientific fact and leave the teaching of religion to the home.

---

* SOURCES: From statements at the *Public Hearings on California Biology Textbooks*, Sacramento, November 9, 1972, and from the *BSCS Newsletter*, November 1972, unless otherwise indicated.
1 John N. Moore, "Evaluation, Creation and the Scientific Method, *American Biology Teacher*, 35 (January 1973); and editorial, *Christianity Today*, 17 (January 1973.)
2 David Ost, "Statement," *American Biology Teacher*, 34 (October 1972), p. 414.
3 Carl Henry, "Theology and Evolution" in R. Mixter (ed.), *Evolution and Christian Thought Today* (Grand Rapids: Eardmans, 1959), p. 218.
4 Ost, *op. cit.*
5 Discussion of the legal issues appears in F. S. LeClercq, "The Monkey Laws and the Public Schools: A Second Consumption ?" *Vanderbilt Law Review*, 27 (March 1974), p. 209-42.
6 Duane Gish, "Creation, Evolution and the Historical Evidence," *American Biology Teacher*, 35(March 1973), p. 140.

munity. It is collegial acceptance that validates one theory and rejects another; acceptance or debate (applause) by those outside the community is irrelevant. Moreover, while science is an open system in terms of social criteria, scientific excellence depends on achievement and rigorous evaluation; indeed, the internal standards of performance in science may run counter to egalitarian notions.[47]

The misconceptions involved in the extension of egalitarian principles to science suggest that in this age dominated by scientific values, the character and content of science are poorly understood by those not directly involved in its practice. Belief in science often remains as fully a matter of faith as a commitment to supernatural explanations. Laymen play a passive role as consumers of science; they may use the language of science but understand little of its methods or underlying assumptions. Nor do they grasp the complex and indirect evidence that take together constitutes the support of certain scientific theories that cannot be verified by direct observation.[48] Without understanding the nature of evidence, they must accept on faith that the theory of evolution is a consistent and plausible explanatory system. Conversely, all too many scientists fail to grasp the differences between the structured, meritocratic processes that operate within science, and the more egalitarian, pluralistic processes outside science. This lack of sensitivity hampers their efforts to effectively communicate science to the public, and contributes to their difficulty in dealing with controversy.

To those whose beliefs are challenged, the scientific merits of a theory that defines man's place in the universe may be less persuasive than its social and moral implications. Evolution theory has had a remarkable impact; it has been described as "a secure basis for ethics" by C. H. Waddington,[49] as "a naturalistic religion" by J. Huxley.[50] It was a comfortable justification for laissez-faire economics for 19th century entrepreneurs who called the growth of large business "merely the survival of the fittest," and industrial competition "merely the working out of a law of nature."[51] The moral implications that can be drawn from the theory, its threat to the concept of absolute ethical values, clearly assume greater importance to creationists than the details of scientific verification.

Scientists, however, are convinced of the rationality of their methods, and expect the validity of their activities to be self-evident. They assume that their careful explanations of scientific theory will be automatically accepted. The persistence of creationism, the MACOS affair and the Kanawha protest, are reminders that people are reluctant to surrender their convictions to the authority of science, that the familiar expectations and beliefs of non-scientists may easily take precedence over scientific explanations. Increased technical information is unlikely to change well-rooted beliefs, for selective factors operate to determine the interpretation of evidence, especiallywhen the nature of such evidence is poorly understood. Creationists avoid, debunk, or disregard information that would repudiate their preconceptions. Where there is powerful and obvious contradictory evidence (as in the case of radio-carbon dating), they question the premises by which such evidence is interpreted. A great deal of social reinforcement helps them maintain their views in the face of repeated frustration, as in a series of abortive Mount Ararat expeditions in the 1970's to find remnants of Noah's Ark.[52]

Despite these characteristics, it is not accurate to dismiss the creationist movement as merely "anti-science." Like many other social movements, creationism is a bizarre combination of religion and science. Theological beliefs are conveyed in a context of diplomas, research monographs and professional societies; for the creationists, modern science is at once the enemy and the model. They are hardly alone in their accommodation to the expectations and images of a scientific age. Some yogis have recommended the use of electroencephalographs to assist in meditation. The popular film, "The Exorcist," portrays a doctor using sophisticated medical technology to exorcize spirits from a possessed child. The Uranius brotherhood advertises "power-packed scientific proof of the continuity of life through the intergalactic confederation." Occultists, astrologers, UFOlogists, and members of other cults often seek scientific validation for their alleged facts.[53] Creationists and many of these groups are a reaction less against science as an activity of professionals than against its image as an infallible source of truth—as a dominant myth of industrial society.[54]

A myth is perpetuated because it satisfies a social need. If it loses credibility ("planet earth is in trouble," the creationists claim), people will grope for more fulfilling constructs. Science threatened the plausibility of non-rational beliefs, but it did not remove the uncertainties that seem to call for such beliefs. Creationism thus fills a social need for its adherents. By using representations that are well adapted to the 20th century, by claiming that their ideas are based on scientific evidence, they offer intellectual plausibility as well as salvation, and the authority of science as well as the certainty of Scripture. Poorly understanding the process of science, they seek to resolve the old conflict between religion and science through popular decision. Democratic values such as freedom of choice, equality and "fairness" become the criteria to judge the merits of science. The creationists have thus sought to bring together science, religious faith, and populist democracy—three venerated pieties of American culture.

REFERENCES

1. Active creationist organizations include: The Institute for Creation Research, the Creation Research Society and the Creation Science Research Center in California, the Bible Science Association in Idaho, the Scientific Creationism Association of New Jersey, Educational Research Analysts in Texas, the Creation Research Science Education Foundation in Ohio, and the Genesis School for Graduate Studies in Florida.

2. California State Department of Education, *Science Framework for California Public Schools* (Sacramento 1970), p. 106.

3. Their demands were defeated by only a one-vote margin by the Board of Education which did, however, require textbook revisions to reduce dogmatism and to consider including creation theory in social science texts. In May, 1975 the Board rejected a proposal to include creationism in social science textbooks, and again the creationists demands were only defeated by one vote.

4. F. S. LeClercq, "The Monkey Laws and the Public Schools: A Second Consumption?",

Vanderbilt Law Review 27, No. 2 (March 1974), pp. 209-42. For most recent rulings see the discussion in the National Association of Biology Teachers' *News and Views*, 19 (April 1975); and *Science*, 188 (May 1975) p. 428.

5. See Irving Zaretsky and Mark Leone (ed.), *Religious Movements in Contemporary America*, (Princeton: Princeton University Press, 1974). Note that creationism also has had wide appeal among a number of well-known aerospace scientists and engineers. Wernher Von Braun once declared his personal support for the "case for design and purpose behind all" and "endorsed the presentation of alternate theories" in the classroom. Wernher Von Braun, Letter to John Ford, published in *Science and Scripture*, (March/April 1973), p. 4. Von Braun later qualified his position, stating that he believed there was "divine intent" behind the processes of nature, but not that all living species were created in their final form 5,000 years ago. The astronauts James Irwin, Frank Borman and Edgar Mitchell have also indicated that they feel the *Genesis* account of creation is an appropriate explanation.

6. Articles on the Kanawha, West Virginia protest appeared in the *New York Times* throughout the fall of 1974.

7. *Willoughby vs. H. Guyford Stever*, U.S. District Court for the District of Columbia, Civil Action No. 1574: 72. Beginning in 1958, the NSF has funded 53 projects for the improvement of pre-college science education. Among these projects was the BSCS at the University of Colorado. Prior to the BSCS series biology textbooks focused on morphology and taxonomy. Teachers and textbook publishers had generally avoided conflicts that they anticipated would follow the teaching of evolution.

8. Objections to MACOS were diverse; critics opposed parts of the course that dealt with natural selection, with animal behavior in its relation to man, and with aspects of Eskimo life and mythology. These were all felt to threaten moral and religious values. *Congressional Record* (April 9, 1975), H2575-H2607. Bibliographies on this issue can be found in the Newsletters 11 and 12 (April and June 1975) and "Early Warning Mailing" No. 1 (May 15, 1975) of the Harvard University Program on the Public Conceptions of Science.

9. Most creationists are trained as physical scientists or engineers, and often at Bible colleges, though some at large state universities as well. See discussion by Moore in this volume.

10. There is now an extensive literature on current concerns with science. See, e.g., J. Ellul, *The Technological Society* (New York: Knopf, 1956); V. Ferkiss, *Technological Man* (New York: Braziller, 1969); and other articles in this volume.

11. Vernon Grose, "Second Thoughts About Textbooks on Sexism," *Science and Scripture*, 4 (January 1974), pp. 14ff. With respect to the increasing concern about secularization, the editor of *Christian Century* suggests that this reflects a "protestant paranoia" developed as America has been transformed from Protestant domination to a much more diverse outlook. See discussion in Will Herberg's "Religion in a Secularized Society," Louis Schneider (ed.), *Religion, Culture and Society* (New York: Wiley, 1964), p. 596.

12. John Klotz, Letter to the Editor, *Christian Century* (March 1, 1967), p. 279.

13. *Acts and Facts* (a publication of the Institute for Creation Research), (April 1974).

14. *The New York Times*, March 10, 1974, p. 49.

15. See discussion in *Science*, 188 (May 1975), pp. 426-8.

16. Personal interviews by the author with creationists in San Diego, California.

17. The relationship between religion and environmental values has provoked much discussion. Some scholars attribute the environmental crisis to the Judeo-Christian tradition that assumes that nature exists to serve man. They argue that western science and technology is "cast in a matrix of Christian theology"; that the creation story in *Genesis* justifies man's subjugation of nature. See Lynn White, Jr. "The Historical Roots of Our

Ecological Crisis," *Science*, 155 (March 1967), pp. 1203ff. The idea that Christian belief necessarily accepts exploitation of nature, however, must consider the many counter examples. The Southern Agrarians, while often fundamental Christians, saw the "Gospel of Progress" as an "unrelenting war on nature—well beyond reason," and advocated harmony with nature. See John Crowe Ransom, "Reconstructed but Unregenerated," in Twelve Southerners', *I'll Take My Stand: The South and the Agrarian Tradition*, (New York: Harpers, 1930), p. 8. Others argue that the creation story implies responsibility and stewardship. See C. F. D. Moule, *Man and Nature in the New Testament*, Philadelphia: Fortress Press, 1967). Moule claims that in *Genesis* the land belongs ultimately to God and man is its trustee. The created order is valuable in itself—to be cared for by man.

18. The perception of scientists and professionals as "arrogant" appears to be a widespread and growing problem. At a recent public meeting of the American Nuclear Society citizens complained of the "arrogance" of scientists from the industry (*Nuclear News*, May 1975, p. 83). The condescension of medical professionals was claimed to influence jurists at the Edelin abortion trial. See articles by Barbara Culliton, *Science*, (January 31, 1975) and (March 7, 1975).

19. Ron Kanigel in the *San Francisco Examiner*, September 16, 1973.

20. Vernon Grose, statement to Board of Education suggesting revisions to *Science Framework* (1969), p. 4.

21. *Awake* (a Jehovah's Witness publication), September 22, 1974.

22. Richard Bube, "Science Teaching in California," *The Reformed Journal*, (April 1973), p. 4.

23. Educational Research Analysts, Florida (handout).

24. *Congressional Record*, (April 9, 1975), H2585-2587.

25. A speaker at a demonstration in Kanawha, cited by Calvin Trillin in "U.S. Journal," *The New Yorker*, September 30, 1974, p. 121. While these protest groups challenge centralized professional authority, they are often strikingly authoritarian in their own ideas as to an appropriate educational style. For example, a prevalent objection to MACOS was its developmental, non-authoritarian method of teaching, involving discussion and group participation. Its critics preferred a much more authoritarian approach to education.

26. Luther Gerlach and Virginia Hine, in *People, Power, Change*, (Indianapolis: Bobbs-Merrill, 1970), attribute the increase in protest to "power deprivation." The reaction against professionals evident in our discussion suggests that these protests do tend to enhance the sense of social power among protesting groups. For general discussion of the reaction to centralization and the appeal of local control, see Leonard Fein, *The Ecology of the Public Schools*, (New York: Pegasus, 1971), and Alan Altshuler, *Community Control*, (New York: Pegasus, 1970).

27. This classic problem for minority sects is discussed in Brian Wilson, "An Analysis of Sect Development," *American Sociological Review*, 24 (February 1959), pp. 3-15.

28. Personal interview by the author with Henry Morris.

29. Letter from Henry Morris to William Mayer, January 1, 1973.

30. George How, statement at hearing before the California State Board of Education, November 9, 1972.

31. Letter from Henry Morris to William Mayer, January 1, 1973.

32. *Federal Register* 10416, (July 1, 1964).

33. This term was used to describe some of the sources of anti-intellectualism by Richard Hofstadter, *Anti-Intellectualism In American Life* (New York: Vintage Books, 1962), p. 23.

34. That the teaching of biology poses a distinct problem for minority religious groups with belief systems incompatible with science has been recognized in court rulings exempting Amish children from school requirements. These decisions recognized that "the values of parental direction of the religious upbringing and education of their children in their early and formative years have a high place in our society." These can outweigh even strong State interest in universal education. See *Wisconsin v. Yoder*, U.S. 205, 213 (1972) discussed in F. LeClercq *op. cit.*

35. See discussion of this point with respect to the pressure from anti-fluoridation groups in Robert Crain *et al.*, *The Politics of Community Conflict* (New York: Bobbs-Merrill, 1969).

36. National Association of Biology Teachers, *News and Views*, 16 (October 1972), p. 2.

37. Quoted in the *Journal of the American Scientific Affiliation*, (June 1972), p. 71.

38. Theodosius Dobzhansky, *Genetics and the Origin of Species* (New York: Columbia University Press, 1951).

39. In *Science*, 186 (October 1974), p. 133.

40. See Philippe Roqueplo, *Le Partage du Savoir* (Paris: éditions du Seuil, 1974), and James Raths, "The Emperor's Clothes Phenomenon in Science Education," *Journal of Research Science Teaching*, 10 (1973), p. 211, for discussion of problems in the communication of science.

41. Stephen G. Brush, "Should the History of Science Rated X ?" *Science*, 183 (March 1974), pp. 1164ff.

42. Strongly worded resolutions opposing the creationists were passed by the National Academy of Sciences, the American Academy for the Advancement of Science, the American Anthropological Association, and other professional societies. In addition 19 Nobel Laureates in California signed a petition appealing to the state board to leave evolutionary explanations intact.

43. William V. Mayer, "The Nineteenth Century Revisited," *BSCS Newsletter*, (November 1972).

44. Statement by David Ost in *American Biology Teacher*, 34 (October 1972), pp. 413-14.

45. S. B. Barnes, "On the Reception of Scientific Beliefs," in Barry Barnes (ed.), *Sociology of Science* (Harmondsworth: Penguin Books, 1972), p. 287.

46. Norman Storer, "The Sociological Context of the Velikovsky Controversy," AAAS Symposium, San Francisco, February 1974.

47. François Hetman, *Society and the Assessment of Technology* (Paris: OECD, 1973), p. 40.

48. See discussion of the relationships between evidence and theory in Harvey Brooks, "Scientific Concepts and Cultural Change," *Dædalus*, 1964.

49. C. H. Waddington, *The Scientific Attitude* (London: Penguin Books, 1941).

50. Julian Huxley, *Evolution in Action* (New York: Harper, 1953), and *Religion Without Revelation* (New York: Harper, 1927). See also Herbert Spencer, "Progress and Its Law and Causes" in *Essays* (New York: Appleton, 1915), for a broad application of the concept of evolution to the social, political, and economic organization of society; and the readings on Darwin's influence in Philip Appleman (ed.), *Darwin* (New York: W.W. Norton, 1970).

51. Andrew Carnegie, cited in Richard Hofstadter, *Social Darwinism In American Thought* Boston, 1955), Ch. II.

52. The Mount Ararat expedition not only failed to produce results but met incredible obstacles. "The men were robbed and beaten by Kurdish outlaws, victimized by city offi-

cials, and fired on in an ambush. And three of them were temporarily incapacitated by the body-rending blows of lightning." *San Diego Union*, (February 23, 1974), p. B-8. Each failure, however, only produced further reinforcement. Leon Festinger's work on cognitive dissonance suggests the factors operating to sustain creationism despite such problems. See *A Theory of Cognitive Dissonance* (Evanston: Row Peterson, 1957).

53. Christopher Evans in *Cults of Unreason* (New York: Farrar, Strauss and Giroux, 1973), discusses how scientism and the UFO cults use scientific apparatus. Marcello Truzzi, "Towards a Sociology of the Occult", in Zaretsky, *op. cit.*, pp. 628-45, suggests that occultists too seek scientific validation for their beliefs.

54. Note the discussion about science as a "myth" in Roszak (this volume) and Ellul, *op. cit.*, The implications of this myth are also developed by Don K. Price, "Science and Technology in a Democratic Society," in *Education for the 21st Century* (Urbana: University of Illinois Press, 1969), pp. 21-36.

AMITAI ETZIONI AND CLYDE NUNN

The Public Appreciation of Science in Contemporary America

THIS PAPER EXPLORES the attitudes of various publics toward science as they are revealed in polls and surveys.[1] What proportion of the American public—considered as a whole and in terms of its subpublics—holds a positive attitude toward the scientific enterprise? On what grounds does it accept or reject scientists or their work? How has the legitimation of science changed in the last fifteen years—a period, it has been said, during which the concept of modernity itself has weakened?

One distinguishing feature of modern society is the high value it accords to an "open-ended" approach to the world, one in which individuals and groups are free to choose whatever instruments will best advance their goals, minimizing most other considerations, such as the normative, emotional, or social status of those means. Indeed, one of Max Weber's central insights was that "rationality," in this instrumental or functional sense, is at the heart of the modern administrative process, the market economy, technological progress, and the scientific enterprise.

Another frequently noted characteristic of modern society is its broad "mobilization base." Institutions are thought to be greatly dependent on mass legitimation, based on a widespread consensus concerning the society's "core values."[2] Our exploration of the legitimation of science in American society will cast some light on how widely institutions embodying rationality are accepted or at least tolerated, and how attitudes vary among different strata of society. Thus, we are interested in the status of science not only as a social institution in its own right, but also as an indicator of the changing status of "rationality," one of the foundations of modernity in our society.

Our discussion here is based only on quantitative data—mainly national public opinion polls and attitude surveys—not on the much larger body of analysis based on the public media, or on in-depth interviews with individuals. We rely, furthermore, on polls and surveys conducted in this country within the last couple of decades. In order to minimize the inherent limitations of data on attitudes, we will pay most attention to data that indicate differences of more than a few percentage points, to distinctions which recur over a period of time, and to points based on two or more sets of data.

Perceptions of the General Status of Science

There is a widely held belief among scientists and nonscientists that appreciation of science in the contemporary United States has declined. Observers vary, however, in the extent to which they see this decline as a "crisis" of legitimation. Robin Clarke ventures the opinion that "science and technology have taken a severe pounding from which they will not recover."[3] Jurgen Schmandt concludes that "a period of faith in science and technology as an engine of social progress has come to an end."[4] *Time* has summarized the public's reaction to science as one of "deepening disillusionment."[5] And, Jerome B. Wiesner is concerned with the "deep mistrust of science and technology . . . expressed by many of our society today."[6] On the other hand, although Kenneth Pitzer does not believe that science can regain the high level of popularity it commanded a decade ago, he thinks that "most people still believe science to be valuable to society."[7]

It is important to note, however, that most quantitative data on public attitudes toward science deal with the period *preceding* the alleged "erosion of faith," and thus they best illuminate what the public attitudes to science were *before* their presumed deterioration.

National data collected in 1957 by Withey and Davis[8] for the Survey Research Center at the University of Michigan revealed that most Americans at that time valued science highly mainly because they saw it as instrumental in achieving goals they valued. A following study six months later, after Sputnik, showed little change.[9] A very high 94 percent agreed that "science makes our lives healthier" and 89 percent agreed that "science makes for rapid progress." The somewhat less openly instrumental statement, "the world is better off with science," also drew support from almost 9 out of 10 Americans—88 percent. Conversely, most people disagree with antiscience statements: less than half (43 percent) agreed that "science makes life change too fast"; and less than a quarter (23 percent) saw science as a force which "breaks down right and wrong."

The study provided lists of specific goals or payoffs and of disutilities that people in the late fifties associated with science. Eighty-one percent felt that the world was "better off" because of science, citing improved health, a higher standard of living, and industrial and technological improvements. Those apprehensive about science also looked upon it primarily as a tool of progress. About one person in ten offered qualifications, the majority viewing armaments, especially atomic weapons, as undesirable. About one person in four said that he paid attention to science because "science may determine whether my family and I, and the world itself, will survive."

In contrast to these relatively positive attitudes toward the instrumentality of science, only about one person in seven saw science as helpful or interesting, and an even smaller minority saw it as "exciting." Such was the status of science in the good old days.

In 1966, Karen Oppenheim pieced together the 1957 Survey Research Center study, the 1958 follow-up study, and national sample data collected in 1964 by the National Opinion Research Center.[10] (Note, however, that all this material precedes the massive falling away from science alleged to have taken place during the late sixties and early seventies.) Her data (see Table 1) suggest that there was an increase, at least in the public's sense of threat from science.

The proportion of people who thought science made life change too fast went up from 43 percent in 1957 to 57 percent in 1964. Moreover, the proportion who saw science as a disintegrating factor nearly doubled, increasing from 23 percent in 1957 to 42 percent in 1964. What, if any, changes in attitude have occurred since

Table 1
Changes Over Time in Agreement with Three Science Attitude Items*

Item	Percent Agreeing		
	1957	1958	1964
Science makes our way of life change too fast	43%	47%	57%
Science breaks down people's ideas of right and wrong	23	25	42
The growth of science means a few people could control us	32	40	—
Number of respondents	1919	1547	923

* Karen Oppenheim, "Acceptance and Distrust of American Adults Toward Science," unpublished Master's thesis, University of Chicago, 1966.

1964? Having, as yet, only limited data on which to base a direct answer, we must rely heavily on an indirect indicator about which we have comparable pre- and post-1965 data—namely, public feeling about scientists (as distinct from science). However, although it might seem obvious that a person who looks with favor upon a discipline will also favor its practitioners and vice versa, this relationship is deceptive. One may have a high regard for politics or religion, but take a jaundiced view of their practitioners. Indeed, the more highly one regards an area, the more apt one is to be critical of its practitioners.

Do people view the institution of science separately from its constituents, the scientists? Survey data from the late fifties show that at that time, a majority of the public was as favorable to scientists as to science; it perceived both as useful, as instrumental.[11] Eighty-eight percent agreed that "most scientists want to work on things that will make life better for the average person"; only 7 percent disagreed. Antiscientist statements, like antiscience statements, drew only minority support: only 25 percent agreed, while 66 percent disagreed, that "scientists always seem to be prying into things they really ought to stay out of."[12] Hence, it seems possible to employ attitudes toward scientists as an indicator of feelings about science, although it is, of course, possible that if and when a disaffection from science set in, feelings about scientists parted company with feelings about science.

Since 1966, both Louis Harris and the National Opinion Research Center have periodically asked the question: "Would you say you have a great deal, only some, or hardly any confidence in those people running the scientific community?"[13]

The data in Table 2 show that confidence dropped 19 percentage points, from 56 percent 1966 to 37 percent in 1973. In fact, it hit a low of 32 percent in 1971, but recouped 5 percent by 1972 and 1973. The value of this particular question, of course, is limited, for it may tap feelings about authority in addition to those about scientists. Scientists themselves may distrust those who run science. Nevertheless, one can conclude, tentatively, that public appreciation of scientists did decline to a degree neither trivial nor monumental, neither reassuring nor alarming. Significantly, Table 2 shows a relatively small percentage (no more than 10 percent) for any given year of what would best be labeled as clearly anti-scientist feelings. The major shift, then, was not from great enthusiasm to great hostility, but from "great confidence" to "only some confidence"—a middling shift by all accounts.

Table 2
Public's Confidence in Those Who Run Science: 1966 vs. 1971-1973°

Confidence in Science	Year of Poll				Change
	1966	1971	1972	1973	1966-1973
Great deal	56%	32%	37%	37%	−19%
Only some	25	47	39	47	+22
Hardly any or none	4	10	8	6	+2
Not sure	15	11	16	10	−5

° Louis Harris and Associates for polls in 1966, 1971, and 1972. National Opinion Research Center, General Survey of the National Data Program for the Social Sciences, 1973.

A Harris poll, taken some five months after the 1973 NORC survey we relied on for data in Table 2, indicated that confidence in most institutions has taken an up-turn. In that poll, all institutions except the executive branch of government (which fell sharply) gained support, some quite substantially.[14] While science was not included, those institutions whose confidence levels most strongly co-vary with those in science posted gains, in some instances of as much as 7 percentage points.

Furthermore, from the Harris and NORC studies, which did include science, we can gauge how much of the long-term loss in confidence is a reflection of a general disaffection from authority and how much is a particular shift away from scientists. An identical "confidence" question was posed concerning several institutional areas, ranging from the military to religion (see Table 3).

In the spring of 1973, science ranked second, with education, in eliciting "great confidence" from the public. Only medicine ranked higher. Thus science gained respect relative to other institutions. In 1972, it ranked third, behind medicine and

Table 3
Percentage of the Public Indicating "A Great Deal" of Confidence in 16 Institutional Areas:
1966 vs. 1971-1973°

Institution	Year of Poll				Change
	1966	1971	1972	1973	1966-1973
Medicine	72%	61%	48%	54%	−18%
Science	56	32	37	37	−19
Education	61	37	33	37	−24
Finance	67	36	39	—	—
Religion	41	27	30	35	−6
Psychiatry	51	35	31	—	—
U.S. Supreme Court	51	23	28	32	−19
Military	62	27	35	32	−30
Retail businesses	48	24	28	—	—
Federal executive branch	41	23	27	29	−12
Major U.S. companies	55	27	27	29	−26
Congress	42	19	21	23	−19
The press	29	18	18	23	−5
Television	25	22	17	19	−6
Labor	22	14	15	15	−7
Advertising	21	13	12	—	—

° Louis Harris and Associates for polls in 1966, 1971, 1972. National Opinion Research Center, General Survey of the National Data Program for the Social Sciences, 1973.

finance; in 1966 and 1971, it ranked fifth, surpassed by the military and education as well as by medicine and finance. The 19 percent decline in support for science was moderate relative to the eclipses suffered by other institutions. In at least three cases, smaller declines may well be statistical artifacts because the institutions involved had such small percentage bases in 1966. Labor, for example, could not decline more than 22 percent since only 22 percent expressed "great confidence" in labor to begin with.

The percentage of respondents who did not know how they felt about a given institution provides a rough estimate of its remoteness from the public's mind. Compared to all other institutions in 1973, science received the *highest* percentage of "don't know's" (10 percent of all respondents and from 15-20 percent of low-status respondents). Major companies ranked next with 7 percent while other institutions scored from 4 percent to less than 1 percent. This suggests that science might stand to gain respect by explaining to the public more fully what it is and what it does. Indeed, this would seem to present an important educational challenge to science, since those for whom science is most remote have similar social backgrounds to those least confident in science.

Certain institutions co-vary more closely with science than others even though confidence in science is positively correlated with all institutional areas included, suggesting that public trust in science varies to some extent with that in other institutions. However, it is most closely linked with the institutions Max Weber cited as prime modern institutions, in terms of their open-ended approaches to the world—namely, science, business and medicine. Not surprisingly, public trust for business and medicine (as well as for the press, which is also open-ended, though Weber does not cite it) is rather closely linked with that for science. Conversely, more traditional, less open-ended institutions, such as the military, the clergy, and labor, correlate less closely with science. Significantly, however, none of the correlations is high, a fact which suggests that the public's views of science are to a considerable degree independent from its views of other institutions, even ones which, in Weber's terms, are also "modern."

Since science seems to command more confidence than many other institutions, particularly government and education, science cannot hope to gain legitimation from public officials and educational leaders. Indeed, government and education may, in future years, turn to scientists for legitimation, since only those who are trusted can lend trust.

Science and Technology

Whereas the goals of science are largely derived from the structure of science itself, those of technology frequently derive from political, economic, and other external spheres. Similarly, whereas science claims a kind of morality and justification intrinsic unto itself, technology is often judged in relation to outside values. Thus, since technology is the aspect of "science" least remote from the public, another way of assessing the changes in public appreciation of science is to focus on attitudes toward technology. Indeed, large segments of the public do not clearly separate science from technology. People tend to see science primarily as a collection of practical tools that is a part of technology. In 1958, 65 percent of the public saw scientists as people interested in getting things done rather than "in knowledge for its own sake." In 1957, only about one person in ten showed interest in science as an intellectual, aesthetic, or "methodological" enterprise.[15]

Most studies on changes in the general public's view of technology are recent, local, and limited in the sense that they used only one or two comparable questions. For example, a 1963 survey found that 85 percent of the public felt that the net effect of computers was to promote a better life.[16] Similarly, in 1972, LaPorte and Metlay reported that 80 percent of a California sample of people believed that the net effect of technology was to make life better rather than worse.[17] Seventy percent of the people interviewed in one 1972 national survey felt that life had changed for the better because of science and technology, while only 8 percent felt that it had deteriorated.[18] In short, the available evidence suggests, as it does in the case of science, that, despite some ambivalence, most Americans are far from rejecting technology in a wrath of Luddite-like sentiments.

Age and Education

"The public" that is for or against science or technology does not exist. Attitudes vary according to a large number of subpublics. And these, in turn, are not monolithically for or against science, as certain grossly oversimplified statements about "the" young and "the" college kids imply.[19] Secondary analysis of the 1973 NORC data shows that people from eighteen to twenty-nine years old, those often believed to harbor strong antiscience sentiment, have more confidence in those who run science than any other age group. This is not to say that they are not divided; 40 percent, substantially more than the national average, had great confidence, 48 percent had "only some" confidence, and 12 percent reported "hardly any" confidence or said that they did not know. Conversely, the oldest group, those sixty years old or more, were least confident: only one out of three reported great confidence.

The data suggest that education is more important than age as a predictor of confidence in scientists. The measure (called the *gamma measure*) of the strength of association between age and confidence in scientists is only 0.09, whereas it is 0.32 between education and trust in scientists. College graduates (and the more educated generally) seem to be the group most favorable to scientists. In her 1957 study, Oppenheim found between education and distrust of science a significant but negative correlation ($-.31$): the more education a person had, the less distrust he tended to feel toward science.[20] Fifty-three percent of the college graduates included in the 1973 NORC data expressed great confidence in scientists, compared to 38 percent of the high school graduates, and 28 percent of those who did not have high school diplomas. The college educated, in other words, were almost twice as likely to be favorably inclined toward science as the least educated group. National data from a *Science Indicators* study in 1972 show a similar pattern.[21] Seventy-four percent of people with at least some college education said science and technology did more good than harm, compared with only 51 percent of those with less than a high school education.

But what about the "rebellious" college students of the 1960's? According to Harris Poll data, gathered from a national sample of college students in the spring of 1965, the scientific community rated more confidence than any other American institution. Of the students polled, 76 percent expressed a great deal of confidence in science,[22] the highest percentage of "great trust" recorded for any social category considered then or at any other time during the years reviewed here. Data collected by Funkhouser and Maccoby from 700 West Coast college students at the height of the student protest movement again found that college youth view science favorably. On a scale where a 7 was the most favorable response, the students' choices averaged out at slightly over 5.[23] College students hold similar attitudes toward technology. In a national survey of youth in 1968, Yankelovitch found that 88 percent of college students agreed that "the problem is not technology—it's what society does with technology."[24] Only 24 percent felt that technology was dehumanizing society.

Assuming from data like these that the more "literate" a person is in science, the more he or she will favor it, scientists might recommend a large-scale campaign of education.

There are two kinds of data available relevant to whether such a course of action would indeed make sense. First, knowledge of science has been found to correlate with trust of science, albeit not very strongly. Oppenheim, for example, reported a negative correlation (-0.31) between scientific knowledge and distrust—that is, the more a person knows about science, the less likely he is to distrust it.[25] Likewise, Taviss' data from a 1970 Boston sample found a strong relationship between a scale measuring technology knowledge and one indicating attitudes toward technology.[26]

Second, there is a positive correlation between one's level of general education and one's positive feelings about science.[27] This could be due to the fact that more educated persons know more science which, in turn, makes them more positive about it. Indeed, Oppenheim reported a 0.47 correlation between levels of education and of scientific knowledge.[28] Schramm and Wade found a similar relationship.[29]

Economic and Geographical Position

Socio-economic factors also seem to have a positive and linear relationship to confidence in scientists. In the 1973 NORC data, those in the lowest decile on occupational prestige ratings had the least confidence (only 26 percent felt great confidence), while those in the highest decile showed the greatest confidence (44 percent felt great confidence). Income levels produced a similar pattern. Of those with annual incomes under $1,000 per annum, only 19 percent had great confidence in science, as compared to 44 percent of those with incomes over $25,000.

Where people live also affects their view of science. Arranged according to what proportion of people said they felt great confidence in scientists, the Middle Atlantic and Pacific Coast regions of the nation, and the geographical heartland from North Dakota and Minnesota to Texas and Louisiana, showed greatest confidence with about 41-42 percent. Another geographical strip running from Montana and Idaho south to Arizona and New Mexico comes next with 38 percent. The east north central states of Wisconsin, Indiana, Illinois, Michigan and Ohio with 36 percent, the New England states with 32 percent, and the southeastern coastal states from Maryland to Florida with 30 percent showed more moderate levels of confidence, while the east south central region of Kentucky, Tennessee, Alabama, and Mississippi showed least confidence with 26 percent.

Although not as large as interregional differences, there were differences in confidence according to whether people lived in large cities, small towns, or rural areas. Of those living in cities with populations of 50,000 or more, 40 percent reported great confidence in scientists as compared with 33 percent among the residents of smaller towns and rural areas.

All of these findings suggest that the major source of distrust for science is not a massive monolithic counterculture group, recruited largely from the middle and upper-middle classes, or college-educated youth, but lower-status, less-educated groups, and people who live in less affluent parts of America. It is perhaps more concentrated among those who found in George Wallace a vocal symbol of their discontent. These groups have been, and continue to be, the main source of antimodernist sentiment in America.

This is not to suggest that these are the only groups who distrust science. It is possible that there is a small and perhaps even a growing enclave among better-educated youth that could be called an antiscience counterculture. It is a plausible hypothesis that they and the "forgotten Americans" are the two prime sources of antiscience sentiment, although the poor are probably apathetic or weak supporters of science, rather than hard-core opponents.

Psycho-Social Dynamics

Robert K. Merton has characterized the scientific ethos as consisting of four institutional imperatives: universalism, communism, disinterestedness, and organized skepticism.[30] In the individual personality, these same qualities find expression in such traits as "open-mindedness," an acceptance of the tenets of rational thinking, a capacity to rise above narrow in-group prejudices for the sake of higher values, and a tendency to subject authority to critical examination before accepting it. In direct contrast, studies of the authoritarian personality describe a type of person who is dogmatic, inflexible, conformist, subservient to authority and intolerant of rationality and, by implication, of science. Although such studies have been criticized for psychological reductionism, measurement problems, and their psychoanalytic view of relatively fixed and unchanging personality patterns,[31] one cannot help noticing that the social groups we have isolated as most distrustful of science are also those often cited as having most members who score high on authoritarian scales.

The original study of the authoritarian personality, conducted in the late forties, utilized groups from various social classes rather than national samples. It reported that working-class men were more authoritarian than middle-class men.[32] More recently Kohn reported that in his national sample there was a positive correlation of 0.38 between social class and authoritarianism.[33] In 1954 Stouffer asked a national sample to what extent they agreed with the statement: "People can be divided into two classes—the weak and the strong." He found clear evidence that more educated people were less likely to give authoritarian replies, and that older people were likely to be more authoritarian than younger people.[34] The kinds of people, then, who distrust scientists are similar to the kinds of people who are more likely to be authoritarian. There is also evidence linking authoritarianism with unscientific beliefs, even though *all* authoritarians are not antiscience. Martin found in a local sample of white adults that authoritarianism and superstitious, pseudoscientific attitudes were highly correlated (0.04 among racially tolerant subjects and 0.58 among prejudiced ones).[35]

In a study designed to distinguish people who hold primarily rational beliefs from those who depend heavily on other kinds of beliefs, Rokeach and Eglash reported high correlations between this measure of rationalism and measures of authoritarianism ($-.71$ and $-.61$ in two samples of college sophomores).[36] It seems that the more authoritarian people are, the less they tend to rely on logical-rational grounds.

Evidence also indicates a close link between authoritarianism and distrust of computers. Lee found not only that the less educated were more distrustful of computers than the more highly educated, but that measures of alienation and intolerance of uncertainty jointly were more powerful as predictors of attitudes toward computers than the joint effects of all seven of the other predictor variables combined, including sex, attitudes toward the new and different, education, dependence on government for a comfortable life, and need for more government control of business.[37]

We should be careful, however, not to conclude that people who are less educated, elderly, poor, rural, or southern are disproportionately distrustful of scientists only because they tend to be more authoritarian. Culturally, persons in these social groups have internalized modern American values only to a very limited degree, as indicated, for example, by Stouffer's 1954 study[38] and confirmed by the 1973 NORC data.[39] They are among the groups most often excluded from the mainstreams of society. Their orientation toward modern American institutions can be affected directly by "structural" factors, such as their place in the stratification system, or their residence in "advantaged" or "disadvantaged" parts of the country or neighborhoods. Whatever the reasons, however, the fact remains that those Americans who are alienated and distrustful of people and institutions in general are most likely to be malcontent with the scientific community as well. The 1973 NORC survey consistently supports this idea. As our compilation of NORC data in Table 4 reveals, adult Americans who perceive their family incomes as below average, who are, relative to others, unhappy with life, in poor health, distrustful of people, and intolerant of atheists, are clearly less likely to have a great deal of confidence in scientists.

A number of factors combine, however, to minimize the likelihood that science will come under severe and damaging attack from this quarter. The kinds of people we have been discussing tend to be politically inactive and unsophisticated, and to command limited political resources. Although, when they are aroused politically, they often tend to gravitate toward extremist movements and charismatic demagogues,[40] their overall record of political participation in the United States is weak and sporadic. They are much less likely than better educated, wealthier, more tolerant urbanites to engage in such traditional political acts as voting, writing letters to their elected representatives, campaigning, and contributing time or money to a political campaign.[41] Observers concerned about what they perceive as

Table 4
General Discontent, Trust in People, and Tolerance of Nonconformity and Confidence in Scientists°

Personality Qualities	Great Deal of Confidence in Scientists	Number of Respondents
Perceived Income Relative to Other Americans		
Below average	31%	335
Average	36%	872
Above average	49%	276
Perceived Happiness in Life		
Not too happy	28%	194
Pretty happy	36%	762
Very happy	41%	536
Perceived State of Health		
Poor	31%	106
Fair or good	35%	908
Excellent	42%	478
Trust in People		
Distrustful	30%	805
Trustful	46%	686
Allow an Atheist to Speak in Community		
No	26%	510
Yes	43%	974

° These data were computed by us from the 1973 NORC General Survey.

a significant disaffection from science among college students, but who overlook the indifference and hostility among low-status individuals, may be very wrong statistically, but they are not so far off base politically. In terms of the negative consequences to society, a small erosion of faith in science in politically active social strata is usually of greater significance than massive disaffection among the geographically, politically, and economically peripheral members of society.

These data might lead one to the conclusion that the crisis of legitimacy is not rooted in science itself. However, a sober second thought about these and related data should produce a quite different response. First, the alienated now make up a sizeable and growing proportion of the population: according to the Harris Index[42] the percentage of the population that is discontent has increased markedly from 29 percent in 1966 to 55 percent in 1973. Furthermore, alienation among the politically active strata is growing at least as fast as it is among skilled labor, rural, southern, and elderly groups. The proportion of people who profess to trust in people has also shown a sizeable decline, from 63 percent in 1954 to 46 percent in 1973.[43] Finally, science is not likely to be excused from blame for a sense of unfulfilled promises among the discontented. More fundamentally, of course, anyone concerned with justice or the educator's mission will want people, regardless of their political potency, to share in the fruits and understanding of science.

Complexity of Individual Conceptions

There is a fundamental sense in which commonly made statements about "the public" gloss over the complexity of reality. Not only are the various segments of the public not of one mind about science, nor quick to change in their appreciation of it, but the *same person* often has a very complex set of feelings about the various facets and meanings of science, and the picture a survey produces—whether attitudes are more or less modern—depends on which facet it mobilizes in each person.

Gerald Holton has delineated seven basic types of popular images of science, and their shortcomings.[44] The first image, whose source he traces to Plato, stresses the unity of theory and action, pure thought, and pragmatic power. One suspects that of all the public images described, this one comes closest to portraying science as scientists would like it to be portrayed. Four of the remaining seven images discussed by Holton can be regarded as largely antiscientific. These view science, respectively, as destructive of established religion; as the fountainhead of ethical perversion; as a tool too dangerous to be entrusted to man; or as potentially disastrous to the normative order. A sixth image is scientism, an addiction to science rather similar to a ritualistic allegiance to bureaucracy. The seventh image is based on scientific ignorance combined with unbounded credulity, and depicts scientists as magicians capable of performing any feat. Holton ventures the opinion that these seven basic images are not mutually exclusive. We agree; in fact, we believe that most people's attitudes toward science and scientists are based on a varying mélange of such concepts, including elements which seem to contrast sharply.

The data already cited suggest that a large number of people approve of science for its usefulness as a technological golden goose, as a source of greater health and wealth, and to some degree as a means for solving social problems.[45] They have much less appreciation for it, however, as an approach to the world, as an exciting or aesthetic experience, or as a great puzzle solver. While it is difficult to obtain data on the scope of public endorsement for noninstrumental aspects of science, there are a few findings which are suggestive. Withey reported from the Survey Research Center study that, in 1958, only about one respondent in ten talked at all about controlled experimentation, the scientific method, measurement, systematic variation, theory, or similar notions.[46] He commented that "probably not more than 12 percent of the adult population really understands what is meant by the scientific approach."[47] A full quarter of the people interviewed freely admitted that they did not know what it meant to study something scientifically. "For about two-thirds," Withey wrote, "science is simply thorough and intensive study, which is, in a way, an adequate label."[48] Moreover, Withey suggests that even those who claim to comprehend science frequently understand it only in a superficial or hollow way. Reading the interviews of those who define science in terms more consistent with a scientist's conception, Withey says, "the sensitive reader . . . is aware . . . of a lack of insight and understanding. It seems as though these people see a stage-set façade, a house front without a house."[49]

The complexity of public understanding of science becomes even more apparent when one considers the correlations among different questions asked of the same individuals Using the 1957 Survey Research Center data[50] dealing with fears that science would lead to rapid change, to control by a few, and to a breakdown of moral values, Oppenheim constructed an index of the threat perceived from science.[51] She found that 34 percent of those who thought science should be limited to that which would produce practical payoffs also feared science intensively. In addition, among those who valued science for less instrumental reasons, although a majority were not threatened by it, a 25 percent minority *were* highly threatened. Thus, even a more noninstrumental appreciation of science is not an unambiguous indicator of a clear-cut pro-science attitude.

Clearly, individual attitudes toward technology are also complex. Robert Lee of IBM found that a large fraction of a nationwide sample simultaneously held attitudes toward computers that could be characterized as both pro and con.[52] Taviss reached a similar conclusion using an occupationally stratified sample from the Boston area.[53] Thirty-three percent of her sample scored high on both pro- and anti-technology statements such as "Machines have made life easier," and "People today have become too dependent upon machines." Another 13 percent were low on both. Thus, while a majority were consistently for or against technology, a large percentage held a more complex mix of attitudes.

The mix of religious and scientific beliefs further serves to illustrate the point. Although science and religion can be seen as representative of rational and nonrational views of the world, studies have revealed that people often believe concurrently in them both. Among those we might least expect to hold religious views—natural scientists, engineers, and natural science graduate students—studies report that a majority hold religious views and commitments.[54] The general public seems to be similarly unconcerned about a conflict between science and religion. Several studies show only small minorities, even among the most orthodox, who consider the conflict very serious or even somewhat serious.[55] Many people, furthermore, offer both scientific and nonscientific explanations for empirical events. For instance, "In the sample of low-income, urban mothers . . . nearly a fourth thought that 'God's Will' was a source of some illness, even though they included in their answers naturalistic sources as well."[56] Finally, the 1957 Survey Research Center study showed that 70 percent of their national sample believed that the world was controlled by God and, at the same time, 83 percent thought the world was better off because of science.[57] To be sure, we need more information on the extent and nature of this seemingly persisting mix of the rational and nonrational, but these examples should at least underscore the point that attitudes of individuals are a complex mix, and great care needs to be taken if we are to understand the public's mind regarding science instead of becoming totally confused.

These figures are best viewed as early approximations that suggest layers upon layers of ambivalence and diversity in the public understanding of science. They suffice, however, to underscore the importance of not assuming that the public is simple-minded about science. If there are simplifications, they are often in the minds of observers, not of those observed.

Conclusion

It is necessary to reiterate that the preceding discussion is based on rather thin data, most of it dated. As we have seen, the last major study of our subject was conducted half a generation ago, and although it is important, it is chiefly descriptive. We know even less about the factors, vectors, and dynamics underlying the status of science in the public mind today.

The available data do support the central thesis that, although science has lost some support, many of the defectors are among politically weaker, less-informed, less-educated groups. Nor can these groups be said to be simply disaffected; our data suggest that a person's attitudes toward science are a complex set, which can be mobilized for or against science, depending on which facet is activated.

If an encompassing, updated, analytic study were done which supported the tentative findings we have presented here, these conclusions would follow for those concerned with the legitimacy of science. Interventions in public affairs tend to be most effective when the problem they seek to correct is of the "middling" size, as the disaffection from science seems to be. A much more severe disillusionment, a true crisis of legitimation, might well be impossible to reverse, and, of course, it would not be worth working to reverse a disaffection of minor proportions. The scientific enterprise seems to be in a state where it could benefit from a major effort to broaden and deepen the public's understanding of science. Of all American institutions, science seems to be the least understood by the wider public. And, spreading science information and educating various publics to its values seem to be relatively effective in improving attitudes toward science. Therefore, a major campaign to inform and educate the public would yield more understanding and support than such campaigns usually yield.

What the theme or themes of such an educational effort should be is difficult to agree upon. On the face of it, informing more people of more scientific facts seems beneficial. Of all the findings, however, this one seems most suspect, because by and large mere cognitive inputs do not change preferences. Nevertheless, the very limited data that we have at present do suggest that, in the case of science, information does increase trust.

Beyond this, matters get more complex. The overwhelming majority of the public seems to confuse science and technology and sees science in a very technological, instrumental light. Should a large effort be made to separate the two realms in the public's mind? Should the public, as part of their general humanistic education, be encouraged to appreciate the internal virtues of science per se in order to relieve the pressure on science to be "productive" and "relevant"? Or should scientists and science educators to some extent accept the public challenge and ask themselves whether indeed all science topics are of equal merit, or whether more attention should be paid to the social consequences of various science projects, in which case they would have to focus more on those which are potentially more socially beneficial? To reach the most indifferent segments of the public, should the goals of science be modified so that they are more relevant socially and economically to "forgotten Americans?"

These questions must be answered by dialogue within the scientific community, and between the scientific community and representatives of the public, rather than by a pair of social scientists like the authors. One thing, however, is clear. Whatever the answers, they will affect the way the public of the future will view science and its legitimation. Although science is unlikely to be regarded as unimportant in the foreseeable future, it *could* lose its status as a prime institution and activity of society and return to what it was just before the onset of modernity—a relatively esoteric unsupported activity.

REFERENCES

1. We are indebted to Pamela Doty, Carol Morrow, and Nancy Castleman for their research assistance in preparing this paper.

2. Edward Shils, "Centre and Periphery," *The Logic of Personal Knowledge* (Glencoe, Ill.: Free Press, 1961), pp. 117-130. See also discussion of rationality in Amitai Etzioni, *The Active Society* (New York: The Free Press, 1968), pp. 264ff; Herbert McClosky, "Consensus and Ideology in American Politics," *American Political Science Review* (June 1964).

3. Robin Clarke, "Technology for an Alternate Society," *New Scientist*, January 11, 1973, p. 66.

4. Jurgen Schmandt, "Crises and Knowledge," *Science*, 174 (October 1971), p. 1.

5. *Time*, April 23, 1973, p. 83.

6. Jerome B. Wiesner, "Technology Is for Mankind," *Technology Review* (May 1973), p. 10.

7. Kenneth S. Pitzer, "Science and Society: Some Policy Changes Are Needed," *Science*, 172 (April 1971), p. 226.

8. Stephen Withey and Robert C. Davis, "The Public Impact of Science in the Media" (Ann Arbor: University of Michigan, Survey Research Center, 1968).

9. Stephen Withey and Robert C. Davis, "Satellites, Science, and the Public" (Ann Arbor: University of Michigan, Survey Research Center, 1959).

10. Karen Oppenheim, "Acceptance and Distrust: Attitudes of American Adults Toward Science," unpublished Master's thesis, University of Chicago, 1966.

11. Withey and Davis, 1957, *op. cit.*

12. *Ibid.*

13. Louis Harris and Associates have kindly permitted us to reanalyze their data poll results on the public's confidence in American institutions. The NORC data are from their 1973 General Survey of the National Data Program for the Social Sciences, National Opinion Research Center, University of Michigan.

14. Harris poll, *The New York Times*, December 3, 1973.

15. Withey and Davis, *op cit.*

16. Robert S. Lee, "Social Attitudes and the Computer Revolution," *Public Opinion Quarterly*, 34 (Spring 1970), pp. 53-59.

17. Todd R. LaPorte and Daniel Metlay, "They Watch and Wonder: The Public's Attitudes Toward Technology," Working Paper #6, 1972.

18. The Harris Survey news release, February 17, 1972.

19. Seymour M. Lipset has recently concluded that the data do not support prevailing characterizations of American youth as a monolithic bloc alienated from our institutions; see *Rebellion in the University* (Boston: Little Brown, 1971).

20. Oppenheim, *op. cit.*

21. *Science Indicators 1972* (Washington, D.C.: U.S. Government Printing Office, 1972).

22. *Newsweek*, "Campus '65," March 22, 1965, p. 45, data from a Harris survey.

23. Ray Funkhouser and Nathan Maccoby, *Communicating Science to Non-Scientists, Phases I and II* (Palo Alto: Stanford University, Institute for Communication Research, 1970).

24. Daniel Yankelovitch, survey reported in *Fortune* (January 1969).

25. Oppenheim, *op. cit.*

26. Irene Taviss, "A Survey of Popular Attitudes Toward Technology," *Technology and Culture*, 13 (1972), pp. 606-621.

27. Oppenheim, *op. cit.*

28. *Ibid.*

29. Wilbur Schramm and Serena Wade, *Knowledge and the Public Mind* (Palo Alto: Stanford University, Institute for Communication Research, 1970).

30. Robert K. Merton, *Social Theory and Social Structure* (Glencoe, Ill.: Free Press, 1957), Ch. 16.

31. Roger Brown, *Social Psychology* (New York: Free Press, 1965), Ch. 10.

32. T. W. Adorno, Else Frankel-Brunswick, D. J. Levinson, and R. N. Sanford, *The Authoritarian Personality* (New York: Harper and Row, 1950), p. 461.

33. Melvin L. Kohn, *Class and Conformity* (Homewood, Ill.: Dorsey Press, 1969), p. 81.

34. Samuel Stouffer, *Communism, Conformity and Civil Liberties* (New York: John Wiley, 1955), p. 96.

35. James Martin, *The Tolerant Personality* (Detroit: Wayne State University Press, 1964), p. 112.

36. Milton Rokeach and A. Eglash, "A Scale for Measuring Intellectual Conviction," *Journal of Social Psychology*, 44 (1956), pp. 135-141.

37. Lee, *op. cit.*

38. Stouffer, *op. cit.*

39. Clyde Z. Nunn, "Tolerance of Nonconformity in the U.S.: 1954-1972," a paper presented at the 1973 meetings of the American Sociological Association, New York City.

40. Seymour M. Lipset, "Working Class Authoritarian," Ch. 4 of *Political Man* (New York: Doubleday, 1960).

41. Lester Milbrath, *Political Participation* (Chicago: Rand McNally, 1965), pp. 114-130.

42. The Harris Survey, *New York Post*, December 6, 1973.

43. Stouffer, *op. cit.* and a recent replication of the Stouffer study by Clyde Z. Nunn, Harry J. Crockett, Jr. and J. Allen Williams, Jr. (forthcoming).

44. Gerald Holton, "Modern Sciences and the Intellectual Tradition," *Thematic Origins of Scientific Thought* (Cambridge, Mass.: Harvard University Press, 1973), pp. 445-460.

45. Withey and Davis, 1957, *op. cit.*

46. *Ibid.*

47. *Ibid.*

48. *Ibid.*

49. *Ibid.*

50. *Ibid.*

51. Oppenheim, *op. cit.*

52. Lee, *op. cit.*

53. Taviss, *op. cit.*

54. Ted R. Vaughn, Douglas H. Smith, and Gideon Sjoberg, "The Religious Orientations of American Natural Scientists," *Social Forces*, 44 (June 1966), pp. 519-526; Andrew M. Greeley, *The Denominational Society* (Glenview, Ill.: Scott, Foresman, 1972), p. 146.

55. Gerhard Lenski, *The Religious Factor* (Garden City, New York: Doubleday, rev. ed. 1961), pp. 281-282; 1973 NORC General Survey, *op. cit.*

56. Clyde Z. Nunn, John Kosa, and Joel J. Alpert, "Causal Locus of Illness and Adaptation to Family Disruptions," *Journal for the Scientific Study of Religion*, 7 (1968), p. 211.

57. Withey and Davis, *op. cit.*

DAVID PERLMAN

Science and the Mass Media

How do journalists who act as the information interface between science and its clients do their job today? We do it, most of us, with varying success. We try to separate science and technology in our writing for the public, and we succeed, I think, more often than not. We try to present news about scientific developments in the context of science as a continuing process, and when space and deadlines permit, at times we succeed. We try to avoid political bias or advocacy when we cover the interaction between science and public affairs, and here we almost always do. We aim for accuracy and try to shun sensationalism, but here our critics say we too often fall short.

Let us look at a recent example of reporting on fundamental science and technology combined: the long voyage of Pioneer 10 to Jupiter, which climaxed December 3, 1973 with the spacecraft's precisely executed fly-by, 81,000 miles above the planet's cloud tops. There was no political controversy here (although there may be doubters among the public who cry "boondoggle" at a $50 million spacecraft measuring fields and particles, and photographing Jovian moons that Galileo spotted more than 350 years ago).

The Pioneer mission was a major event in planetary astronomy, a superb feat of technology, a collector of fundamental scientific data about important physical phenomena, and a forerunner of other interplanetary missions that will tell us much about how our solar system was formed.

For science writers, Pioneer offered an opportunity to treat the public to something quite unique: a glimpse of science in its moment of pursuit. As the spacecraft's radio data poured in to the Missions Operations Center at the Ames Research Center of the National Aeronautics and Space Administration in Mountain View, California, the thirteen principal investigators joined reporters to engage in daily colloquies of considerable depth. More than 130 reporters, photographers and television crew members were registered in the special press room set up at Ames. At least two dozen reporters—many of them specialized science writers from major newspapers, magazines and wire services—attended every briefing by the scientific teams. Those briefings began a week before Pioneer's rendezvous with Jupiter, and continued each day until a week after encounter. Some of the discussions were relatively brief hour-long progress reports on the functioning of Pioneer's array of instruments; others lasted up to three hours, and offered detailed interpretations of complex data.

Among the scientists some, like James Van Allen of the University of Iowa, who was investigating Jovian radiation, were veterans of press encounters; others, like Tom Gehrels of the University of Arizona, who directed Pioneer's imaging photopolarimetry experiments, had rarely, if ever, met experienced science writers before. During the briefings, the reporters pressed for details on what the data from Pioneer meant. The scientists, elated at the mission's continuing success and the near-perfect functioning of their instruments, were eager to answer. They covered blackboards with speculative diagrams; they built magnetic field models out of wire, and pushed the wires into new shapes with each new set of readings from Pioneer. They interpreted S-band occultation readings on the spot. They likened the huge planet's magnetic field to a soggy, leaky, bulging doughnut. They debated whether it rotated with the planet, or not; whether its dipole center was offset from Jupiter's equator or not; whether its trapped radiation extended 4 million miles out, or only 2 million miles. Energetic electron fluxes rose predictably; proton fluxes less so. With every new batch of data, the experimenters devised new models, or altered old ones.

This was "instant science," as the reporters dubbed it, and somehow, we felt, we were able to capture for our readers the legitimate excitement of the instant in terms of its purely scientific content. We were able to help our readers share the sense of wonder which Victor Weisskopf contends is implicit in science as "the greatest cultural achievement of our time."[1] We didn't have to write about plastic flags on the moon, or talking satellites, or photograph astronauts floating upside-down like the men of *2001*. We could report a purely scientific mission to explore man's home solar system, and the new concepts it was generating. We could try to convey in dramatic terms the magnitude of the universe as Pioneer headed out beyond Jupiter, beyond the solar system, to fly, perhaps, forever—"If," as one journalist reminded his readers, "there is such a thing as forever."

This sense of excitement, of discovery, of questioning, imbued much of the Pioneer news coverage with a flavor that commanded front-page attention. Newspapers all over the country featured the Pioneer story prominently for days, and even science writers who were not sent personally to California to cover the story wrote interpretative commentaries as Jupiter yielded up its secrets. The television networks carried news reports on the mission every day, with interviews, charts and models to explain the data. The Johnny Carson Show featured Carl Sagan for half an hour speculating on pre-biotic chemicals in the Jovian atmosphere. *Time* and *Newsweek* both published detailed articles on Jupiter, its moons, and its magnetic field.

Editorial comment in major newspapers accurately identified the mission's meaning:

Pioneer 10 is an extension of mankind, a probing of the unknown. We treasure the facts we gather, but mankind's glory is the questions it asks.

Houston Chronicle, December 14, 1973

Pioneer 10 is a triumph of technology, but it is more than that, it is another successful attempt by endlessly curious man to puzzle out the secrets of the endless universe and then, some day, turn them to his own use.

Los Angeles Times, December 5, 1973

Now the data from Pioneer 10 inaugurate a qualitatively new stage in man's knowledge of this giant planet and in his understanding of how its physical and other properties relate to the larger problem of the origin of the solar system and of the universe.

New York Times, December 4, 1973

Yet despite occasions such as these, when a scientific event captures the public imagination, it seems clear that only a small fraction of the public understands science as scientists or science reporters might hope they would. Amitai Etzioni reports, elsewhere in this volume, that while public confidence in the institutions of science has declined in recent years, the decline has not reached crisis proportions. Perhaps not. But there is no evidence that public confidence is growing.

A gifted, mystical friend recently sent me a poster bearing a picture, a tarot card of The Fool. The poster's message is this:

> To arrive at the simplest truth, as Newton knew and practiced, requires years of contemplation. Not activity. Not reasoning. Not calculating. No busy behavior of any kind. Not reading. Not talking. Not making an effort. Not thinking. Simply bearing in mind what one needs to know.

There is no indication here that any thought has been given to how "simple truth" is really reached: no thought of Newton's calculus, of the desperately painstaking observations by Brahe, of Darwin's years aboard *Beagle* amidst the armadillos and iguanas, of Michelson's and Morley's mirrors upon mirrors that sought to detect the luminiferous ether, of the x-ray crystallography (and high-spirited chicanery, too, if we are to believe Watson) that pinned down the double helix of molecular genetics. What seems to be new in science's problem of public understanding is an unexpectedly swift shift in public attitudes as reflected not only in Congressional budgets and Nixonian impoundments, but also in fundamentalist textbook rewriters; growing bands of seekers after instant satori; and growing numbers of citizens who seem to equate all physics with nuclear holocaust, all biology with DDT, all psychiatry with mind control, all electronics with Big Brother and the end of privacy.

This development has led to a response from the scientific community. The American Association for the Advancement of Science and the National Science Foundation are at this moment intensifying their programs aimed at increasing the public understanding of science. Experimental conferences, briefings, and community education ventures are underway. The efforts are promising, and in my judgment soundly directed. Some will suggest, as Daniel Greenberg does in *Science and Government Reports*, that this activity is a Pavlovian response to the drying up of real dollars in federal support for science. Greenberg contends that these new programs are "essentially a public relations campaign designed to allay public fears of science and drum up support for increased spending on research."[2] Even if this were the case—and I believe Greenberg oversimplifies it—there is an urgent need to increase the general public's understanding of the processes, the motives, and the results of scientific inquiry.

In 1868, Thomas Henry Huxley offered this statement to an audience of ordinary English workingmen in Norwich:

> I weigh my words well when I assert that the man who should know the true history of the bit of chalk which every carpenter carries about in his breeches pocket, though ignorant of all other history, is likely, if he will think his knowledge out to its ultimate results, to have a truer, and therefore a better conception of this wonderful universe and of man's relation to it than the most learned student who is deep-read in the records of humanity and ignorant of those of nature.[3]

It is this kind of quasi-missionary sense about the search for the universe's truths that the communicators of science—be they scientists or reporters—believe is fun-

damental to a public understanding of science today. Communication is essential
not because Huxley's "humanity" and "nature," or C. P. Snow's "Two Cultures"
are antithetical, but because they co-exist and are interdependent.

My colleagues and I, as science writers, believe the public's understanding of
science can and must be improved for reasons that should be self-evident. The scientific enterprise is expensive; it requires financial support and knowledgeable oversee-
ing by the public. Science yields practical consequences that require public decision-
making: to fluoridate or not to fluoridate; to finance dialysis centers or not to finance
them; to build breeder reactors, or fund fusion research, or both; to engineer genes,
abort after amniocentesis, or screen universally for sickle-cell trait. And above all,
science is and has been mankind's greatest intellectual adventure, as much a part of
our culture as music or art or literature. Surely the mass media have as much business
reporting and interpreting science as they do ballet or baseball.

In a newspaperman's terms science can be a Good Story. My own editors, for example, have no particular bias in favor of science, yet they have headlined on our
front pages such varied subjects as the discovery of distant pulsars; the paleo-
magnetic evidence for polar reversal; the concept of a sea-floor spreading and
continental drift; the historical emergence of theories in molecular evolution, from
Oparin to Ponnamperuma; and the synthesis of biologically active viral DNA.

Recently, the comet Kohoutek made page-one news, first because it was coming
in all its brilliance as "the comet of the century," and later because it failed mis-
erably to live up to its advance billing. Was there a backlash from Kohoutek? Was
science generally, or astronomy in particular, diminished in the public mind as a
result? Did the astronomers or the press fail the public? I think not.

David Cudaback, of the University of California at Berkeley, concedes that he
and his fellow astronomers probably erred in their early public discussions of
Kohoutek by not noting clearly enough that predictions of comet magnitude are
notoriously uncertain, and that Kohoutek itself might, in fact, barely attain naked-
eye brightness at all.

Nevertheless, the backlash appears to have been minimal. There have been few
letters to the editor ridiculing the fallibility of science, and no editorial denunciations
of astronomy. The only real ridicule, in fact, has been against cultists and astrologers
who saw Kohoutek as the portent of the century. Kohoutek T-shirts lay in shops un-
sold, and a paperback book, *The Comet Kohoutek: Greatest Fiery Chariot of All
Time,* which proclaimed that "the harbinger of God is coming!" sold briskly for a
while, then vanished from bookstores.[4]

II

That not all those people "out there" beyond the laboratory are cultists or Archie
Bunkers is evident from the high level of public interest in science. The two major
American newsmagazines, *Time* and *Newsweek,* devote significant space to their
sections on Science and Medicine. Jerry Bishop, the *Wall Street Journal*'s perceptive
chief science writer, frequently publishes long articles on scientific topics far
removed from the day-to-day interests of stockbrokers or investors. Meetings of the
American Physical Society, the Federation of Societies for Experimental Biology, the
American College of Surgeons, the American Chemical Society, the American

Geophysical Union and the American Association for the Advancement of Science are widely covered by the press. Reports in scientific journals are regular sources of news, and often lead to personal interviews with investigators.

Nor are the mass media the only avenues for communicating science to the public. Huxley has been updated with enormous success. For example, in the summer of 1972, a thirteen-week lecture series entitled, "Cosmic Evolution: Man's Descent from the Stars" was held in San Francisco.[5] Its sponsors rented a 500-seat auditorium and worried about filling all the seats, even though the series was opened by Ray Bradbury and featured such luminaries as Geoffrey Burbidge, Melvin Calvin, Sherwood Washburn, Freeman Dyson, and Philip Morrison. On the first night 3000 people showed up. Bradbury gave his lecture twice, the audience was crammed into every aisle, and the overflow listened from the lobby and the floor of the "Exploratorium," a science museum next door. A thousand people still had to be turned away, and the same thing occurred at the next five lectures. The total audience for the series was more than 26,000. The following summer a similar lecture series, called "The Next Billion Years: Our Future in a Cosmic Perspective," was even more successful. This time a larger hall was rented in San Francisco, and the series was repeated in Los Angeles, San Diego and Cupertino, a suburb of San Francisco with a junior college. Total attendance was over 100,000, and educational television stations in each city broadcast the entire series from videotapes. A third series is projected for the summer of 1974, focusing more specifically on the interaction of science and society in solving major problems during the coming century.

Much of the attraction of the series, I am convinced, has lain in the intelligent public's eagerness to understand the processes by which science asks questions of nature. All the speakers shared with the public their own curiosity about the universe and man's place in it, including both the insights and the ignorance that mark scientific knowledge as of today.

III

Science journalists face many problems as they confront audiences on one side, scientists on another, and the traditions of the media in the middle.

The current directory of the National Association of Science Writers lists some 400 active members, engaged, by definition, primarily in the dissemination of science information directly to the public. "Public" is a broad term, for active members include writers and editors at such specialized publications as *Science, Chemical and Engineering News, Medical World News, Physics Today,* and the *Journal of the American Medical Association.* There are also many free-lance writers, a tiny handful of television workers, and a few staffers on general magazines. But among the nation's 1750 daily newspapers and two major wire services, there are fewer than seventy-five full-time science writers all told. The *New York Times,* by latest count, has a science and medical staff of eleven, including two in Washington. *The Boston Globe* has four. The *Washington Post* and *Los Angeles Times* have three each. A handful of other papers, no more than a dozen, boast two science reporters, who usually divide their work so that one covers the biomedical sciences and health care, while the other does the "hard" science and technology, including everything from physics to manned space flight. Lately, some papers have assigned specialized reporters to cover the "environment" beat, the "energy" beat, and "consumer affairs," each of which includes significant components of science and technology.

Until the 1930's, science writing was a virtually unknown specialty in journalism. The NASW was conceived in 1934 in the bar of a Philadelphia hotel during the annual meeting of the American Philosophical Society; its three founders were David Dietz, science editor of the Scripps Howard Newspapers, William Laurence of the *New York Times*, and Robert D. Potter of the *New York Herald Tribune*. A more formal meeting in Washington during that year's meeting of the National Academy of Sciences launched the organization. The first eleven active members included virtually all the full-time science writers in the country. By 1938 there were 18; by 1945 there were 61; by 1963 there were 200; and today there are about 400.

Very few science journalists have any specialized training in the sciences; fewer than one-quarter were science majors in college. Most are college-educated in the liberal arts, and, according to a 1973 survey, nearly half majored in either English or journalism and spent several years in more general newspaper reporting before specializing in science. Younger science writers today are beginning to come from the ranks of undergraduate science majors, but there is little formal training in the field. A few university journalism departments offer seminars or single courses in science writing; one or two provide "internships" with experienced science writers in major metropolitan centers. But, by and large, entry into the field is unsystematic at best. I am not sure this is all bad. I, for one, am without formal scientific training, but I think of my own rather typical job in science writing as a full-time, perpetual fellowship to a graduate school with an endlessly varied and endlessly challenging curriculum. And all I have to produce each year, instead of a dissertation with footnotes, is a file of 100 to 150 newspaper articles, most of them written under a daily deadline, and each running in length from 500 to 1200 words!

Most newspaper editors, it is fair to say, still cling to the journalistic tradition that names make news, that "human interest" is essential, and that any good reporter can cover any story. So specialists like science writers are not always considered team players. They demand expensive reference books and shelf room for journals. They disappear on out-of-town trips to meetings, rocket-launchings, and laboratories. They insist that their stories cannot be written in less than two columns. They balk at explaining what a proton or a molecule is every time they use such words. They argue—all too often in vain—with the copy editors who incorrigibly top their carefully qualified stories with headlines heralding "cure" and "breakthrough."

Science writers are not always loved by scientists, either. And if we are inaccurate, or we sensationalize, we deserve our lumps. But Edward Shils, for example, is also concerned that the press airs controversies where, as he puts it, scientists publicly attack their own colleagues for such sins as "not serving society." Shils has said:

Journalists, especially those involved in reporting and commenting on science policy, pick up all these criticisms. They delight in finding the "establishment" of science in the wrong. . . . Their animosity against government makes them critical of governmental science and of the scientists who perform it. They like to catch them in contradiction or impugn the veracity of their statements.[6]

We specialized journalists who cover science and science policy consider ourselves reporters with a particular mission. We are in business to report on the activities of the house of science, not to protect it, just as political writers report on politics and politicians. At times these activities may seem ignoble. If "politically adroit and accomplished senior academicians" edge out promising young investigators in the scramble for limited grant funds, as the President of the American Federation for Clinical Research has alleged,[7] that is legitimate news in a story analyzing the funding problems of science.

In 1971, research teams at the University of Southern California and Georgetown University simultaneously announced the discovery of what the *New York Times* called "two candidates for the ominous title of human cancer virus."[8] Neither group had as yet published its report in a refereed journal, but each group summoned a press conference for its announcement. Six months earlier a group at the M.D. Anderson Tumor Institute in Houston had also reported a possible "candidate" virus, publishing in *Nature* and subsequently in a press release.[9] It seems to me fully appropriate that a perceptive and knowledgeable science reporter should comment on the background of such races for priority and note: "The development of the controversy also reflects the pressure of the present political maneuverings to legislate a cure for cancer."[10]

Surely it is an obligation for science writers to report most carefully on the qualifications of those who claim scientific expertise, and at times to seek to balance their controversial statements. An eminent physicist may expatiate on his theories linking race and I.Q. When he does, however, the conscientious science writer will note the physicist's track record in genetics, and place his conclusions in the context of the ongoing controversy over race as a definable concept and I.Q. as a measurable culture-free entity; the reporter may even solicit comments on the physicist's message from geneticists known to oppose his views.

No formalized code of journalistic ethics covers this kind of reporting. But reporters who specialize in any field—be it science, politics, labor, or education—are expected on most newspapers to interpret the background of complex issues in their field. Their responsibility goes beyond the traditional "who-what-when-where" of elementary police-court reporting.

Science writers, in short, are now exercising "clinical judgment" on many of the stories they cover. A virologist, for example, reported in 1971 that virus-like particles had been found in the breast milk of nursing mothers with a family history of breast cancer. "Look," he said in response to questions during a press conference at the National Academy of Sciences, "if a woman has a familial history of breast cancer in her family and if she shows virus particles, and if she was my sister, I would tell her not to nurse the child." Here was a story, from a most eminent source, in a vitally important area of human interest, that could have alarmed millions of women. How did the press handle it? In the *New York Times* Harold Schmeck discussed the research that led to the finding of the virus particles; he mentioned the scientist's reservations about nursing in the fourth paragraph of his story, and immediately added: "He and other scientists emphasized, however, that the particles had not been proved to be viruses related to the cause of breast cancer in humans." In the *New York Daily News* Edward Edelson wrote this interpretive paragraph: "There is no definite proof that the suspected virus causes breast cancer. Even if the milk agent does cause breast cancer, there is no definite proof that transmission of the particle in human milk is responsible." And for the Associated Press Frank Carey wrote this lead: "A

Columbia University cancer researcher said Tuesday new findings suggest a conceivable though wholly unproven danger that some breast-nursing mothers may
transmit a potential for breast cancer to their female babies."[11]

Whether we science reporters are backgrounding the technological problems of
the energy crisis, the biological activity of synthetic DNA, the cosmological
questions of quasars, or the Hunting of the Quark, we try never to forget our basic
role as journalists. Our stories must be compellingly told, for if they are not, neither
our editors nor our readers will accept them. We try to point out how even the most
fundamental and arcane research has its relevance to human purposes. Yet, since the
rubric of "science news" covers such an unbelievably broad range of topics, it is no
wonder that science writers often feel frustrated by their inability to remain familiar
with it all, to translate it all, to convey the nature and implications of scientific
progress promptly yet accurately.

There are aids and obstacles to covering the science beat. The organizers of most
major meetings have long since learned to retain experienced public information
staff to distribute texts or abstracts of papers to the press in advance, and to help
arrange interviews and press conferences when science writers request them or the
organizers feel that particular papers are especially newsworthy. The major scientific
institutions—government agencies, universities, medical centers, "think tanks," and
technically based corporations—employ professional public relations staffs. Most
science writers read or scan regularly as many major scientific journals as they can.
And all science writers, wherever they are based, try to enlist the best scientists in
their neighborhoods to help them simplify formidable journal language and focus
appropriately on the significance of new developments. I do not often encounter a
scientist who, when he understands the kind of help I need, refuses to discuss his
work or to help me interpret the work of others in his field. If I am reporting results
of a scientist's own research, I am perfectly willing, when asked, to hold up my story
until the research has been published in a journal or presented at a meeting.
"Refereeing" is at least tentative assurance that the story is valid. If the story is complex and technically difficult (and deadlines permit), I am eager to have a scientist's
help in checking the facts in my account. I may show him my piece, or read it over
the phone, but always with the most careful prior understanding that only the facts
are at issue, never my personal interpretative comments, my emphasis, or my writing
style.

After a few years of experience, most science writers find they have developed
personal contacts among the professionals in the fields they cover. Walter Sullivan of
the *New York Times* is virtually a card-carrying geophysicist by now, he has written
so often on the subject; because of his personal interest in it, he has developed a wide
circle of scientific friendships ever since he first covered the International
Geophysical Year in 1957-1958. Among my own personal friends—first encountered
because I covered news stories involving them—I can number a radio astronomer, a
neurophysiologist, a cardiac surgeon, a biophysicist, a biochemist, a nuclear
physicist, a radiologist, a population biologist, and a number of environmental activists who also happen to be good scientists. Any one of these men and women is
willing to help me when I need a guide—anonymous or not—through some scientific
thicket.

Many scientists, however, still flatly refuse to cooperate with reporters—some because they have in fact been misquoted or misinterpreted in the past; others because they mistrust the press generally; still others because they really do cling to the outmoded idea that science is none of the public's business. Similarly, those who cover the interface between science and political affairs from Washington often encounter public information officers in federal agencies who withhold information or distort it through mistaken zeal.

In some scientific institutions—a few grant-hungry universities and high-technology corporations are the worst offenders—public relations writers exaggerate the achievements of their scientists and proclaim "breakthrough" in wholesale press releases. I recall one handout from a university in Texas, dealing with a very minor development in accelerator design, where the writer did not quite have the nerve to use the word "breakthrough." The development, he wrote, was a "major ripple" in the world of physics!

IV

There are, of course, always the problems of competition—among the mass media for "exclusive" stories, and among the varied happenings world-wide that mark each day's news budget. More than a million words a day will pour onto the desks of a major newspaper's editors; only a fraction can be printed. In the scramble for space or air time, does the overthrow of parity outrank the overthrow of a Latin American government? Does the synthesis of growth hormone overshadow *The French Connection?* Are Kohoutek and Pioneer 10 more spell-binding than the aftershocks of Watergate? The science drama in the skies—despite Kohoutek's poor performance—achieved prominence last winter partly because many editors decided with canny news judgment that it would provide welcome front-page relief from a depressing spate of aircraft hi-jackings, White House tape erasures, stock market plunges, and energy brown-outs. The public learned a lot of astronomy last winter, as well as politics. In a less ominous time the science might not have been as welcome.

There are no orderly rules for this kind of decision-making in journalism, for journalism deals with news, and news is ad hoc; it happens and it must be reported *now*. True, there are Sunday sections, and a few papers feature weekly science pages where background articles, sometimes written by scientists themselves, are purveyed. There is a strong suspicion among many reporters, however, that these special pages and sections, because they are dull, wind up pinned to high school bulletin boards, largely ignored.

Newsmagazines have time to be more reflective, and their science and medicine sections frequently produce extremely well-backgrounded accounts of new developments. But this country does not boast a single broadly based popular magazine of science with a wide circulation. *Science News*, a sixteen-page weekly magazine, is published in Washington by the nonprofit Science Service, Inc.; according to E. G. Sherburne, the magazine's publishers, it has a circulation of about 100,-000, made up largely of "scientists and engineers, science teachers, students and interested citizens," and is now in "pretty good shape" financially. *The New Scientist*, published in England, fares somewhat better; it has a circulation of 70,000, only half of which is in the United Kingdom. It is part of the profitable International

Publishing Corporation and is self-supporting.)

Scientific American is superb in its rather special field, that of communicating primarily to scientists, engineers, executives in technical management, and scientifically literate laymen. Since it crosses all disciplines, however, its publisher, Gerard Piel, argues that every reader of the magazine is a "layman" in every field the magazine covers except his own. *Popular Science* and even *Popular Mechanics* occasionally publish articles about scientific and technological developments, but hardly do much to convey the true spirit of science or the nature of scientific work. The *National Geographic* ranges now and then into astronomy, geophysics and oceanography, but not much further. The women's magazines present clear and accurate information at times on developments in clinical medicine, but their style is breathless and their focus is heavily biased toward nutrition, contraception, sexuality and new approaches to breast and cervical cancer. *Harper's*, the *Atlantic Monthly* and *The New Yorker* carry occasional excellent science articles, but these magazines have such small circulations that "mass media" becomes a misnomer.

V

Television, the newest of the media, has been called "the chief popular form of discourse in a technological society" by Michael Ambrosino, producer of science documentaries for WGBH, the Boston educational TV station. But in terms of continuing discourse between scientist and citizen, American commercial television is the most bankrupt of the mass media. Except for major developments that make the front pages of virtually every newspaper, TV networks pay little attention to science news; they seem to save their talents for the spectacular.

When Apollo 11 carried three astronauts to the moon, for example, commercial television truly shone: for nearly two weeks the American public was saturated with coverage every step of the way. Lunar scientists, armed with magnificent mockups and models and animated drawings, discoursed on maria and craters, meteoroids and volcanism, gravity and the whole history of the solar system. Every network dedicated itself to the enterprise. At CBS, no fewer than 1000 employees were involved in the Apollo coverage; behind Walter Cronkite and Eric Sevareid stood 37 correspondents, 9 researchers assigned from the regular CBS news research staff, 6 special scientific and technical consultants, 22 producers, 13 directors, 16 engineers and 65 guest discussants who ranged all the way from Lyndon Johnson and Spiro Agnew to Sir Bernard Lovell, Harold Urey, James D. Watson, Buckminster Fuller and Gloria Steinem! Four corporate sponsors—Western Electric, General Foods, International Paper, and Kellogg's—helped pay for the effort, although the network will not disclose cost and revenue figures. And all that was only CBS. The efforts of NBC and ABC were equally valiant. Apollo 11 was covered like the great human adventure it truly was; and if anyone in America failed to learn at least a little about space flight, selenology and Newtonian physics along the way, it was not television's fault.

But what of the prime-time hours since? Television news programs do indeed cover new developments in science from time to time; occasional interviews on talk shows do try to elucidate complex technical controversies. But Ambrosino has noted that during the 1972-1973 TV season—leaving out coverage of the Apollo 17 flight—fewer than 25 of 4,368 prime-time commercial network hours were devoted to science documentaries which explored topics in depth. There is, then, virtually no biology, no behavioral science, no physical science on everyday television.

According to records from the networks, when Pioneer 10 flew past Jupiter last December, ABC television carried three three-minute reports on three successive evening news programs. CBS News presented five successive spots of two to three minutes each. NBC did better: the *Today Show* featured two discussions of Pioneer's flight results—one six minutes long, the other nine minutes. At encounter time, NBC News described the event in five reports both morning and night, while NBC's syndicated service fed daily stories of two to three minutes each to some 200 independent TV stations. United Press International, whose *Telenews* service feeds film to 700 client TV stations, covered the Jupiter encounter with two three-minute news spots.

From personal experience I can testify how less spectacular science news may typically reach the public. A major meeting takes place in a large city—a meeting, say, on geophysics. A press conference is held at which scientists from Columbia University, California Institute of Technology, and the U.S. Geological Survey discuss efforts at earthquake prediction. They explain a new theory of crustal dilatation, and describe how they and Soviet geophysicists have used such parameters as changes in strain gauges, changes in well-water levels, and changes in the radioisotope content of ground water to foretell earthquakes. They contend that a few quakes have been predicted—although not yet with precision as to time, magnitude or exact epicenter—and that new analyses of old records before the 1971 San Fernando quake show how that devastating temblor might have been forecast.

This press conference lasts, perhaps, an hour. Science writers for a handful of newspapers and the two wire services take notes, ask questions, and prepare to write articles that may run between a few hundred words and a thousand. Their deadlines will come in a very few hours, but some reporters will later write more reflective, interpretive, longer feature articles based on further interviews and fresh background reading.

At the end of the press conference, the television reporters will seek out the scientist whose explanations have seemed the most colorful and articulate, and ask him if he can please explain the whole matter in two or three minutes. The scientist, if he is adroit and cooperative, may take four or five minutes. On the six o'clock news that night his explanation will be cut down to thirty seconds or a minute, with another minute for the TV reporter's summary, often rewritten from a press release prepared earlier by the public relations office of the scientist's institution. Nor is such coverage surprising when one considers that a major local television station in a big city could announce with pride recently that its new chief newscaster "is well known for his services as master of ceremonies at a wide variety of banquets and civic functions, and has served as singing 'emcee' at 70 beauty pageants. . . ." "The News Department's catch-phrase is 'never a dull moment'—and they mean it," says this station.

In England, where the BBC is entirely financed by a tax on radio and television receivers, the quality of science coverage is very different. BBC's science department is advised by a nine-member committee of scientists (six are Fellows of the Royal Society) which meets twice a year with the network's science producers and writers. The BBC's Features Group, a close-knit and independent team with a $20 million yearly budget, produces some 600 programs a year in 32 different series covering every area of interest. Ten of these series involve science, and three, notably the *Horizon* science documentary series, have been among the finest ever made. Several *Horizon* programs have been purchased and shown in the United States.

The most recent BBC science series, and one of the most successful, is *The Ascent of Man*, a highly personal view of the history of man's continuing discovery about his world and his universe, conceived and narrated by Jacob Bronowski. The budget for its 13 installments was $3 million. As of last February, Time-Life Films, its American distributors, had not yet sold it to a commercial American network, although Bronowski's book with the same title has already been published and the films have been released for educational use.[12] Michael Ambrosino says of the series: "It is filled with passion and visual excitement. It's a personal, idiosyncratic view of science. It's Bronowski's view, and a noble one."

Except for a few outstanding films on nature like those of Jacques Yves Cousteau and Jane van Lawick-Goodall, it is outrageous that commercial American television should be so bereft of material in an area that can produce so much visually satisfying, entertaining, and enlightening information on a most vital aspect of human culture. Fortunately, the outlook for more effective TV communication of science, at least via the public channels, is improving right now. In 1973-1974 two major projects were launched. The more ambitious is that of Ambrosino's "Science Program Group for Public Television" at WGBH in Boston. Its initial series of 13 weekly documentaries, entitled *Nova*, opened in March, 1974, over the 234 stations of the Public Broadcasting Network. The series deals imaginatively with varied aspects of science, research, and technology: the origins of life in the solar system, intelligence in cetaceans, a stone-age Brazilian culture, the history of anesthesia, reactors and the prospects of fusion power, astrophysics and the Crab Nebula in history. Some of the programs were produced at WGBH, some were made jointly by the BBC and the Boston group; and some were purchased from other documentary film makers.

The Boston group's initial planning was financed by a $40,000 grant from AAAS. Funds for the first year's films, totaling $1.5 million, came from the Carnegie Corporation of New York, the Corporation for Public Broadcasting, the National Science Foundation and the Polaroid Corporation. For the future, given adequate funding, the Science Group plans a separate series in the behavioral sciences; a series for children involving activities that can lead to scientific questions ("we would climb a mountain, build a flying machine, predict the weather, study a cubic yard of dirt, excavate a site, plant a garden," says Ambrosino); a series on archeology, anthropology, linguistics, and history; and a special film project to provide expert up-to-date backgrounding of topical and controversial issues arising from science and technology.

The second science project of interest this year has been a five-part series, also on public television, called the *Killers*, financed by the Bristol-Myers Company and produced by David Prowitt at WNET in New York. Prowitt enlisted a team of twenty-three medical consultants to assure the accuracy of programs that dealt with research, clinical advances and prevention in five medical problem areas: heart disease, cancer, inborn genetic defects, trauma, and pulmonary disease.

VI

Surveys of newspaper content show that far more science information and science news is being conveyed to the public today, and with greater accuracy and sophistication, than was the case in the years before Sputnik symbolized the opening of a space age. Newspaper journalism, with its primary focus on immediate events, is not designed as an educational venture; by and large, newspaper publishers are in business to record the day's news and to make a profit doing so—not to serve as an extension of school and college. Even the best of newspaper science articles are ephemeral, though they can still be informative and interesting without sensationalism.

Yet I see real possibilities that science writers and the scientific community can, together, bring about major improvements in the reporting of science news. We science writers, I believe, can and should be more aggressive in persuading our managements that the scientific enterprise merits even fuller coverage because of its drama, mystery, human relevance, successes, failures and newsworthiness. We can, in our writing, show our readers more effectively that science is a continuing process by which the laws and workings of the universe are uncovered step by step, and that each step, however small, is potentially crucial. We need to develop more skill in reporting the full social significance of technological developments. We should show how these developments flow from the basic science that preceded them, and we should more energetically enlist thoughtful scientists who can point out, as quickly and fully as possible, what the social effects—both intended and unintended—are likely to be of new science and technology.

We should report more thoroughly than we do now on the political institutions of science and technology—on the complex processes of public funding, the internal operations of the great science-supporting agencies like the National Institutes of Health, the National Aeronautics and Space Administration, and the Atomic Energy Commission. We should be covering the ebb and flow of power among the scientists who advise, or should advise, the White House and the governors of the states. *Science* and Daniel Greenberg's *Science and Government Reports* do this effectively for their limited audiences, and occasionally their reporting is picked up by the more general press. But many a *Science* story on an AEC reorganization or an impoundment of HEW funds has direct repercussions on academic communities around the nation, and we "provincial" science writers should become far more alert to these issues in our own regions.

If this kind of depth reporting is to be done thoroughly, accurately, fairly, and interestingly, however, it will require a new level of assistance from scientists and technological leaders. These people will have to make themselves far more accessible to the public, not as apologists or propagandists, but as explainers and interpreters.

I see no reason why every component organization in the house of science, from astronomers to zoologists, should not develop cadres across the country to serve as guides when science news needs translation or local scientific developments warrant public notice. The Scientists Institute for Public Information (SIPI) has developed an effective nation-wide network for communication on environmental issues. But where are the physicists? the immunologists? the molecular biologists? Would it be undignified if an astronomer at a local state college were to telephone the news director of a local television station and volunteer to spend a few minutes explaining why Kohoutek fizzled, and to provide some good photographs of earlier comets that didn't? Would an assistant professor of physiology risk refusal of tenure if he invited a newspaper reporter to watch the operation of a scanning electron microscope, and made available some photomicrographs of virus particles for a Sunday feature article? For five years, from 1966 to 1971, Joshua Lederberg wrote his personal interpretations of scientific developments in a column syndicated by the *Washington Post*. Right now two faculty members at the University of California School of Public Health in Los Angeles contribute a weekly column on scientific nutrition to the *Los Angeles Times*, and Jean Mayer's *Boston Globe* column is widely syndicated. This kind of effort by scientists who like to write would be welcomed by scores of newspapers in small cities everywhere. It requires only three elements: the ability to write simply and clearly; the ability to sense what interests the public as well as the scientific community; and a commitment to publish for the public as well as to "publish or perish" for the rewards of tenure and promotion.

Scientists may not always succeed, however, when they try to improve communication with the public. Editors may trim their contributions to meet space requirements. Headline writers may miss their major points. In interviews, scientists may find themselves misunderstood and sometimes bruised by encounters with journalists and the public (although the bruising will prove healthy if it impels scientists to become activists for science education in their local school districts). They may often find their work described in the press with less than academic completeness, and even at times, inaccurately; reporters, after all, can make errors, although they do not knowingly commit malpractice any more than physicians would.

Happily, the American Association for the Advancement of Science and the National Science Foundation are at this moment intensifying their programs aimed at the public understanding of science. Experimental conferences between local government officials and scientists have been held to underscore the contributions science and technology can make to community problems in states and cities. Community education ventures have begun which bring scientists into personal contact with small local communities. On several university campuses in recent years, editors, news executives, and other "opinion leaders" have been brought face to face with scientists in small working conferences that explore both the processes of basic science and the relevance of applied science.

The Council for the Advancement of Science Writing, a small offshoot of the National Association of Science Writers, whose members include both scientists and science writers, sponsors a most valuable series of briefings for journalists—usually three a year—in the physical sciences, health and medicine, and the social sciences. These briefings attract fifty to seventy-five science writers from all over the country, and recruit leaders in the various scientific fields for programs that last from two to five days. Costs are borne by research institutions, foundation grants, and technological industries whose sustaining memberships keep CASW functioning.[13]

To do the urgent job of science writing more and more effectively, to assure an increasingly constructive interaction between the scientific community and the public, my colleagues and I need an unimpeded flow of information. We need scientists who are willing to explain their work, to interpret its significance, to defend its relevance without special pleading. Scientists, in turn, have every right to expect that reporting in the mass media be accurate, and to refuse to have anything to do with reporters whose work is careless or sensationalized. But scientists cannot expect reporting in the mass media to equal the technical detail of journal articles (nor, thank goodness the turgidity!). At the same time, the public has a right to depend on scientists as unbiased technical advisers when public policy issues arise from science, and on science writers to report the issues fairly and objectively.

All scientific inquiry must ultimately serve society, for it is the whole of society that endows science with its charter. The services science performs may be as practical as creating the transistor, or as intellectually exciting as investigating the neurotransmitters of the brain. But science can serve society only if it is healthy, and responsibly independent; and these qualities depend most critically on an informed public. The mass media are the public's principal channels to timely information. They too require health and freedom.[14]

REFERENCES

1. Victor F. Weisskopf, *Knowledge and Wonder* (New York: Doubleday, 1962).

2. *Science and Government Reports*, December 1, 1972, p. 5.

3. Thomas H. Huxley, *On a Piece of Chalk*, with an Introduction by Loren Eiseley (New York: Charles Scribner's Sons 1967).

4. Joseph Goodavage, *The Comet Kohoutek* (New York: Pinnacle Books, 1973).

5. This series came into being under the initial stimulus of Dr. Hans Mark, Director of the Ames Research Center. Both this series and the one given the following summer were co-sponsored by the National Aeronautics and Space Administration, the Astronomical Society of the Pacific, and local educational institutions.

6. Edward Shils, *The Religion of Science*, unpublished manuscript.

7. Kenneth L. Melmon, "New Challenges for the American Federation for Clinical Research: A Progress Report," *Clinical Research*, 22, No. 2 (February 1974).

8. Harold Schmeck, Jr., "Two Viruses Found in Cancer Studies," *New York Times*, December 6, 1971.

9. Elizabeth S. Priori, Leon Dmochowski, Brooks Myers, and J. R. Wilbur, "Constant Production of Type C Virus Particles in a Continuous Tissue Culture Derived from Pleural Effusion Cells of a Lymphoma Patient," 232, *Nature New Biology*, July 14, 1971.

10. Nicholas Wade, "Race for Human Cancer Virus," *Science*, 173 (September 24, 1971).

11. Nicholas Wade, "Scientists and the Press: Cancer Scare Story that Wasn't," *Science* (November 12, 1971).

12. Jacob Bronowski, *The Ascent of Man* (Boston: Little Brown, forthcoming in 1974).

13. The Council's 1974 membership includes: Alton Blakeslee, Science Editor, Associated Press, New York; Melvin Calvin, Professor of Chemistry, University of California, Berkeley; Victor Cohn, Science Reporter, *Washington Post*; I. W. Cole, Dean, Medill School of Journalism, Northwestern University, Evanston, Illinois; William J. Cromie, Director of Research and Development, Field Enterprises Educational Corporation, Chicago; René Dubos, Professor Emeritus of Microbiology, The Rockefeller University, New York; Nancy A. Hicks, Science Reporter, the *New York Times*; Donald C. Kirkman, Science Writer, Scripps-Howard Newspapers, Washington, D.C.; Samuel L. Kountz, M.D., Professor and Chairman, Department of Surgery, Downstate Medical Center, State

University of New York, Brooklyn, N.Y.; Martin Mann, Text Director, Time-Life Books, New York; Philip E. Meyer, National Correspondent, Knight Newspapers, Washington, D.C.; Glenn Paulson, Staff Scientist, Natural Resources Defense Council, New York; David Perlman, Science Editor, *San Francisco Chronicle*, San Francisco; Judith E. Randal, Science Writer, *Washington Star-News*, Washington, D. C.; Albert Rosenfeld, Science Editor, *Saturday Review/World*, New Rochelle, New York; Peter H. Rossi, Chairman, Department of Social Relations, The Johns Hopkins University, Baltimore; Carl Sagan, Professor of Astronomy, Cornell University, Ithaca, New York; Earl Ubell, Director, New York television news, NBC News, New York; Pierre Fraley, former Science Writer of the *Philadelphia Bulletin*, and now CASW Executive Director.

14. The writer gratefully acknowledges the opportunity to prepare this manuscript during his appointment as Regents Professor at the University of California in San Francisco.

GEORGE BASALLA

Pop Science: The Depiction of Science in Popular Culture

IN ONE OF Kurt Vonnegut's novels a young writer persistently and critically questions a research laboratory director about scientists and their ways. The director becomes angry and accuses the writer of believing "that scientists are heartless, conscienceless, narrow boobies, indifferent to the fate of the rest of the human race, or maybe not really members of the human race at all." "Where did you ever get such ideas ?", he continues, "From the funny papers ?"[1]

My answer to the last query would be in the affirmative. The writer conducting the interview, along with many other Americans, is likely to have derived his ideas of science and scientists from a source to which he was introduced in early youth and which he continued to read faithfully in adulthood. It is through the comics and other forms of mass media that a large majority of Americans get their impressions of the scientific community and its work. This portrayal of science in popular culture is not to be confused with what is generally called "popular science". In fact, it is so radically different that I prefer to call it "pop science" and separate it from the more restricted and sophisticated region of "popular science".

Popular science is science presented not to the masses but to a highly educated, and thereby limited, segment of the population; in many cases it is merely one group of scientists writing for another group of scientists and technicians. The advertising material printed in a popular science periodical such as *Scientific American* is a good indication of the journal's readership. The devices, books, and services offered in its pages are obviously intended for academic and/or corporate-technical readers and not for the mass pop science audience that is my primary concern.

Pop science is an aspect of the popularization of science that has never been systematically studied.[2] It deserves our special attention because it reveals some basic American attitudes towards science, technology, and the intellect. By presenting these attitudes in a popular medium and attractive form, the creators of popular culture perpetuate and strengthen them. There exists a feedback loop between widely-held American ideas of science and their popular artistic representation in comic strips, television shows, and feature films.

Scientists are rarely the heroes in the current world of popular culture. More likely, one encounters the pop scientist as a villain who uses his knowledge to destroy or thwart the hero who has the public's sympathy. In comic strips the villainous scientist is recognized by his title of Doctor or Professor, his peculiar features and personality, his well equipped laboratory, his intellectual brilliance, and his nefarious schemes. Cartoonist Jules Feiffer described him as

G. Holton and W. A. Blanpied (eds.), Science and its Public, 261–278. All Rights Reserved.
This article Copyright © 1976 by D. Reidel Publishing Company, Dordrecht-Holland.

an elderly man, with bad eyesight and posture, who clutches a test tube in his hairy hand or leafs though a thick book "doing research on a secret formula to rule the world".[3]

Earlier, the American poet Stephen Vincent Benet suggested a similar portrait of the typical scientist. In 1935 Benet wrote a letter to Paramount Pictures in Hollywood, outlining a scenario for a projected screen play based on Jules Verne's novel *From the Earth to the Moon* and offering short descriptions of the film's main characters. Science was to be represented by one "John Latham...a white-haired, amiable but ruthless scientist, perfectly willing to sacrifice his daughter or anyone else to his ideas". As might be expected, this unpleasant and dangerous man did not live to share the happiness the other characters found as their reward at the film's conclusion.[4]

The negative stereotype of the scientist is so blatant and pervasive in popular culture that the Franco regime prohibited publication of some episodes of the *Superman* strip. Spanish censors claimed that scientists were depicted as eccentric, absent-minded, ambitious, traitorous, and evil, and that they were often drawn to resemble wizards and alchemists.[5] Any close reader of the comics would agree with their contention, if not with their decision to resort to censorship.

Superman first gained popularity in the 1940's; Benet was adapting a science fiction story for the cinema audience of the thirties; Jules Feiffer was recalling the mad and wicked scientists of his boyhood days during the golden age of comic books, 1938-1954. Perhaps, one might wonder, these three examples reflect an outmoded conception of the scientist. We know that the comics underwent a significant transformation in the 1960's. Responding to a changing social and intellectual climate, comics became more sophisticated and developed a social conscience of sorts. Comic books, once read chiefly by boys in their early teens and by nearly illiterate adults, found new and discriminating readers on college campuses.[6] This revolution in popular art and taste was fostered in large measure by Stan Lee with his new Marvel Comics. What view of science and its practitioners, then, does one find in the pages of Marvel Comics?

One of Stan Lee's most popular creations was The Hulk, "a scientist transformed into a raging behemoth by a nuclear accident". Dr Bruce Banner, a famous American nuclear physicist working on a new "G-Bomb", was accidently exposed to the weapon's gamma radiation and consequently became The Hulk, a huge, green-skinned figure with the physical features of Frankenstein's monster and the split personality of Dr Jekyll and Mr Hyde. Half-man and half-monster, The Hulk is a dangerous creature possessing superhuman strength coupled to a "dull and sluggish" mind.[7]

The Hulk at least has a human form; not so another Marvel creation: Man-Thing. Brilliant chemist Ted Sallis was working on a secret government project to produce a super-soldier when he was forced to inject himself with a potion he had developed in the course of his research. When the chemist's car crashed into a swamp his body and mind were transformed by a "chain reaction" between the super-soldier injection and the swamp ooze. He became Man-Thing, a hideous "creature of root and muck and slime...unclean, malodorous, misshapen". "He belongs", we are told, "wherever humans are not, for he is not human ... but the macabre Man-Thing".[8]

Let me also introduce the one-armed scientist Dr Curtis Connors. Although a devoted family-man and friend of the forces of good, he has a penchant for changing himself into "the loathsome Lizard, whose lust for violence and domination over mankind" taxes the powers of Marvel Comics' greatest hero, Spider-Man.[9]

Stan Lee's imaginative powers have produced in the last few years a new set of comic heroes and monsters, but his portrayal of the scientist is remarkably consistent with an older tradition that saw the scientist as a dangerous figure who tended toward mental instability and social irresponsibility. In weighing the possible public impact of the pop scientists of Marvel Comics one should compare the 134 million copies of Marvel Comics distributed annually with the 6.85 million copies of *Scientific American* and recall, once again, the difference in audience.

Sinister scientists are by no means exclusively confined to comic strips and books. They are stock figures in most science fiction and horror films where their experimentation generates the terror that distinguishes that genre of cinema. Critic Dwight MacDonald[10] notes that the familiar white coat of the mad scientist in a horror film is every bit as blood curdling a sight as Count Dracula's black coat, while *New York Times* columnist Russell Baker writes:

Obviously a lot of us fear and dislike science, but don't dare admit it. The movies know it, however. They give us the mad scientist whom we can fear and despise without feeling bigoted and anti-intellectual, even though, truth to tell, his work is often saner than that of sane scientists.[11]

Beyond comics and the cinema, the evil scientist inhabits children's television cartoons where he battles the forces of good on weekend mornings. These animated cartoons are usually poor copies of the superior artwork found in better comic strips and books, but they are faithful to their originals in depicting scientists as distorted individuals bent upon disrupting the lives of peaceful citizens. But the parent who limits his children's television viewing to the educational channels will not protect them from meeting evil pop scientists on public television. The highly respected *Electric Company*, a children's series that has pioneered in challenging the stereotyped views of minority groups and women, continues to offer its viewers skits featuring mad scientists. Granted, there is a difference; the insane, bushy-haired scientist and his assistant are played by *black* actors who recklessly experiment upon a white, supine monster strapped to a laboratory table!

Adult television offers a more subtle version of the scientist as villain. Gone —except of course in the late movie—is the wild-eyed maniac raving in his dungeon-laboratory. He has been replaced by a figure who is more human than caricature but no less dangerous to the innocent. Nuclear physicist Dr Jerome Cooper, who appeared in a *Mission:Impossible* show, might be mistaken for a prosperous businessman; he was well-dressed and had an attractive wife and home. However, he was a scientist and therefore not to be trusted. Dr Cooper had hidden a 50-megaton hydrogen bomb under Los Angeles and was prepared to kill all of its inhabitants if certain demands he made upon the

President of the United States were not met. Only the quick intervention of the *Mission: Impossible* team narrowly rescued the city from the dangerous machinations of yet another scientist.

When the pop scientist is not portrayed as a malevolent figure he is likely to appear as an individual who is easily manipulated, or dominated, because of some flaw in his character. If he is personally ambitious, he will lose sight of the human consequences of his research, and society will suffer. If he is exceedingly naive when dealing with powerful political or business interests, his scientific discoveries will be wrested from him and put to some perverted use by others. If he is free of these flaws, then he watches in horror, powerless to act, as his research program results in disastrous consequences which were wholly unforeseen or unintended. In all of these cases scientific discoveries are seen as a threat to human life and happiness. Despite his initial good intentions, the end results of this pop scientist's work are identical with those of his wicked or insane colleagues.

Mary Shelley, in her novel *Frankenstein*, created one of the best-known stories of the harmful consequences of seemingly beneficent scientific research. Victor Frankenstein was a dedicated scientist operating from the highest motives. Yet in his search for the secret of life he unleashed a monster that threatened him, his family, and his community. The monster, nameless in the novel, soon came to be called Frankenstein; the scientist and the monstrous product of his research became one. With the confusion of master and creation the scientist Frankenstein was increasingly pictured as deranged, anti-social, and inherently evil.

Shelley's novel was an immediate success and its story has been retold countless times since the book appeared in 1818. Throughout the nineteenth century one could read one of the many reprints or translations of *Frankenstein* or see it enacted in several dramatic versions in England, on the Continent, and in the United States. In the early days of moving pictures Thomas Edison produced a short film entitled *Frankenstein* (1910), but it was the James Whale version of 1931, with Boris Karloff as the monster, that spawned hundreds of films and animated cartoons on the theme. Most recently, Mel Brooks and Andy Warhol have been inspired to offer their own interpretations of the legend.

The Frankenstein story has formed the basis for radio dramas, adult and children television features, puppet shows, and comic books. The monster's patched, gruesome face has appeared in printed advertising material, television commercials, and a long list of children's toys. Without a doubt, Doctor Frankenstein is better known in America today than any other scientist, living or dead. And all who recognize the name associate it with the story of science out of control, of science releasing a murderous monster upon an innocent public.[12]

In the twentieth century Kurt Vonnegut, Jr. produced a modern variation on the Frankenstein theme in *Cat's Cradle*. There one meets Dr Felix Hoenikker, a mild-mannered scientific genius who gained fame as one of the "fathers" of the atomic bomb. Nuclear physicist Hoenikker had isolated a new form of water, *ice-nine*, to be used in freezing swamp terrain for easy foot passage by military forces. At the novel's conclusion, through a series of bizarre circumstances, *ice-nine* has turned the earth with all its life into a single, frozen, dead mass.

Ice-nine is one of Vonnegut's zany ideas; the atomic bomb is not. Stanley Kubrick, in a film that mixed fact with fiction (*Dr Strangelove*), depicted the final development in man's long search to learn and use the secret of the atom. In attempting to unravel the mystery of matter, nuclear physicists unwittingly laid the theoretical foundations upon which the Dr Strangeloves and their military accomplices could build a Doomsday Machine with the capability of destroying every living organism on the earth's surface. As Vonnegut said of his fictional physicist Dr Hoenikker, "...how the hell innocent is a man who makes a thing like an atomic bomb?"[13]

Kubrick, Vonnegut, and Shelley are all making the same point. Good intentions are not sufficient to insure that scientific activity will not have catastrophic consequences for mankind. Even the pure in heart may create monsters in their single-minded pursuit of scientific truths.

II

In the midst of the villains and the well-intentioned but nevertheless dangerous scientists one finds some good "pop" scientists. They are laboratory technicians carrying out routine analyses for the hero, computer experts with access to information necessary for the successful completion of the hero's mission, or, less likely, members of a research team. The lab men work in criminological laboratories attached to police departments, where they help the hero detective in his solution of crimes. Hundreds of such men have served Dick Tracy down through the years. Their names are rarely mentioned and they receive little recognition for their contribution to Tracy's success in closing still another criminal case. Dick Tracy's notorious enemies, men like Flat Top and Prune Face, will be remembered long after we have forgotten these anonymous, nondescript lab men who stay at their benches and have little opportunity to engage in heroic adventures. [14]

Within the past few years a new pop scientist-technician has emerged along with the computer. Someone has to push the buttons that will activate the bewildering series of dancing, irrelevant lights that sparkle on the consoles of computers conceived by television stage-set designers. Someone has to read the computer print-out for the hero. That person is the white-coated computer expert who has appeared in *Mission: Impossible, Search, UFO, Six Million-Dollar Man* and other television adventure dramas.

Actor Burgess Meredith directed a team of such computer specialists, called Probe Control, in *Search*. Their job was to back up television star Hugh O'Brian who spent his time slugging it out with the bad guys. The computer people, content to watch their oscilloscopes and to mouth meaningless numbers, were lackluster individuals. Their chief, however, had a wit and personality of his own, and engaged in banter with the hero. On occasion he revealed the true nature of the pop scientist, as he did when the latter, on purely rational grounds, wanted to drop the search for O'Brian's lost friend. He was overruled by the hero who read him a lesson in fundamental ethics and then went on with the job. That is one of the main faults of the pop scientists; they are apt to put logic and reason above humane considerations.

The positive portrayal of the lab man, or computer expert, as opposed to the distrust shown for the brilliant but misguided scientist, was reflected in the treatment of the technician in *Mission: Impossible*. Barney (Greg Morris), the electronics expert who produced an arsenal of ingenious gadgets each week, was one of the most important and popular members of the team.[15] The *Mission: Impossible* scientist, however, was characterized differently. On one show, entitled "Project Meningitis", he was Dr Oswald Beck, a bacteriological warfare scientist replete with beard, eyeglasses, ominous looks, and a white laboratory coat. Not only did Beck perform experiments on unwilling human subjects in order to demonstrate the killing power of his newly-discovered germ strain, but he was anxious to sell his research secret to the country offering the highest bid. Dr Beck met the end of most evil scientists—he was killed outright.

III

Generally, the scientist in popular culture is seen as a Faust-like figure who has great power over natural forces gained from his knowledge of the physical universe. Because of his evil or deranged nature, or because of circumstances beyond his control, the power of science is likely to be used against mankind. When the scientist is depicted as making positive contributions to social welfare, he is nevertheless presented as a pale, somewhat eccentric or unpleasant indiviudal who is ruled by logic and deficient in the human passions.

With all his other faults the pop scientist (almost always male) also tends to have sexual problems—he has difficulty in relating to the women he encounters. If he is entirely perverse and possesses hideous physical features, then he may kidnap young virgins, whisking them off to his dungeon-laboratory where he carries out his scientific and sexual experiments at leisure. Otherwise, he is the kid who stayed home reading or working with his chemistry set while the neighborhood boys were learning about girls. By the time this scientific recluse reaches maturity, his intellectual proclivities and his lack of romantic experiences make him unattractive to young women. That is why so many scientists in motion pictures are shown as divorced, widowed, or as bachelors. The pattern is for an older, wifeless scientist to have a young daughter, or to have an attractive female assistant who is romantically involved with some one other than the scientist.

The central problem in the pop scientist's sexual life was nicely summarized in a set of cartoons appearing in the semi-pornographic *World's Best Cartoons*[16] under the title "The Wild, Sexy Scientist". The voluptuous wife of a young scientist is shown complaining to her father-scientist: "Oh, Dad, Tom is so dedicated to science that our sex life suffers." She recounts, in lurid detail, her futile attempts to arouse her husband's sexual interests and her father promises to develop a sexual stimulant to resolve the problem. Our frustrated heroine puts the stimulant into her husband's supper, changes into a transparent nightie, and manages to distract her husband from the laboratory for one evening. In the final scene the couple are pictured making love but the wife is still displeased. Even at this tender moment her husband has attached their bodies to a huge machine so that he can measure their activity and thrills. The moral is obvious: the measured, rational scientific life has no place for love and emotional fulfillment.

Whether he is good or evil, the pop scientist is easily identifiable by his clothing and personal appearance. He is usually shown in the comics as an older man, wearing of course his white coat and a pair of eyeglasses, working in a laboratory. His physical features tend to the extremes: either he is small and stout or tall and thin, either he is completely bald or he is bearded and wearing a halo of tangled hair in the style of Albert Einstein. Some familiar pop scientists so pictured are Dr Elbert Wonmug of *Alley Oop*, Mr Homer Sapiens of *Little Orphan Annie*, Dr Huer of *Buck Rogers*, and the comic scientists who go through their capers for the juvenile audience of *Electric Company* or who fill the pages of cartoon books.

IV

In 1957 Margaret Mead and Rhoda Métraux published a study of the American high-school student's image of the scientist. The composite portrait they drew from their research is identical with the foregoing description of the pop scientist. According to the students, a scientist is:

...a man who wears a white coat and works in a laboratory. He is elderly or middle-aged and wears glasses. He is small, sometimes small and stout, or tall and thin. He may be bald. He may wear a beard, may be unshaven and unkempt. He may be stooped and tired.[17]

Almost two decades have passed since the Mead-Métraux study was made, and during that time America has experienced some spectacular scientific and technological advances, and her youth have been subjected to the efforts of an illustrious band of reforming science educators. Despite all of this, the image of the scientist in popular culture has changed little. One of the surprising results of my systematic analysis of comic strips was that the essential characteristics of science and the scientist remained stable throughout the period 1945-1975. When I began my researches I expected to find a rising curve of interest in, and comprehension of, the scientist in popular culture. What I found was a portrait of the scientist and his activities that received minor embellishments through time but underwent no major alteration. I suspect that this stability extends to a much earlier date than the one I chose to begin my study. Professor Radium[18], the scientist of a 1908 English comic strip, has much in common with his modern counterpart in the latest generation of comics, say Gary Trudeau's *Doonesbury*.

The standard uniform of the pop scientist, the white laboratory coat, deserves another comment. It is so much a part of the pop scientist's costume that he wears it not only in the laboratory but also when he is at home, in a taxicab, or speaking before a scientific society. The white lab coat is required dress for computer technicians even though they are not likely to soil their street clothes near a computer. A comic strip character who participates but for a short time in scientific activity also assumes the accouterments of his new role. When young Bruce Wayne was preparing himself for his career as Batman he found it necessary to become a "master scientist". He is pictured dressed in a white laboratory outfit gazing thoughtfully at an uplifted test tube. Wayne is wearing eyeglasses, an ocular aid he never needed before and one he was never to use as Batman. As a scientist, though, the glasses were as important as the white coat. By wearing them he served notice that he was an intellectual who needed them to read his many books.

A *Steve Canyon* comic strip episode provides us with the rare opportunity to see the kind of transformation a regular guy must undergo before he can become a scientist. In the strip we meet Hogan, a nonchalant, rugged adventurer and notorious free agent, who is prevailed upon by Colonel Canyon to assume temporarily the role of scientist in order to work closely with visiting computer expert Anna "Brain" Payne who holds a doctorate in "Data processing—computers, cybernetics". A female scientist, needless to say, is a strange creature indeed. Dr Payne is logical, efficient, and without emotion; she is described as never having stopped working in the laboratory long enough to discover whether she was a woman.

Hogan prepares himself for Dr Payne's arrival by reading a computer manual overnight, and by discarding his casual dress. When the transformed "Dr Hogan" meets the young and attractive, though spectacled, lady scientist, he is wearing glasses, a coat and tie, and a blank, stupid expression on his face. Without a doubt he is a scientist. But lurking behind the disguise is he-man Hogan who soon catches "Brain" Payne off-guard by kissing her and using his manliness to conquer her despite her intellectual superiority.

We are not to be misled by Miss Payne's feminine beauty. As a scientist she is perfectly capable of committing one of the many anti-social or dangerous acts that typify her kind. Shortly after meeting "Dr Hogan" she attempts to use her computer expertise to blackmail the United States into a nuclear war with the U.S.S.R. The threat she poses to international peace only ends with her timely death.[20]

Apart from his personal effects, the pop scientist can also be identified by his laboratory surroundings. According to the Mead-Métraux report, high-school students pictured the scientist as working with "test tubes, bunsen burners, flasks and bottles, a jungle gym of blown glass tubes and weird machines with dials."[21] Electrical equipment, especially computers, are gaining popularity, and now one sometimes finds the glass tubing directly and inexplicably connected to the electronic gadgetry.

This is the sort of equipment used by physical scientists, in many cases chemists. Chemistry and physics are the typical pre-occupations of the pop scientist; biology comes next in line. An occasional earth scientist makes his appearance in the comics—there is geologist Buster Stone in the lithic environment of *Alley Oop*—and pith-helmeted archeologists, always useful as guides to hidden treasures and lost civilizations, are frequently encountered. Social and behavioral scientists are rarely found in pop science, outside of the few realistic medical strips, and then usually in the form of a psychiatrist or psychologist.

V

In contrast to the portrayal of the pop scientist, consider the image of the scientist in science fiction. The genre has experienced several fluctuations in popularity in the past few decades and today it is once again in vogue. Despite its current popularity, however, science fiction is read by a relatively small, and select, group of people. There are many more readers of comic strips and viewers of feature films and television shows.

The science fiction audience is drawn from college-educated technical and professional people and from technically-oriented high-school students. Scientists and engineers are among the regular readers of science fiction, and a goodly number of scientists have gained recognition as authors of science fiction novels and short stories.[22] To name several of the best known: biochemist Isaac Asimov, astronomer Fred Hoyle, and anthropologist Chad Oliver.

A literary genre that is both read and written by scientists is likely to depict them in a favorable light. So it is not surprising to find in a Hoyle novel a small group of scientists who, when faced with the imminent destruction of all terrestrial life, manage to save it by circumventing ordinary political processes. The scientists are the heroes while the politicians are dismissed as "an archaic crowd of nitwits".[23] Nor is it surprising to read Isaac Asimov's claim that science fiction, because it presents science so favorably, should be used to battle anti-intellectualism and to recruit bright young students for scientific careers.[24]

The matter is not quite so simple as these two examples might indicate. A thorough analysis of the image of the scientist in science fiction made by sociologist Walter Hirsch does not substantiate Asimov's hope. According to Hirsch, the naive worship of the "omnipotent and omniscient" scientist is fast disappearing from the genre. Nevertheless, upon comparing pop science with science fiction it is clear that the scientist fares much better in the latter.[25]

Science fiction is found in its purest form in books and magazines. It is adulterated with elements of popular culture when any attempt is made to prepare it for the mass audience of the cinema. Science fiction fans have long been critical of what they believe to be the corruption of the genre as it is adapted for the feature film. It is the intrusion of pop science that disturbs them.

Critic Susan Sontag has written a perceptive essay on science-fiction films entitled: "The Imagination of Disaster". In this essay she accurately records the results of the amalgamation of science fiction with pop science. The cinema scientist, she tells us, is portrayed "as satanist and savior", as the bringer of both evil and good because of his mastery over natural forces. The "satanist" is, in most instances, the lone individual who neglects family or fiancée, pursuing his "daring and dangerous experiments" while holed-up in his laboratory. Conversely, the scientist as a "loyal member of a team...is treated quite respectfully".[26] But whatever the portrayal of the scientist, *science* is invariably presented to the cinematic audience in the context of overwhelming social disaster and chaos.

Television, like the cinema, merges pop science and science fiction attitudes in the portrayal of the scientist. A good case in point is the popular television series *Star Trek*. Science in *Star Trek* is personified by Mr Spock, who is listed as

"Science Officer" of the spaceship U.S.S. Enterprise. In a spaceship made up of earthlings of several national cultures, Spock is unique because of his extra-terrestrial origin. He has pointed ears, a yellowish complexion, green blood, a pulse rate of 242 beats per minute, and a mating urge that comes upon him once every seven years. In accord with his scientific background, he operates solely upon logical grounds, expresses virtually no emotion, and firmly rejects female advances, preferring to amuse himself by playing three-dimensional chess with the ship's computer. He represents a familiar attempt to make a science-fiction figure recognizable to the mass audience of television.[27]

Whether in space or on earth, the scientist in popular culture undertakes many tasks, but he is readily identifiable. He spends his time seeking the knowledge that will consolidate his dominion over nature, and therefore over mankind. Or he may be carrying out the more routine work associated with scientific crime-fighting or computer technology. But whatever his duties, you are certain to find this eccentric individual isolated from the mainstream of society.

The eccentricity, social isolation, and irresponsibility of the pop scientist is documented in the popular science-fiction novel (and film), *The Hephaestus Plague*, by Thomas Page. When large fire-producing, carbon-eating insects emerged from a fissure in the earth, most Americans were horrified, but not Dr James L. Parmiter, entomologist, who succeeded in breeding the fire-bugs with cockroaches, thus creating a horde of urban insect-arsonists. Here is Page's description of the scientist at work:

Deep in his rank basement, cut off from humanity at large, the Bainboro [College] entomologist playing with his nervous insects had enough eccentricity for greatness. But even he himself did not suspect this until the following day when the roaches still raged through the cities of the country, burning cities and towns and woods at an even more desperate rate.[28]

VI

It seems to me there are at least four reasons why the modern scientist has been transformed into the pop scientist I have been discussing.

First, the scientist suffers from a more general bias against intellectuals in this country. He must bear the onus of our "ingrained contemporary mistrust of the intellect", Susan Sontag writes.[29] Those who work in classroom, study, or laboratory have traditionally been viewed as "pale bloodless shadows" of real men.[30] They have deliberately excluded themselves from the outdoors that is valued in American mythology, and settled for the world of books and ideas. Jules Feiffer recalls that the two stock evil characters of the old comic books— the mad scientist and the Oriental mastermind—were customarily shown as bookworms.[31] Other men of intellect—writers, thinkers, poets, artists—undergo somewhat the same fate as the scientist when they are depicted in popular culture. There is this difference, however: other intellectual types are comical figures or weaklings; scientists are dangerous because they use their minds to control nature.

The scientist-as-intellectual is also suspect because it is believed that he does not share the simpler tastes of his fellow Americans. Here is a United States

Senator revealing this suspicion at a Congressional Committee hearing:

I notice in the Orphan Annie strip they now have the atom bomb down to hand size and you can tote your own private demolition. But as a scientific man, you wouldn't know about the Little Orphan Annie strip.[32]

The distrust of the scientist as intellectual was a standard feature of the science fiction films of the 1950's. The scientist was portrayed as a man who, because of his intellectual interests, was directly or indirectly responsible for some social or physical catastrophe. In the classic film *The Thing* (1951) a vegetable monster arrived on a flying saucer and threatened the lives of a group of military personnel stationed near the North Pole. Even though the monster was thirsting for fresh human blood he was protected by the scientist. Convinced that "knowledge is more important than life", the scientist saw "no enemies in science, only phenomena to be studied." Finally, United States Air Force officers intervened, subdued the scientist, and killed the monster. The finale is typical for these films, the heroes were the men of brawn—soldiers, policemen, adventurers—who physically stopped the scientist and literally saved the earth.

At first glance it might appear that Clarke and Kubrick's *2001: A Space Odyssey* should not be classed with this older type of film. Astronauts, and not scientists, are the leading figures in the film and the classic confrontation between irresponsible scientists and the saviors with good sense and powerful muscles is not evident. A close study of the plot, however, will reveal the old distrust of the intellectual.

Aboard the spaceship in the film are five crew members, three of whom are in hibernation. Astronauts Bowman and Poole are awake, they operate the ship with the help of a sixth "crew member" HAL, a Heuristically programmed ALgorithmic computer. HAL is the personification of the intellect in *2001*. He is familiar with every aspect of the ship's operation; he is incapable of making an error; he alone was trusted with the knowledge of the ship's mission and destination.

HAL, the super-intellectual, deliberately murders four of the astronauts. Commander Bowman, the sole survivor, is able to stay alive because he crawls into the interior of HAL's brain to perform a lobotomy on the computer. HAL is to be trusted only when he is reduced to an idiotic state. He is deprived of all of his knowledge save for the words of an old song: "Daisy, Daisy give me your answer do." Now at last the danger is past, the intellect has been destroyed.[33]

The story of HAL is illustrative of one of the two prevailing images of the computer. The machine as symbol of the super intellect is a threat to man in *2001*, the television film *The Forbin Project*, and a host of science fiction tales. The computer as just another instrument, with its intellectual powers closely guarded by technicians, is the supplier of information useful to adventurous heroes.

Research work done by the scientist is not really "visible" to the public and that is the second reason he is misunderstood. Members of other professions, medical doctors and lawyers, noticeably interact with the public in our social and political systems. Rex Morgan, M.D. can truthfully be shown in a comic strip as a faithful servant of society. On the other hand, it is impossible to imagine a television series featuring a physicist or chemist which could compete with *Marcus Welby, M.D.*, *Medical Center*, or *MASH*. Nor is it possible to imagine that a scientist would ever be described in as flattering terms as was the physician in the best-selling medical novel *Not As a Stranger*:

> there's something about a doctor, learned, wise, clad in white wisdom, the ancient robe of the seer—healing in his hands—intellect in his eyes, the eyes of the world on him, pleading, hoping against hope, and the doctor is the last resort, awful as Jove, the right hand of God, to give and to take, the last appeal.[34]

The public has had few opportunities to meet with scientists or seek their direct help on a problem. Little wonder then that scientists are usually referred to as a remote "they". "They" work hidden from public scrutiny. "They" speak a language that is incomprehensible to a large segment of the nation, educated or not. "They" are grasping for the secrets of the universe. Can "they" be trusted with such knowledge, these men who work in secret and with whom one cannot hold extended discourse? At best the answer is an uneasy yes; in popular culture it is more often no.[35]

The feature film *Escape From the Planet of the Apes* gives vivid testimony to the different treatment accorded the scientist and the physician in popular culture. Science is represented by the American President's science advisor, physicist Dr Otto Hasslein (German for "little hate"), who is young, coldly handsome, and, as it turns out, a brutal murderer. Medicine takes the form of a pleasant male-female medical team doing its best to protect the human-like ape parents and their newborn offspring. In the film's final scene Dr Hasslein, defying the orders of the President, pursues the apes onto a cargo ship, murders mother and child, and in turn is killed by the father. As the dead physicist's body fell overboard into the water, the audience in the theatre where I saw the film spontaneously broke into a cheer. It seemed a fitting end for a scientist, but a most unlikely one for a physician.

The third reason for the negative portrayal of the scientist in popular culture stems from a widespread confusion between science and its technological applications. Because science is so readily identified with practical results, and because science is not seen as an effective means of aiding men to understand complex natural phenomena, the pop scientist is characterized as lacking in human qualities and engaged in anti-social activity.

Science in popular culture is not presented as an intellectual adventure, nor as the disinterested search for truths about man and the physical universe. Instead, the technological fruits of science are stressed to such a degree that basic scientific research merges with engineering, and straight-forward technological achievements are called scientific. Science becomes a source of products and power, not a means of helping men better to comprehend himself and his surroundings.

The classic popular-culture embodiment of this confusion between science and its applications is found in the *Dick Tracy* strip. At one time Dick Tracy was pursuing the criminal Piggy Butcher. Piggy was very elusive but the case was finally solved with the use of a Voice-o-graph, an instrument capable of making voice comparisons. When policewoman Lizz learned that Piggy's voice prints were identical with those of an hitherto unknown murderer she exclaimed; "As the feller says, Science marches on".

Among comic readers the *Dick Tracy* strip has the reputation of being "scientific". By that is meant that he uses ring cameras, two-way wrist radios and TV's, tranquillizer guns, atomic lights, magnetic space coupes, laser goldmining machines, atomic-powered television cameras, and the like. Many of these devices were created by his friend, the industrialist Diet Smith.[36]

The blame for misunderstanding the nature of science is not to be placed upon Chester Gould who draws *Dick Tracy*, or upon the other creators of popular culture. This misunderstanding has long been part of the American way of thinking about such matters.[37] In 1883 a president of the American Association for the Advancement of Science complained that his countrymen called "telegraphs, electric lights, and such conveniences by the name of science."[38] In the twentieth century the Walt Disney film *Son of Flubber* testifies to the persistence of this misunderstanding when a news commentator in it reads a list of the greatest scientists of all time: Sir Isaac Newton, James Watt, Thomas Edison, Albert Einstein, and Professor Brainard, inventor of Flubber. Gravitational theory, steam engines, light bulbs, relativistic physics, and levitating Flubber all become SCIENCE.

The public reacts ambiguously towards technology. They acknowledge its many benefits and yet they are personally aware of the problems it has engendered in our society. Therefore, when the scientist is not depicted as a disinterested seeker of knowledge but as the dispenser of machines that pollute the environment or promote nuclear warfare, he is held responsible for the deleterious effects of advanced technology.

The fourth source of the caricaturization of the pop scientist is the scientist himself. Without doing so knowingly, he often contributes to the negative image of the practitioner of science. For example, by overemphasizing the practical results of his work, especially when seeking public funds, he contributes to the existing national confusion between science and technological application and opens himself to criticism that might better be directed against engineers, managers, and industrialists.[39] Second, by deliberately cultivating the image of the scientist as a remote, superbly dedicated, logical, and humorless individual he creates an easy target for cartoonists and popular satirists. The second contribution is evident in the response of a number of men of science who were scandalized by James Watson's *The Double Helix*, in which scientists were portrayed with all-too-human foibles and science was depicted as, at times, a haphazard enterprise.

The soberness of the scientist's self-image is exposed in their unqualified praise for the way their portraits were drawn by Sinclair Lewis in *Arrowsmith*.[40] Here is Lewis's description of the typical scientist:

He had never dined with a duchess, never received a prize, never been interviewed, never produced anything the public could understand, nor experienced anything since his school boy amours which nice people could regard as romantic. He was, in fact, an authentic scientist.[41]

And again, Lewis writing in praise of the dedicated medical research scientist:

He was so devoted to Pure Science, to art for art's sake, that he would rather have people die by the right therapy than be cured by the wrong. Having built a shrine for humanity, he wanted to kick out of it all mere human beings.[42]

It is a short step from Lewis's dedicated scientist to the pop scientist of the funnies. If working scientists wish to counteract the pop science caricature, *The Double Helix* would serve as a better model than *Arrowsmith*.

Underlying all of the contemporary explanations for the image of the scientist in popular culture is an explanation that would trace it back to the Renaissance, or even earlier. When anthropologist Claude Levi-Strauss first heard about the American high-school student's image of the scientist he remarked that it was ˙ikingly similar to the ambiguous attitude primitive peoples had towards metal workers in their society.[43] I would refine that observation by suggesting that it was similar to the response to such Renaissance figures as the wizard, alchemist, or magician.

The wizard was a man of intellect, literally a wise man, whose magical studies had given him influence over natural and supernatural forces. He was usually elderly, having devoted his entire life to acquiring the esoteric knowledge that made him superior to his fellow men. The extraordinary powers of the wizard, and his predilection to work apart from the rest of society, made him a figure held in awe by the populace. However, he paid a price for his special status. Excessive study distorted his physical features and personality, perhaps even his mind, and caused him to lose his moral sense. In his passionate search for knowledge he was often driven, as was Faust, to make a pact with Satan.

If innocent bystanders were not harmed by the evil actions of an immoral wizard they might still become victims of one of the mistakes of the wizard's —or sorcerer's—apprentice. The apprentice's error, made in his clumsy attempt to gain control of nature, could not always be rectified by his master, and mankind suffered as a result.

Obvious parallels are immediately apparent between the pop scientist and the wizard as described here, including the ubiquitous young apprentice to the scientist who appears in countless science fiction films. Furthermore, wizard and wizardry are words often used in connection with modern science from the time of Thomas Edison—the wizard of Menlo Park, New Jersey—to the television era: Mr Wizard with his magical scientific experiments for children. In 1879 *Harper's Weekly* published this account of wizard Edison in his laboratory:

At an open red brick chimney, fitfully outlined from the darkness by the light of fiercely smoking lamps, stands a rough clothed gray-headed man ... His eager countenance is lighted up by the yellow glare of the unsteady lamps, as he glances into a heavy old book lying there, while his broad shoulders keep out the gloom that lurks in all the corners and hides among the masses of machinery. He is a fit occupant for this weird scene; a midnight workman with supernal forces whose mysterious phenomena have taught men their largest idea of elemental power; a modern alchemist, who finds the philosopher's stone to be made of carbon, and with his magnetic wand changes every-day knowledge into the pure gold of new applications and original uses. He is THOMAS A. EDISON, at work in his laboratory, deep in conjuring of Nature while the world sleeps.[44]

I submit that this gothic picture of a nineteenth-century inventor could serve as the inspiration for the opening scene of a modern horror or science fiction film, or for the first pages of a comic book.

The Renaissance wizard's interest in alchemy could account for the antiquated chemical glassware that appears on the pop chemist's laboratory bench. The flasks and retorts filled with bubbling colored liquids have few analogues in the modern chemical laboratory, but they are close in spirit to the equipment of the alchemist.

There is an air of magic about many aspects of pop science. Pop scientific experiments, for example, are not formal, methodical, and often frustrating interrogations of nature. Instead, they are almost any trial or study that happens to touch upon natural phenomena. Thus Diet Smith conducted what he called an "interesting experiment" when he orbited his necktie and cigar from a spaceship. The pop experiment is easily understood; it is short and simple, and its results are immediately apparent without further need to refine or extend the original investigation.

Many pop experiments are run by chemists searching for the "key equations" and "new formulas" that will yield substances with miraculous powers. This reliance upon a simple substance, usually a chemical element, with wide-ranging effects can be traced back to the philosopher's stone of the alchemist which in turn can be explained as a more general human yearning for a panacea capable of resolving all phyiscal problems. In any case, no matter how difficult a problem is to be solved, there exists a pop element offering a simple, universal solution. Following is a list of some of the better known pop elements:

(1)   Eonite—named after discoverer Eli Eon;
according to Little Orphan Annie: "Leapin' Lizards! With Eonite anything is possible!"

(2)   Caltechium—mentioned by Buck Rogers;
powerful enough to blast the earth out of its solar orbit.

(3)   Kryptonite—possesses power to stop Superman.

It is entirely possible that at a popular level, and even beyond, there lingers remnants of archaic attitudes towards magic and the wizard-alchemist-magician that had been transferred to science and the scientist when they first appeared in history. If this is true, then we must admit the stubborn strength of irrational ideas in a rational society and acknowledge the failure of efforts of science educators and the recognized popularizers of science to act sufficiently effectively at that level of human understanding. The current popularity of astrology, divination, and witchcraft is certainly one indication of the extent of that failure.

Conclusion

I hope this excursion into the world of popular culture will hereafter make us more careful when we think about popular science and the popularization of science. It is all too easy for the educated community to believe that what they call popular science reaches the general populace when it only touches a small part of it, and then that part that probably needs it least. We must always ask ourselves how popular is the popular science we are producing. Most popular scientific journals have monthly circulations in the 150 thousand range, with the exception of *Scientific American* which has a monthly circulation of about 570,000. Probably the most popular general magazine carrying an occasional article with scientific content is *Reader's Digest*, with over 19 million monthly subscribers. Now evaluate their impact on a nation of 211 million people. Even if the circulation figures are doubled or tripled to indicate readers as opposed to purchasers, a large majority of Americans are not being reached by the usually recognized popularizers of science who write for magazines, or publish books with even smaller sales.

In contrast to these numbers, we know that an estimated 100 million Americans read at least one comic strip in the Sunday papers and that 90 million regular readers follow their favorite comic characters in daily episodes.[45] When a newspaper temporarily suspends publication the loudest complaints come from the comic fans.

As for television, 97% of American homes have at least one television set. Between 50% and 60% of the American population is watching some television program on a given day. The television audience, as well as the comic readers and moviegoers, are receiving whatever message they contain about science and often in an attractive and entertaining form. Television does have its non-fiction documentaries on scientific topics now and then, but they do not generate Nielsen ratings comparable to those earned by the popular television series. Jacob Bronowski's *Ascent of Man* and the *Nova* series on Public TV can never compete, in popularity ratings, with *Marcus Welby, M.D.* and *Star Trek*. Those concerned with the popularization of science need to remember how far we have to go before we can really claim to be bringing science to the general public.

Public opinion, it should be remembered, is an elusive phenomenon; and I would hesitate to argue for a direct correlation between the image of science in popular culture and public understanding of science or public willingness to support scientific education and research. Nevertheless, when one reflects upon the persistent distortion science undergoes in popular culture, one must consider the effect of pop science upon the public's response to science. I conclude that the individuals who are in a position to reach substantial numbers of Americans with their conception of science are therefore not science writers and journalists, not the winners of the Arches of Science Award and the Kalinga Prize who write for an audience which already accepts their outlook, and not the producers of film or television documentaries of a scientific nature. It is rather the persons who produce popular television shows and feature films, along with the cartoonists of the favorite comic strips—in short the creators of popular culture— from whom the wide American public receives its portrayals of science and scientists.

REFERENCES

1. K. Vonnegut, Jr., *Cat's Cradle* (New York, 1963), pp. 41-42.

2. O. Handlin, Dædalus, 94 (Winter 1965) pp. 156-70. For the British scene, see M. Millhauser, Nineteenth Century Fiction, 28 (December 1973), pp. 287-304.

3. J. Feiffer, New York: The Sunday Herald Tribune Magazine, January 9, 1966, p. 6.

4. S. V. Benet, *From the Earth to the Moon* (New Haven, 1958).

5. A. A. Villar, *Revista Española de la Opinión Pública* 6 (1966), p. 217.

6. S. Braun, *New York Times Magazine*, May 2, 1971, p. 32; L. van Gelder, *New York*, October 19, 1970, p. 36.

7. S. Lee, *Origins of Marvel Comics* (New York, 1974), p. 75.

8. *The Man-Thing* (New York, 1975).

9. S. Lee, *Origins of Marvel Comics* (New York, 1974), p. 138.

10. D. MacDonald in *Mass Culture: The Popular Arts in America*, B. Rosenberg and D. M. White, eds. (New York, 1957), pp. 68-9.

11. R. Baker, *The New York Times Magazine*, September 22, 1974, p. 6.

12. M. W. Shelley, *Frankenstein*, J. Rieger ed. (New York, 1974), pp. xi-xliii; D. F. Glut, *The Frankenstein Legend* (Methuen, N. J., 1973).

13. K. Vonnegut, Jr., *Cat's Cradle* (New York, 1963), p. 63.

14. C. Gould, *The Celebrated Cases of Dick Tracy* (New York, 1970).

15. R. C. Carpenter, *Journal of Popular Culture*, 1 (Winter 1967) p. 286, J. Klemesrud, *New York Times*, May 11, 1969, p. D19.

16. *World's Best Cartoons*, August 1975.

17. M. Mead and R. Métraux in *The Sociology of Science*, B. Barber and W. Hirsch, eds. (Glencoe, Ill., 1968), pp. 237-38, R. Plank, *The Emotional Significance of Imaginary Beings* (Springfield, Ill., 1968), p. 152.

18. G. Perry and A. Aldridge, *The Penguin Book of Comics* (Harmondsworth, Great Britain, 1967), p. 61.

19. J. Feiffer, *The Great Comic Book Heroes* (New York, 1965), pp. 26-27, p. 70.

20. This episode was published in April 1968.

21. M. Mead and R. Métraux in *The Sociology of Science*, B. Barber and W. Hirsch, eds. (Glencoe, Ill., 1962), p. 238.

22. S. E. Finer, *Sociological Review*, 2 (1954), 239-246, K. Amis, *New Maps of Hell* (New York, 1960), pp. 48-52; A. S. Barron, *Bulletin of the Atomic Scientists*, 13 (1975) 62.

23. F. Hoyle, *The Black Cloud* (New York, 1965), p. 92; D. K. Price, *The Scientific Estate* (Cambridge, Mass., 1965), p. 9.

24. I. Asimov, *Smithsonian*, 1 (May 1970), pp. 41-47.

25. W. Hirsch in *The Sociology of Science*, B. Barber and W. Hirsch, eds. (Glencoe, Ill., 1962), p. 265.

26. S. Sontag, *Against Interpretation* (New York, 1969), pp. 220, 225.

27. S. E. Whitfield and G. Roddenberry, *The Making of Star Trek* (New York, 1968), pp. 2, 24, 26, 223-38, 280.

28. T. Page, *The Hephaestus Plague* (New York, 1973), p. 119.

29. S. Sontag, *op. cit.*, p. 220.

30. L. Gurko, *Heroes, Highbrows and the Popular Mind* (Indianapolis, 1962), pp. 38-40, 54-62;
    R. Hofstadter, *Anti-Intellectualism in American Life* (New York, 1963), pp. 11-12, 24-51.

31. J. Feiffer, *New York: The Sunday Herald Tribune Magazine*, January 9, 1966, p. 6.

32. Quoted in H. S. Hall, *The Sociology of Science*, B. Barber and W. Hirsch, eds. (Glencoe, Ill.,
    1962), p. 279.

33. A. C. Clarke, *2001: A Space Odyssey* (New York, 1968).

34. M. Thompson, *Not As a Stranger* (New York, 1954), p. 96.

35. H. Cantril, *The Invasion from Mars* (New York, 1966), p. 158.

36. Science Research Associates, *The Sunday Comics* (New York, 1956), p. 136; M. Sheridan,
    Comics and Their Creators (Boston, 1944), p. 123.

37. O. Handlin, *Dædalus 94* (Winter, 1965), p. 156.

38. H. A. Rowland, *The Physical Papers of Rowland* (Baltimore, 1902), p. 594.

39. R. Dubos, *The Dreams of Reason* (New York, 1961), p. 156

40. J. R. Pierce, *Science*, 113 (1951), p. 431, B. Glass, *The Scientific Monthly*, 85 (1957), 288.

41. S. Lewis, *Arrowsmith* (New York, 1961), p. 121.

42. S. Lewis, *Arrowsmith* (New York, 1961), p. 120.

43. G. Charbonnier, *Conversations with Claude Levi-Strauss* (London, 1969), p. 46.

44. Anonymous, *Harper's Weekly*, 23 (August 1879), p. 607.

45. D. M. White and R. H. Abel in *The Funnies: An American Idiom*, D. M. White and R. H.
    Abel, eds. (Glencoe, Ill. 1963), pp. 3-4.

Notes on Contributors

GEORGE BASALLA, born 1928, is associate professor of the history of science and technology at the University of Delaware. His prime research area is the social implications of science and technology, with a special interest in the presentation of science, medicine, and technology in popular culture.

DAVID Z. BECKLER, born in 1918, is Assistant to the President of the National Academy of Sciences. He served as Executive Officer of the President's Science Advisory Committee (including its predecessor committee) from 1953 to 1973, and as principal assistant to all six Presidential science advisers from 1957 until the termination of the White House post in 1973.

WILLIAM A. BLANPIED, born in 1933, is director of communications for the American Association for the Advancement of Science. He was formerly associate professor of physics at Case Western Reserve University, served for two years as a member of the U.S. National Science Foundation's liaison staff in New Delhi, and later became the first executive director of the Harvard University Program on the Public Conceptions of Science. His publications include numerous articles on science education, elementary particle physics, and the history of science in India.

EMILIO Q. DADDARIO, born in 1918, is director of the Office of Technology Assessment. He has served as Mayor of Middletown, Connecticut, judge of Middletown Munipal Court, and Democratic nominee for Governor of Connecticut. From 1959 to 1971, he was a member of the U.S. House of Representatives, where he served as a member of the House Committee on Science and Astronautics, and of the Manned Space Flight Subcommittee. In addition, he was chairman of the Subcommittee on Science, Research and Development, and of the Special Subcommittee on Patents and Scientific Inventions.

AMITAI ETZIONI, born in 1929, is professor of sociology, Columbia University, and director of the Center for Policy Research. Among his many publications are *Genetic Fix* (1973), *The Active Society* (1968), *Modern Organizations* (1964), and *A Comparative Analysis of Complex Organizations* (1961). He has served on the editorial board of *Science*, the NSF Science Information Council, and the AAAS Committee on Public Understanding of Science.

GERALD HOLTON, born in 1922, is professor of physics and associate of the history of science department at Harvard University. From 1957 to 1963, he was editor of the American Academy of Arts and Sciences and, until 1961, editor of *Dædalus*. Among his book publications are *Thematic Origins of Scientific Thought: Kepler to Einstein* (1973), *Introduction to Concepts and Theories in Physical Science* (1952, 1973), and, as editor, *The Twentieth-Century Sciences: Studies in the Biography of Ideas* (1972), and *Science and Culture* (1965).

SALLY GREGORY KOHLSTEDT, born in Michigan in 1943, has taught American social history at Simmons College, Boston, and is currently a faculty member of the history department in the Maxwell School, Syracuse University. She has published several essays on scientific institutionalization in the nineteenth century and her study of the early American Association for the Advancement of Science will be published by the University of Illinois Press in early 1977.

ANDRÉ MAYER, born in 1946, is a doctoral candidate and Dean's fellow in history at the University of California, Berkeley. He is currently working on a study of early American science.

JEAN MAYER, born in 1920, is professor of nutrition, lecturer on the history of public health, and master of Dudley House at Harvard University. He is the author of *Human Nutrition: Its Physiological, Medical Social Aspects* (1972), and *Overweight: Causes, Cost, and Control* (1968), and of numerous articles. He is also the editor and chief author of *Health* (1974) and *U.S. Nutrition Policies in the Seventies* (1972). He served as Special Consultant to the President of the United States in 1969–1970 and as Chairman of the First White House Conference on Food, Nutrition and Health in 1969. He is presently organizing a Conference on National Nutrition Policies in the U.S. Senate.

RUSSELL MCCORMMACH, born in 1935, is associate professor of the history of science at Johns Hopkins University. He has published numerous articles on the recent history of physics, and is editor of the series, *Historical Studies in the Physical Sciences*.

JOHN A. MOORE, born in 1915, is professor of biology at the University of California, Riverside. He is the author of *Readings in Heredity and Development* (1972), *A Guidebook to Washington* (1963), *Heredity and Development* (1963, 1972), *Biological Science: An Inquiry into Life* (1960, 1961, 1963, 1968, 1973), and *Principles of Zoology* (1957).

DOROTHY NELKIN, born in 1933, is associate professor in the Program on Science, Technology, and Society and the department of city and regional planning at Cornell University. Among her book publications are *Nuclear Power and Its Critics* (1971), *The University and Military Research* (1972), *Methadone Maintenance — A Technological Fix* (1973), and *Jetport: The Boston Airport Controversy* (1974).

CLYDE NUNN, born in 1934, is senior research associate at the Center for Policy Research. He is a co-author of *Achievement in American Society* (1969), and the author of numerous articles. He is currently engaged in a NSF-supported study of tolerance and nonconformity.

DAVID PERLMAN, born in 1918, is science editor of the *San Francisco Chronicle*. In 1974, he was on leave of absence to serve as Regents Professor of Human Biology at the University of California, San Francisco. He is Past President of the National Association of Science Writers and a member of the Council for the Advancement of Science Writing and the AAAS Committee on Public Understanding of Science. A former foreign correspondent and political reporter, he has published articles in most major American magazines.

DON K. PRICE, born in 1910, is dean of the John Fitzgerald Kennedy School of Government at Harvard University. He is the author of *The Scientific Estate* (1965) — which was awarded the Faculty Prize of the Harvard University Press — and *Government and Science* (1954), and the editor of *The Secretary of State*. He has served as president of the American Association for the Advancement of Science, vice-president of the Ford Foundation, and deputy chairman of the Research and Development Board of the Department of Defense.

MARC J. ROBERTS, born in 1943, is associate professor of economics at Harvard University. He is the author of a number of articles on the economics of public policy. He is currently chairman of the AAAS Committee on Environmental Alteration.

DAVID J. ROSE, born in 1922, is professor of nuclear engineering at the Massachusetts Institute of Technology. He is the author of a great many papers and co-author of *Plasmas and Controlled Fusion* (1961).

THEODORE ROSZAK, born in 1933, is chairman of general studies and professor of history at the California State University at Hayward. During the forthcoming year he will be a visiting professor at the University of British Columbia. He is the author of *Pontifex: A Revolutionary Entertainment* (1974), *Where the Wasteland Ends* (1972), and *The Making of a Counter Culture* (1969); the editor of *Sources* (1970); and co-editor, with Betty Roszak, of *Masculine/Feminine* (1970).

EDWARD SHILS, born in 1911, is professor of sociology and social thought at the University of Chicago, a Fellow of Peterhouse, Cambridge University, and editor of *Minerva*. His books include *The Intellectual Between Tradition and Modernity* (1961) and *The Intellectuals and the Powers* (1972).

STEVEN WEINBERG, born in 1933, is Higgins Professor of Physics at Harvard University and Senior Scientist at the Smithsonian Astrophysical Observatory. He is the author of numerous articles on the theory of elementary particles, cosmology, and arms control. He has served as consultant to the U.S. Arms Control and Disarmament Agency and the Institute for Defense Analyses; he is currently chairman of the Academy Committee on Research Funds, and a member of the Council on Foreign Relations, the National Academy of Sciences, the Council of the American Physical Society, and the American Mediaeval Academy.

Index

SYNTHESE LIBRARY

Monographs on Epistemology, Logic, Methodology,
Philosophy of Science, Sociology of Science and of Knowledge, and on the
Mathematical Methods of Social and Behavioral Sciences

Managing Editor:
JAAKKO HINTIKKA (Academy of Finland and Stanford University)

Editors:
ROBERT S. COHEN (Boston University)
DONALD DAVIDSON (The Rockefeller University and Princeton University)
GABRIËL NUCHELMANS (University of Leyden)
WESLEY C. SALMON (University of Arizona)

15. C. D. BROAD, *Induction, Probability, and Causation. Selected Papers.* 1968, XI + 296 pp.
16. GÜNTHER PATZIG, *Aristotle's Theory of the Syllogism. A Logical-Philosophical Study of Book A of the Prior Analytics.* 1968, XVII + 215 pp.
17. NICHOLAS RESCHER, *Topics in Philosophical Logic.* 1968, XIV + 347 pp.
18. ROBERT S. COHEN and MARX W. WARTOFSKY (eds.), *Proceedings of the Boston Colloquium for the Philosophy of Science 1966–1968,* Boston Studies in the Philosophy of Science (ed. by Robert S. Cohen and Marx W. Wartofsky), Volume IV. 1969, VIII + 537 pp.
19. ROBERT S. COHEN and MARX W. WARTOFSKY (eds.), *Proceedings of the Boston Colloquium for the Philosophy of Science 1966–1968,* Boston Studies in the Philosophy of Science (ed. by Robert S. Cohen and Marx W. Wartofsky), Volume V. 1969, VIII + 482 pp.
20. J. W. DAVIS, D. J. HOCKNEY, and W. K. WILSON (eds.), *Philosophical Logic.* 1969, VIII + 277 pp.
21. D. DAVIDSON and J. HINTIKKA (eds.), *Words and Objections: Essays on the Work of W. V. Quine.* 1969, VIII + 366 pp.
22. PATRICK SUPPES, *Studies in the Methodology and Foundations of Science. Selected Papers from 1911 to 1969,* XII + 473 pp.
23. JAAKKO HINTIKKA, *Models for Modalities. Selected Essays.* 1969, IX + 220 pp.
24. NICHOLAS RESCHER et al. (eds.), *Essays in Honor of Carl G. Hempel. A Tribute on the Occasion of his Sixty-Fifth Birthday.* 1969, VII + 272 pp.
25. P. V. TAVANEC (ed.), *Problems of the Logic of Scientific Knowledge.* 1969, XII + 429 pp.
26. MARSHALL SWAIN (ed.), *Induction, Acceptance, and Rational Belief.* 1970, VII + 232 pp.
27. ROBERT S. COHEN and RAYMOND J. SEEGER (eds.), *Ernst Mach; Physicist and Philosopher,* Boston Studies in the Philosophy of Science (ed. by Robert S. Cohen and Marx W. Wartofsky), Volume VI. 1970, VIII + 295 pp.
28. JAAKKO HINTIKKA and PATRICK SUPPES, *Information and Inference.* 1970, X + 336 pp.
29. KAREL LAMBERT, *Philosophical Problems in Logic. Some Recent Developments.* 1970, VII + 176 pp.
30. ROLF A. EBERLE, *Nominalistic Systems.* 1970, IX + 217 pp.
31. PAUL WEINGARTNER and GERHARD ZECHA (eds.), *Induction, Physics, and Ethics, Proceedings and Discussions of the 1968 Salzburg Colloquium in the Philosophy of Science.* 1970, X + 382 pp.
32. EVERT W. BETH, *Aspects of Modern Logic.* 1970, XI + 176 pp.
33. RISTO HILPINEN (ed.), *Deontic Logic: Introductory and Systematic Readings.* 1971, VII + 182 pp.
34. JEAN-LOUIS KRIVINE, *Introduction to Axiomatic Set Theory.* 1971, VII + 98 pp.
35. JOSEPH D. SNEED, *The Logical Structure of Mathematical Physics.* 1971, XV + 311 pp.
36. CARL R. KORDIG, *The Justification of Scientific Change.* 1971, XIV + 119 pp.
37. MILIČ ČAPEK, *Bergson and Modern Physics,* Boston Studies in the Philosophy of Science (ed. by Robert S. Cohen and Marx W. Wartofsky), Volume VII. 1971, XV + 414 pp.
38. NORWOOD RUSSELL HANSON, *What I do not Believe, and other Essays,* (ed. by Stephen Toulmin and Harry Woolf), 1971, XII + 390 pp.
39. ROGER C. BUCK and ROBERT S. COHEN (eds.), *PSA 1970. In Memory of Rudolf Carnap,* Boston Studies in the Philosophy of Science (ed. by Robert S. Cohen and Marx W. Wartofsky), Volume VIII. 1971, LXVI + 615 pp. Also available as a paperback.
40. DONALD DAVIDSON and GILBERT HARMAN (eds.), *Semantics of Natural Language.* 1972, X + 769 pp. Also available as a paperback.

41. YEHOSHUA BAR-HILLEL (ed.), *Pragmatics of Natural Languages*. 1971, VII + 231 pp.
42. SÖREN STENLUND, *Combinators, λ-Terms and Proof Theory*. 1972, 184 pp.
43. MARTIN STRAUSS, *Modern Physics and Its Philosophy. Selected Papers in the Logic, History, and Philosophy of Science*. 1972, X + 297 pp.
44. MARIO BUNGE, *Method, Model and Matter*. 1973, VII + 196 pp.
45. MARIO BUNGE, *Philosophy of Physics*. 1973, IX + 248 pp.
46. A. A. ZINOV'EV, *Foundations of the Logical Theory of Scientific Knowledge* (*Complex Logic*), Boston Studies in the Philosophy of Science (ed. by Robert S. Cohen and Marx W. Wartofsky), Volume IX. Revised and enlarged English edition with an appendix, by G. A. Smirnov, E. A. Sidorenka, A. M. Fedina, and L. A. Bobrova 1973, XXII + 301 pp. Also available as a paperback.
47. LADISLAV TONDL, *Scientific Procedures*, Boston Studies in the Philosophy of Science (ed. by Robert S. Cohen and Marx W. Wartofsky), Volume X. 1973, XII + 268 pp. Also available as a paperback.
48. NORWOOD RUSSELL HANSON, *Constellations and Conjectures*, (ed. by Willard C. Humphreys, Jr.), 1973, X + 282 pp.
49. K. J. J. HINTIKKA, J. M. E. MORAVCSIK, and P. SUPPES (eds.), *Approaches to Natural Language. Proceedings of the 1970 Stanford Workshop on Grammar and Semantics*. 1973, VIII + 526 pp. Also available as a paperback.
50. MARIO BUNGE (ed.), *Exact Philosophy – Problems, Tools, and Goals*. 1973, X + 214 pp.
51. RADU J. BOGDAN and ILKKA NIINILUOTO (eds.), *Logic, Language, and Probability*. A selection of papers contributed to Sections IV, VI, and XI of the Fourth International Congress for Logic, Methodology, and Philosophy of Science, Bucharest, September 1971. 1973, X + 323 pp.
52. GLENN PEARCE and PATRICK MAYNARD (eds.), *Conceptual Chance*. 1973, XII + 282 pp.
53. ILKKA NIINILUOTO and RAIMO TUOMELA, *Theoretical Concepts and Hypothetico-Inductive Inference*. 1973, VII + 264 pp.
54. ROLAND FRAÏSSÉ, *Course of Mathematical Logic – Volume 1: Relation and Logical Formula*. 1973, XVI + 186 pp. Also available as a paperback.
55. ADOLF GRÜNBAUM, *Philosophical Problems of Space and Time*. Second, enlarged edition, Boston Studies in the Philosophy of Science (ed. by Robert S. Cohen and Marx W. Wartofsky), Volume XII. 1973, XXIII + 884 pp. Also available as a paperback.
56. PATRICK SUPPES (ed.), *Space, Time, and Geometry*. 1973, XI + 424 pp.
57. HANS KELSEN, *Essays in Legal and Moral Philosophy*, selected and introduced by Ota Weinberger. 1973, XXVIII + 300 pp.
58. R. J. SEEGER and ROBERT S. COHEN (eds.), *Philosophical Foundations of Science. Proceedings of an AAAS Program, 1969*. Boston Studies in the Philosophy of Science (ed. by Robert S. Cohen and Marx W. Wartofsky), Volume XI. 1974, X + 545 pp. Also available as paperback.
59. ROBERT S. COHEN and MARX W. WARTOFSKY (eds.), *Logical and Epistemological Studies in Contemporary Physics*, Boston Studies in the Philosophy of Science (ed. by Robert S. Cohen and Marx W. Wartofsky), Volume XIII. 1973, VIII + 462 pp. Also available as paperback.
60. ROBERT S. COHEN and MARX W. WARTOFSKY (eds.), *Methodological and Historical Essays in the Natural and Social Sciences. Proceedings of the Boston Colloquium for the Philosophy of Science, 1969–1972*, Boston Studies in the Philosophy of Science (ed. by Robert S. Cohen and Marx W. Wartofsky), Volume XIV. 1974, VIII + 405 pp. Also available as paperback.
61. ROBERT S. COHEN, J. J. STACHEL and MARX W. WARTOFSKY (eds.), *For Dirk Struik*.

Scientific, Historical and Political Essays in Honor of Dirk J. Struik, Boston Studies in the Philosophy of Science (ed. by Robert S. Cohen and Marx W. Wartofsky), Volume XV. 1974, XXVII + 652 pp. Also available as paperback.

62. KAZIMIERZ AJDUKIEWICZ, *Pragmatic Logic*, transl. from the Polish by Olgierd Wojtasiewicz. 1974, XV + 460 pp.
63. SÖREN STENLUND (ed.), *Logical Theory and Semantic Analysis. Essays Dedicated to Stig Kanger on His Fiftieth Birthday*. 1974, V + 217 pp.
64. KENNETH F. SCHAFFNER and ROBERT S. COHEN (eds.), *Proceedings of the 1972 Biennial Meeting, Philosophy of Science Association*, Boston Studies in the Philosophy of Science (ed. by Robert S. Cohen and Marx W. Wartofsky), Volume XX. 1974, IX + 444 pp. Also available as paperback.
65. HENRY E. KYBURG, JR., *The Logical Foundations of Statistical Inference*. 1974, IX + 421 pp.
66. MARJORIE GRENE, *The Understanding of Nature: Essays in the Philosophy of Biology*, Boston Studies in the Philosophy of Science (ed. by Robert S. Cohen and Marx W. Wartofsky), Volume XXIII. 1974, XII + 360 pp. Also available as paperback.
67. JAN M. BROEKMAN, *Structuralism: Moscow, Prague, Paris*. 1974, IX + 117 pp.
68. NORMAN GESCHWIND, *Selected Papers on Language and the Brain*, Boston Studies in the Philosophy of Science (ed. by Robert S. Cohen and Marx W. Wartofsky), Volume XVI. 1974, XII + 549 pp. Also available as paperback.
69. ROLAND FRAÏSSÉ, *Course of Mathematical Logic – Volume II: Model Theory*. 1974, XIX + 192 pp.
70. ANDRZEJ GRZEGORCZYK, *An Outline of Mathematical Logic. Fundamental Results and Notions Explained with All Details*. 1974, X + 596 pp.
71. FRANZ VON KUTSCHERA, *Philosophy of Language*. 1975, VII + 305 pp.
75. JAAKKO HINTIKKA and UNTO REMES, *The Method of Analysis. Its Geometrical Origin and Its General Significance*. Boston Studies in the Philosophy of Science (ed. by Robert S. Cohen and Marx W. Wartofsky), Volume XXV. 1974, XVIII + 144 pp. Also available as paperback.
76. JOHN EMERY MURDOCH and EDITH DUDLEY SYLLA, *The Cultural Context of Medieval Learning. Proceedings of the First International Colloquium on Philosophy, Science, and Theology in the Middle Ages – September 1973*. Boston Studies in the Philosophy of Science (ed. by Robert S. Cohen and Marx W. Wartofsky), Volume XXVI. 1975, X + 566 pp. Also available as paperback.
77. STEFAN AMSTERDAMSKI, *Between Experience and Metaphysics. Philosophical Problems of the Evolution of Science*. Boston Studies in the Philosophy of Science (ed. by Robert S. Cohen and Marx W. Wartofsky), Volume XXXV. 1975, XVIII + 193 pp. Also available as paperback.
80. JOSEPH AGASSI, *Science in Flux*. Boston Studies in the Philosophy of Science (ed. by Robert S. Cohen and Marx W. Wartofsky), Volume XXVIII. 1975, XXVI + 553 pp. Also available as paperback.

SYNTHESE HISTORICAL LIBRARY

Texts and Studies
in the History of Logic and Philosophy

Editors:

N. KRETZMANN (Cornell University)
G. NUCHELMANS (University of Leyden)
L. M. DE RIJK (University of Leyden)

1. M. T. BEONIO-BROCCHIERI FUMAGALLI, *The Logic of Abelard.* Translated from the Italian. 1969, IX + 101 pp.

2. GOTTFRIED WILHELM LEIBNITZ, *Philosophical Papers and Letters.* A selection translated and edited, with an introduction, by Leroy E. Loemker. 1969, XII + 736 pp.

3. ERNST MALLY, *Logische Schriften,* ed. by Karl Wolf and Paul Weingartner. 1971, X + 340 pp.

4. LEWIS WHITE BECK (ed.), *Proceedings of the Third International Kant Congress.* 1972, XI + 718 pp.

5. BERNARD BOLZANO, *Theory of Science,* ed. by Jan Berg. 1973, XV + 398 pp.

6. J. M. E. MORAVCSIK (ed.), *Patterns in Plato's Thought. Papers arising out of the 1971 West Coast Greek Philosophy Conference.* 1973, VIII + 212 pp.

7. NABIL SHEHABY, *The Propositional Logic of Avicenna: A Translation from al-Shifā: al-Qiyās,* with Introduction, Commentary and Glossary. 1973, XIII + 296 pp.

8. DESMOND PAUL HENRY, *Commentary on De Grammatico: The Historical-Logical Dimensions of a Dialogue of St. Anselm's.* 1974, IX + 345 pp.

9. JOHN CORCORAN, *Ancient Logic and Its Modern Interpretations.* 1974, X + 208 pp.

10. E. M. BARTH, *The Logic of the Articles in Traditional Philosophy.* 1974, XXVII + 533 pp.

11. JAAKKO HINTIKKA, *Knowledge and the Known. Historical Perspectives in Epistemology.* 1974, XII + 243 pp.

12. E. J. ASHWORTH, *Language and Logic in the Post-Medieval Period.* 1974, XIII + 304 pp.

13. ARISTOTLE, *The Nicomachean Ethics.* Translated with Commentaries and Glossary by Hypocrates G. Apostle. 1975, XXI + 372 pp.

14. R. M. DANCY, *Sense and Contradiction: A Study in Aristotle*. 1975, XII + 184 pp.

15. WILBUR RICHARD KNORR, *The Evolution of the Euclidean Elements. A Study of the Theory of Incommensurable Magnitudes and Its Significance for Early Greek Geometry*. 1975, IX + 374 pp.

16. AUGUSTINE, *De Dialectica*. Translated with the Introduction and Notes by B. Darrell Jackson. 1975, XI + 151 pp.